Urban Sustainability

The Urban Sustainability Book Series is a valuable resource for sustainability and urban-related education and research. It offers an inter-disciplinary platform covering all four areas of practice, policy, education, research, and their nexus. The publications in this series are related to critical areas of sustainability, urban studies, planning, and urban geography.

This book series aims to put together cutting-edge research findings linked to the overarching field of urban sustainability. The scope and nature of the topic are broad and interdisciplinary and bring together various associated disciplines from sustainable development, environmental sciences, urbanism, etc. With many advanced research findings in the field, there is a need to put together various discussions and contributions on specific sustainability fields, covering a good range of topics on sustainable development, sustainable urbanism, and urban sustainability. Despite the broad range of issues, we note the importance of practical and policy-oriented directions, extending the literature and directions and pathways towards achieving urban sustainability.

The series will appeal to urbanists, geographers, planners, engineers, architects, governmental authorities, policymakers, researchers of all levels, and to all of those interested in a wide-ranging overview of urban sustainability and its associated fields. The series includes monographs and edited volumes, covering a range of topics under the urban sustainability topic, which can also be used for teaching materials.

Baojie He · Yupeng Wang · Ali Cheshmehzangi
Editors

Urban Climate and Urban Design

Springer

Editors
Baojie He
School of Architecture and Urban Planning
Chongqing University
Chongqing, China

Yupeng Wang
School of Human Settlement and Civil Engineering
Xi'an Jiaotong University
Xian, China

Ali Cheshmehzangi
School of Architecture,
Design and Planning
The University of Queensland
Brisbane, QLD, Australia

ISSN 2731-6483 ISSN 2731-6491 (electronic)
Urban Sustainability
ISBN 978-981-96-1520-9 ISBN 978-981-96-1521-6 (eBook)
https://doi.org/10.1007/978-981-96-1521-6

This Springer imprint is published by the registered company Springer Nature Singapore Pte Ltd.
The registered company address is: 152 Beach Road, #21-01/04 Gateway East, Singapore 189721, Singapore

Baojie He dedicates this book to all his collaborators and colleagues at Centre for Climate-resilient and Low-Carbon Cities at Chongqing University, China. Also dedicated to all contributors and scholars who attended the 1st, 2nd, 3rd, and 4th International Conference on Urban Climate and Urban Design (UCUD).

Yupeng Wang dedicates this book to researchers in the field of urban climate and urban design, dedicated to the editors on the board of international journal of Urban Climate and dedicates to the partners and friends who have supported the 4th International Conference on Urban Climate and Urban Design (UCUD) at Xi'an Jiaotong University in May 2024.

Ali Cheshmehzangi dedicates this book to climate and environmental activities worldwide, as well as the academics and participants of our UCUD events in recent years.

Acknowledgements

We would like to sincerely thank all authors and contributors for their hard work and dedication in writing their chapters. While we met some of them online in recent months, we hope we get the opportunity to meet all of them in person in the near future. Their support, dedication, and continuous efforts are recognised, genuinely valued, and highly appreciated.

Baojie He acknowledges the support from the National Natural Science Foundation of China (Grant No. 42301339), the Fundamental Research Funds for the Central Universities (Grant No. 2024CDJXY014), Research on the Establishment and Application of Green and Low Carbon Habitat Environment in Chongqing East Railway Station Area (Grant No. 2022-KJ-JY-20 and CSTB2022TIAD-KPX0195), the China Meteorological Administration "Research on value realization of climate ecological products" Youth Innovation Team Project (Grant No. CMA2024QN15), and Chongqing Natural Science Foundation Project (Grant No. CSTB2024NSCQ-MSX0670). Baojie He also appreciates the tremendous support from the Chongqing University, Chongqing Jiaotong University, Xi'an Jiaotong University, Northwest A&F University, and Chang'an University for holding the International Conference on Urban Climate and Urban Design.

About This Book

Due to climate change and urbanization, urban climate change is increasingly prominent with significantly environmental, economic, social, and health consequences. Addressing climate-related risks, threats and disasters in cities to a significant extent means creating resilient, healthy, safe and inclusive built environments. Such an assumption is much truer since cities are the main human settlements of human beings. Typically, urban design is an implementable and tangible pathway to practically deal with the city-climate interactions, while there should be a large transformation from unsustainable urban design patterns towards sustainable ones.

Urban Climate and Urban Design is a comprehensive collection of theoretical perspectives and global case study examples focused on four core areas of (i) Urban Climate Monitoring, Assessing and Forecasting, (ii) Mitigation and Adaptation Strategies, (iii) Advanced and Emerging Design Models and Tools, and (iv) Action Plans and Policy Formulation. This edited volume provides theoretical and methodological references for urban climate research and generates practical implications for mitigation and adaptation capacity improvement.

This book will be of interest to a range of researchers from earth and planetary sciences and environmental sciences to engineering, architecture, and urban planning. Beyond them, this book will enlighten policy makers, practitioners, and developers how to properly regulate urban climate through design interventions. We believe this book is promising to narrow the gap in the transition towards climate-resilient cities, and enhance the understanding of new ideas, methods, strategies, and policies.

Contents

Nexus of Urban Climate and Urban Design 1
Baojie He, Yupeng Wang, and Ali Cheshmehzangi

Urban Climate Monitoring, Assessing and Forecasting

Monitoring, Assessing, and Forecasting Urban Climate Issues and Challenges 19
Tengku Nilam Baizura Tengku Ibrahim, Nur Azalina Suzianti Feisal, Wai Yan Cheah, Nurzawani Md Sofwan, Nur Faseeha Suhaimi, Nursalfarina Abdul Samat, and Noor Haziqah Kamaludin

Heatwave Resilience: What Do We Know About Its Monitoring, Assessing and Forecasting? 49
Chenyu Yang, Junqing Tang, Pengjun Zhao, and Jing Wang

Modeling Natural Ventilation in Urban Environments: A Review of Existing Techniques and Tools 73
Humera Mughal, Anber Rana, and Salma Sherbaz

Mitigation and Adaptation Strategies

From Vernacular to Vernomimicry: Vernacular Design Principles for Resilient Communities 87
Ahmed Khoja

Navigating Nearly Zero-Energy Strategies for Urban Climate Change Adaptation and Mitigation 101
Osama Omar

Enhancing Climate Resilience Through Urban Landscape Design: Strategies for Green Infrastructure and Adaptation 117
Reza Farhadi, Hadi Soltanifard, and Behdad Alizadeh

Prioritizing Energy Efficiency and Low-Carbon Approaches: Climate Adaptation Strategies in Urban West Asia 129
Akram Ahmed Noman Alabsi

Advanced and Emerging Design Models and Tools

Cooling Pakistan: Exploring Historical Wisdom and Modern Strategies for Passive Cooling in Pakistan 147
Shanzae Naeem, Aisha Aziz, and Baojie He

The Urban Design Climate Workshop: Integrating Mitigation and Adaptation Strategies to Confront Urban Climate Change 167
Jeffrey Raven, Mattia Federico Leone, and Giovanni Nocerino

Understanding Urban Heat Islands in Tunisian City Centers: An Integrated Approach 181
Messaoudi Abir

The Rehabilitation of Riverbanks in Beijing's Urban Area in the Light of an Ambitious Policy to Enhance Nature and Quality of Life for Residents 197
Stéphanie Boufflet and Vassilia Pirierros

Action Plans and Policy Formulation

Approaches to Urban Climate Change Policymaking: Challenges and Way Forward 213
Kirti Kusum Joshi and Ajay Chandra Lal

Building Climate-Resilient Cities Through Community Engagement and Participatory Design 233
Amir Tayyebi

Analysing Policies for Public Green Spaces of Indian Cities 253
Piyush Pandya, Gagan Jain, and Shweta Prasad

Teaching Practice: Analyzing and Transmitting the Wisdom of Climate-Adaptive Construction of Traditional Heritage Settlements 297
Xu Li, Pengcheng Liu, and Baojie He

Editors and Contributors

About the Editors

Baojie He is a Professor of Urban Climate and Sustainable Built Environment at School of Architecture and Urban Planning, and the leader of the Centre for Climate-Resilient and Low-Carbon Cities, Chongqing University. Baojie is also the Adjunct Associate Professor at University of Queensland, Australia, and the Research Fellow at Network for Education and Research on Peace and Sustainability, Hiroshima University, Japan. Baojie He was a Postdoc Research Fellow, Research Associate, Ph.D. researcher, at UNSW Sydney, Australia. Baojie is working on the technologies of Cool Cities and Communities and Net Zero Carbon Built Environment. Baojie has published more than 170 peer-reviewed papers in high-ranking and delivered more than 40 invited talks in reputable conferences. Baojie has a H-index of 51 (Scopus). Baojie has been involved in several large research projects on energy and built environment in Australia. Baojie has been invited to act Editor-in-Chief, Leading Guest Editor, Associate Editor, Editorial Board Member, Conference Chair, Sessional Chair, Scientific Committee of a variety of reputable international journals and conferences. Baojie received the Highly Cited Researcher Title (Clarivate) in 2022, 2023, and 2024, the Sustainability Young Investigator Award in 2022, the Green Talents Award (Germany) in 2021, and National Scholarship for Outstanding Foreign Students (China) in 2019. Baojie was ranked as one of the Single-year & Career 100,000 global scientists (2%) by the Stanford University in 2020, 2021, 2022, 2023, and 2024.

Yupeng Wang is a Professor and the Associate Dean at the School of Human Settlements and Civil Engineering, Xi'an Jiaotong University, China. He obtained his Ph.D. degree from the Faculty of Environmental Engineering, the University of Kitakyushu, Japan. He has the working experience as an urban planner in NIKKEN SEKKEI, the largest architecture and planning firm in Japan, and the research experience in the Heat Island Group at the Department of Building, Civil and Environmental Engineering of the Concordia University for sustainable urban planning, funded by the

Natural Sciences and Engineering Council of Canada. He has a wide range of research experiences including environmental modeling and evaluation, sustainable urban planning, urban vegetation and climate change mitigation, and published more than 30 journal papers. Because of his interdisciplinary educational background, research experiences and work experiences, he has received a "Hundred Talents Project" of Shaanxi Province, China in 2017, "Person of the Year" from Scientific Chinese magazine in 2018, "AIUE-O Prize" from Asia Institute of Urban Environment in 2019, and "Award of Excellence Young Scholar" from The International Society of the Built Environment in 2020. He currently serves as an Editorial Board Member for Indoor and Built Environment, Guest Editor of Buildings, International Journal of Environmental Research and Public Health, and Sustainability.

Ali Cheshmehzangi is the World's top 2% field leader, recognised by Stanford University. He is a Professor and Head of the School of Architecture, Design and Planning (ADP) at the University of Queensland. He is among the top 30 global scholars in the urban sustainability research area. Prior to joining UQ, Ali held several strategic leadership and senior managerial roles, such as Vice-President for International Engagement and Global Partnership, Founding Director/Head of the Center for Innovation in Education and Research, Head of the Department of Architecture and Built Environment, Founding Director of the Urban Innovation Lab, Director of a university-wide Teaching and Learning platform, Director of International Research Network for Rural and Urban Development, Head of Research Group for Sustainable Built Environment, co-director of university-wide research priority areas, Director of Center for Sustainable Energy Technologies, and Interim Director of Digital Design Lab. So far, Ali has published over 500 journal papers, articles, conference papers, book chapters, and reports. He also has 33 other academic books, three of which have received awards at the national, provincial, and municipal levels. He also has received international awards and recognition for his research on urban resilience studies and sustainability research, as well as a Vice-Chancellor's award for his impactful contribution to higher education. Ali is also globally known for his extended work on urban sustainability research, and particular as the Editor-in-Chief of Springer's Urban Sustainability (US) book series (Link: https://link.springer.com/bookseries/16930).

Contributors

Nursalfarina Abdul Samat Department of Environmental Health, Faculty of Health Sciences, MAHSA University, Jenjarom, Selangor, Malaysia

Messaoudi Abir Research Laboratory Governance, Ambiance and Development of Cities (GADEV Lab), ENAU, University of Carthage, Tunis, Tunisia

Akram Ahmed Noman Alabsi College of Architecture and Urban Planning, Fujian University of Technology, Fuzhou, China

Behdad Alizadeh Department of Horticulture and Landscape Architecture, Ferdowsi University of Mashhad, Mashhad, Iran

Aisha Aziz University of Engineering and Technology, Lahore, Pakistan

Stéphanie Boufflet Department of Theories and Practices of Architectural and Urban Design (TPCAU), National School of Architecture of Paris Val-de-Seine, Paris, France

Wai Yan Cheah Centre of Research in Development, Social and Environment (SEEDS), Faculty of Social Sciences and Humanities, Universiti Kebangsaan Malaysia, UKM Bangi, Selangor Darul Ehsan, Malaysia

Ali Cheshmehzangi School of Architecture, Design and Planning, The University of Queensland, Brisbane, Australia

Reza Farhadi Department of Human Geography and Planning, Faculty of Geography, University of Tehran, Tehran, Iran

Nur Azalina Suzianti Feisal Department of Diagnostic and Allied Health Science, Faculty of Health and Life Sciences, Management and Science University, Shah Alam, Selangor, Malaysia

Baojie He Centre for Climate–Resilient and Low–Carbon Cities, School of Architecture and Urban Planning, Key Laboratory of New Technology for Construction of Cities in Mountain Area, Ministry of Education, Chongqing University, Chongqing, China;
School of Architecture, Design and Planning, The University of Queensland, Brisbane, Australia;
Institute for Smart City, Chongqing University, Liyang, Jiangsu, China;
CMA Key Open Laboratory of Transforming Climate Resources to Economy, Chongqing, China;
Institute for Smart City of Chongqing University in Liyang, Chongqing University, Liyang, Jiangsu, China

Gagan Jain School of Architecture and Planning, Woxsen University, Telangana, India

Kirti Kusum Joshi Institute of Applied Sciences, Madan Bhandari University of Science and Technology, Chitlang, Nepal

Noor Haziqah Kamaludin Occupational Health and Safety Risk Management (OHSeRM) Research Initiative Group and Center of Environmental Health and Safety, Faculty of Health Sciences, Universiti Teknologi MARA, Puncak Alam, Selangor, Malaysia

Ahmed Khoja Department of Architecture, Munich University of Applied Sciences, Munich, Germany

Ajay Chandra Lal Institute of Engineering, Tribhuvan University, Lalitpur, Nepal

Mattia Federico Leone Department of Architecture, Università di Napoli Federico II, Napoli, Italy

Xu Li Centre for Climate–Resilient and Low–Carbon Cities, School of Architecture and Urban Planning, Key Laboratory of New Technology for Construction of Cities in Mountain Area, Ministry of Education, Chongqing University, Chongqing, China

Pengcheng Liu Centre for Climate–Resilient and Low–Carbon Cities, School of Architecture and Urban Planning, Key Laboratory of New Technology for Construction of Cities in Mountain Area, Ministry of Education, Chongqing University, Chongqing, China

Nurzawani Md Sofwan Faculty of Health Sciences, Universiti Teknologi MARA, Kepala Batas, Pulau Pinang, Malaysia;
Environmental Health Program, Faculty of Health Sciences, Universiti Teknologi MARA, Kota Samarahan, Sarawak, Malaysia

Humera Mughal College of Architecture and Design, Prince Sultan University, Riyadh, Saudi Arabia

Shanzae Naeem University of Engineering and Technology, Lahore, Pakistan

Giovanni Nocerino Department of Architecture, Università di Napoli Federico II, Napoli, Italy

Osama Omar Associate Professor of Architecture, Department of Architecture and Interior Design, College of Engineering, University of Bahrain, Manama, Kingdom of Bahrain

Piyush Pandya School of Architecture and Planning, Centre of Excellence-Built Environment and Urban Resilience, Woxsen University, Telangana, India

Vassilia Pirierros School of Architecture, Tianjin University, Tianjin, China

Shweta Prasad School of Architecture and Planning, Woxsen University, Telangana, India

Anber Rana Department of Architecture, University of Washington, Seattle, USA

Jeffrey Raven School of Architecture and Design, New York Institute of Technology, New York City, USA

Salma Sherbaz Research Center for Modeling and Simulation, National University of Sciences and Technology, Islamabad, Pakistan

Hadi Soltanifard Faculty of Geography and Environmental Sciences, Hakim Sabzevari University, Sabzevar, Iran

Nur Faseeha Suhaimi Department of Environmental and Occupational Health, Faculty of Medicine and Health Sciences, Universiti Putra Malaysia, Serdang, Selangor, Malaysia

Junqing Tang School of Urban Planning and Design, Shenzhen Graduate School, Peking University, Shenzhen, China

Amir Tayyebi Architecture and Planning Department, Jundi-Shapur University of Technology, Dezful, Iran

Tengku Nilam Baizura Tengku Ibrahim Faculty of Health Sciences, Universiti Teknologi MARA, Kepala Batas, Pulau Pinang, Malaysia

Jing Wang School of Urban Planning and Design, Shenzhen Graduate School, Peking University, Shenzhen, China

Yupeng Wang School of Human Settlements and Civil Engineering, Xi'an Jiaotong University, Xi'an, China

Chenyu Yang School of Urban Planning and Design, Shenzhen Graduate School, Peking University, Shenzhen, China

Pengjun Zhao School of Urban Planning and Design, Shenzhen Graduate School, Peking University, Shenzhen, China

Nexus of Urban Climate and Urban Design

Baojie He, Yupeng Wang, and Ali Cheshmehzangi

Abstract Cities are challenged by many problems, where problems related to urban climates are very critical. However, existing efforts in both academia and industries are not enough to close the gaps among urban climate science, mitigation and adaptation strategies, and planning and design practices. This pushes relevant stakeholders to deepen basic and applied research on incorporating urban climate into urban design. This chapter, a brief outline of the book titled Urban Climate and Urban Design, is designed to demonstrate how researchers and stakeholders from different fields are working together to promote the incorporation of urban climate sciences and advanced and emerging technologies into the practices of urban planning and design. This chapter charts the aim, objectives, and structures of the book. Afterwards, the key information of all chapters in four parts including (i) Urban Climate Monitoring, Assessing, and Forecasting, (ii) Mitigation and Adaptation Strategies, (iii) Advanced and Emerging Design Models and Tools, and (iv) Action Plans and Policy Formulation are presented.

B. He (✉)
Centre for Climate–Resilient and Low–Carbon Cities, School of Architecture and Urban Planning, Key Laboratory of New Technology for Construction of Cities in Mountain Area, Ministry of Education, Chongqing University, Chongqing 400045, China
e-mail: baojie.unsw@gmail.com

B. He · A. Cheshmehzangi
School of Architecture, Design and Planning, The University of Queensland, Brisbane 4072, Australia

B. He
Institute for Smart City, Chongqing University, Liyang 213300, Jiangsu, China

CMA Key Open Laboratory of Transforming Climate Resources to Economy, Chongqing 401147, China

Y. Wang
School of Human Settlements and Civil Engineering, Xi'an Jiaotong University, Xi'an 710049, China

B. He et al. (eds.), *Urban Climate and Urban Design*, Urban Sustainability,
https://doi.org/10.1007/978-981-96-1521-6_1

1 Nexus of Urban Climate and Urban Design and the Solutions

Urbanization has significantly shifted cities as the main location of human settlements. However, the way urban planners, designers, and managers adopt to modify spatial landscapes and force urban operation is far from sustainable. As a result, cities are challenged by numerous problems, among which problems related to urban climates are very critical. The major components of urban climate problems can be divided into two parts. The first part is the climate change at the global scale, associated with a series of disasters such as extreme heat, drought, intense precipitation, and storm [10, 15]. The second part is the different behaviours of radiative, aerodynamic, moisture, hydrological, thermal, and reflective properties of built-up areas at the local scale, from those of natural areas [12, 16]. The critical problems include air pollution, heat island effects, uneven precipitation, and associated hazardous weathers. Furthermore, there are possible synergies between the global and local climates, with much stronger impacts and threats [1, 5].

The history of human civilization is a process of human beings seeking healthy, safe and comfortable living environments. For thousands of years, people have actively responded to, adapted to and utilized climate characteristics, forming urban planning and architectural design theories, methods and paradigms for creating a green, low-carbon, healthy and liveable environment [3]. However, over the past nearly 100 years, climate change and urbanization have caused urban climate conditions to break the bottom line of liveability, health, and safety, causing great impacts on human production, survival, and life. Humans must urgently reflect on climate-adaptive design methods to first make sure the climate resilience of built environments and then to maintain the built environment to continue to provide suitable housing for humans [8]. For instance, many houses can no longer provide cooler shelters for citizens during overheating days; Instead, the indoor overheating of these houses has significantly threatened occupant's health and safety [9, 19]. It is critical to increase the resilience of these houses to extreme heat before achieving the expectation of cooling down and escaping the heat.

A system of applicable and effective mitigation and adaptation strategies, technologies, and schemes are particularly important, to reshape theories and practical paradigms of urban planning and design [2]. Scholars from disciplines such as urban planning, architecture, landscape architecture, civil engineers, transportation engineering, geography, climatology, medicine, and management have contributed to build an innovative knowledge system of urban climate and urban design [7]. Existing technologies, strategies, plans, and actions are far from enough to merge the gaps among urban climate science, mitigation and adaptation strategies, and planning and design practices. The impacts of climate change at global and local scales are unprecedented, fast, and far-reaching that it far exceeds the responsive speed of human cognition, knowledge, and action. People are not prepared to address urban climate problems, leading to many barriers such as lack of industry professionals, insufficient policy support, financial investment shortage, and low public awareness and participation [6].

To close these gaps, it is essential to conduct basic and applied research on incorporating urban climate into urban design. In the viewpoint of urban planning and design, deciding planning and design needs and targets, among different stakeholders, with the accurate analyses and assessment of urban climate problems, is a prerequisite. Development of an assessment indicator system in environmental, economic, social, and health dimensions are highly preferable to balance the diverse functions of urban spaces. For instance, extreme heat problems in cities are associated with microclimatic indicators (e.g., airborne temperature, surface temperature, wind speed, relative humidity), health indicators (e.g., morbidity and mortality), urban quality indicators (e.g., heat island intensity, public outdoor participation), and economic indicators (e.g., electricity use, water use, and economic losses) [17]. Moreover, understanding of climate-related impacts and impact mechanisms should be tailored to urban spatial form and morphological characteristics, to pinpoint the locations and strategies to reshape urban spaces. For instance, heat island effects in both spatial and temporal aspects have been explored at different scales (e.g., regional, city, and block) in terms of 2D surface cover and 3D spatial and landscape pattern characteristics [4, 20].

Facing urban climate change, both mitigation and adaptation are critical to protect people from climate-related threats and risks in the short term and to alleviate climate-related impacts in the long run. On the climate change, decarbonisation solutions should be implemented in different fields (e.g., buildings, transportation, and industry) and every stage of urban planning, design, and management. For instance, the design of a neighbourhood is expected to enable low-carbon transportation, low embodied carbon and operational carbon in buildings, the uptake of renewables, and high carbon sinks via vegetation, and the adoption of carbon capture, utilisation and storage techniques [18]. The direct response to a specific urban climate problem is often related to urban climate adaptation. For instance, the reduction of air pollution is possibly achievable via the replacement of energy sources from fossil fuels to renewables and the reduction of energy needs. The increase of ventilation efficiency by changing spatial and morphological characteristics, by developing ventilation corridors and modifying building height, density, and street properties, is also feasible [13].

Though many mitigation and adaptation techniques and strategies have been developed, there are still large gaps in their effective application in urban planning, design, and management. The ability of different techniques and strategies to mitigate and adapt to specific urban climatic problems is generally influenced by climate and geographic factors, urban development patterns, construction culture, demographics, and health and safety requirements [2]. Furthermore, there are unanticipated repercussions that come along with strategies and solutions. A good incorporation of urban climate mitigation and adaptation into planning and design, namely the improvement of transformation capacity, should be aided by novel planning and design methods and models at the technical aspect. By accommodating various stakeholder needs, integrating divergent priorities, demonstrating outcomes under various conditions, and promoting both independent and cooperative actions, decision support tools can aid in closing these gaps [14]. It can be used to translate potential methods and approaches into adjustments and improvements in operations and behaviours. Furthermore, models and tools that are ready for implementation should be demonstrated and delivered to different stakeholders by participatory practical activities [2].

Beyond urban climate sciences, mitigation and adaptation strategies, and design models and tools, the policy formulation, economic investment, and social participation are also important to enhance the implementation capacity of climate-responsive planning and design [6]. The policy formulation in terms of governmental policies, laws, standards, rules and regulations often offers strategic guidance and insights to mitigate and adapt to urban climate change. The national and provincial plans for climate change mitigation and adaptation are stringent to shift institutional atmosphere and then the planning and design activities of all stakeholders. In order to improve planners' and designers' ability to mitigate and adapt, as well as their intention and willingness to change, social participation is also given priority. Propaganda and education are frequently required to accomplish this through recognizing effects, recognizing warning signs and symbols, and identifying potential strategies [11]. Collaboration among governments, enterprises, non-governmental organizations, and owners is imperative in shaping a sound economic circle through economic investments. When it comes to attracting investment, allocating funds, and initiating incentive programs are essential to enhance the engagement of pertinent businesses, as well as to raise awareness and recognition of non-governmental organizations, businesses, and owners.

2 The Aim and Objectives of the Book

This book aims to the latest research outcomes on urban climate and urban design. This book demonstrates the chances and possibilities to bring researchers and stakeholders from different fields together to effectively incorporate urban climate sciences and advanced and emerging technologies into the practices of urban planning and design. From the planning and design angle, this book showcases the assessment and identification of critical urban climate problems to pinpoint the planning and design goals and targets; presents the mitigation and adaptation strategies and techniques to determine the solutions that urban planners and designers can adopt; demonstrates advanced and emerging planning and design models and tools that can eliminate the technical barriers in generating planning and design schemes; and advances the understanding of action plans and policy formulation to provide urban planners, designers, and managers with sufficient policy, economic, and social support for implementation capacity improvement. Given this, the objectives of this book include:

– Presenting emerging, innovative and comprehensive techniques and methods to monitor, assess, and forecast urban climate issues and challenges

- Deepening the understanding of traditional, regional, advanced, and emerging techniques and strategies for urban climate change mitigation and adaptation
- Showcasing the development of planning and design models and tools that can enhance the implementation of mitigation and adaptation strategies in different contexts
- Exploring the policy, economic, and social drivers to improve the societal implementation capacity of mitigation and adaptation strategies.

Overall, this book is by now the first one to integrate urban climate sciences, mitigation and adaptation strategies, design models and tools, and policy formulation together to show a comprehensive roadmap to promote the transformation from urban climate science to planning and design practices. This book is also advantaged to bring scholars and stakeholders in the fields of urban planning, architecture, landscape architecture, civil engineers, transportation engineering, geography, climatology, medicine, and management together to address urban climate change collaboratively through planning and design interventions. Furthermore, this book can provide implications for developed and developing cities through showing theoretical and practical results in both Global North and Global South.

3 Structure of the Book

The book is intentionally divided into four parts, addressing key aspects related to the climatic facts, implementable strategies, design methods, and enabling policies, following a chain to promote the transformation from urban climate sciences and technologies into planning and design practices to address urban climate problems. For each part, we have invited contributors from around the globe, where they present their latest study findings and case study work to chart a comprehensive, advanced, tangible, and implementable scheme to achieve urban climate change mitigation and adaptation in urban planning and design (Fig. 1). These four parts are structured as:

(1) Urban Climate Monitoring, Assessing, and Forecasting;
(2) Mitigation and Adaptation Strategies;
(3) Advanced and Emerging Design Models and Tools; and
(4) Action Plans and Policy Formulation.

PART I: Urban Climate Monitoring, Assessing and Forecasting

The first part investigates methods and approaches to record, estimate, and track the facts of urban climate, as well as assessment techniques and tools to identify critical urban climate problems. This part holistically defines the key challenges of urbanization and climate change, urban heat island effect, climatic impacts on air pollution, extreme weather events, water pollution and scarcity, impact on human health, and extreme precipitation and flood, and explores the state-of-the-art of the

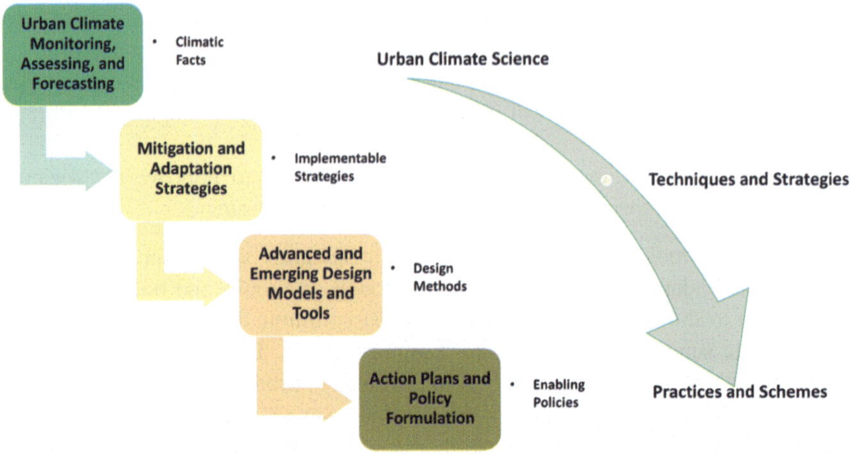

Fig. 1 Pathways to achieve the transformation from urban climate science and technologies to planning and design practices and schemes

methods adopted to monitor, assess, and forecast urban climate issues and challenges. Below is the summary of the three chapters included in this part.

Chapter 2: Monitoring, Assessing, and Forecasting Urban Climate Issues and Challenges

By: Tengku Nilam Baizura Tengku Ibrahim, Nur Azalina Suzianti Feisal, Wai Yan Cheah, Nurzawani Md Sofwan, Nur Faseeha Suhaimi, Nursalfarina Abdul Samat, and Noor Haziqah Kamaludin

The urgency of climate change demands the development of sustainable urban systems. Although advanced modelling holds significant potential for designing such systems, current tools have two primary limitations: lack of user-friendliness, which hampers accessibility for a broad range of stakeholders, and they struggle to integrate up-to-date climate data, especially regarding extreme weather events. Accurate forecasting is essential for protecting life and property through climate adaptation, mitigation actions, and disaster risk reduction strategies. Dynamic approaches that view cities as evolving systems are crucial for enabling humanity to adapt, survive, and thrive amidst volatile shocks and stresses. This chapter identifies a significant research gap in creating user-centric urban climate modelling tools that seamlessly incorporate the latest urban climate, state-of-the-art for the monitoring and assessing urban climate issues, data integration, sensing, and visualization. The chapter also discusses challenges and provides recommendations for addressing urban climate issues, aiming to formulate effective climate adaptation and mitigation strategies, ultimately benefiting human health and societal well-being.

Chapter 3: Heatwave Resilience: What Do We Know About Its Monitoring, Assessing and Forecasting?

By: Chenyu Yang, Junqing Tang, Pengjun Zhao, and Jing Wang

In the studies of climate change, the intensity, frequency, and duration of heatwave situations have escalated and are projected to continuously worsen in coming years, leading to detrimental effects on ecosystems, infrastructure, public health, and social life. To effectively mitigate the adverse effects of heatwaves and foster sustainable development, it is essential for urban areas to prioritize the consideration of the concept of "heatwave resilience". However, the current approaches to addressing heatwave resilience, including its assessment, monitoring, and forecast, are still in the early phase and need further development. In this chapter, we structured the content into six sections. We begin by introducing the background and significance of heatwave resilience and elaborating on related concepts and their significance. Next, we illustrate the impacts of heatwaves on modern cities. We then delve into the assessment framework and methods of heatwave resilience based on its key determinants, followed by a discussion on the monitoring methods, tools, and platforms available. We also survey the forecasting methods, highlighting various predictive model frameworks that enhance preparedness for heatwave impacts. Lastly, we discuss future directions and possibilities for building heatwave resilience. We have put forth a set of suggestions and implications for policymakers to address heatwave resilience, which serves as a novel paradigm for appraising prospective policy shifts, government grants, and community education engagements. This holds significance for maintaining sustainable urban development as well as public health and social justice.

Chapter 4: Modelling Natural Ventilation in Urban Environments: A Review of Existing Techniques and Tools

By: Humera Mughal, Anber Rana, and Salma Sherbaz

The urban climate faces numerous challenges stemming from climate change and escalating populations. The built environment and lifestyle patterns exacerbate these challenges, resulting in an intensified urban heating effect. These challenges include poor air quality, heat island effects, and increase risks of floods/droughts with change in land use cover. Addressing many of these challenges hinges on effective urban ventilation strategies. Urban ventilation can be evaluated and improved using different tools and techniques. This study's primary objective is to thoroughly review these tools and techniques used in modelling urban ventilation, with a specific focus on identifying common challenges in modelling and charting future directions for improvement. The study commences by providing an overview of the need to evaluate urban ventilation. A brief overview of these tools and techniques is followed by the most prevalent issues and essential research avenues for improving urban ventilation assessment. The outcomes of this study hold considerable value in fostering sustainable urban environments, as they furnish guidance for enhancing existing techniques and tools. Specifically, this work serves as a valuable resource for the modelling community and newcomers to the field of research.

PART II: Mitigation and Adaptation Strategies

The second part concentrates more on the development and advancement of mitigation and adaptation techniques and strategies that can alleviate urban climate problems or at least avoid the impacts on citizens. This part demonstrates four key collections of mitigation and adaptation strategies by exploring the vernacular design principles, nearly zero-energy strategies, urban landscape design, and energy efficiency and low-carbon approaches. Below is the summary of the four chapters included in this part.

Chapter 5: From Vernacular to Vernomimicry: Vernacular Design Principles for Resilient Communities

By: Ahmed Khoja

Climate change poses increasing threats to our built environment and the debate is gradually shifting toward its magnitude and not its occurrence. As we embrace for a new climatic era, designers and planners are forced to produce architectural design that can address the dual challenge of mitigation and adaptation. This requires incorporating a design paradigm that is "with" nature and not "despite" it. Vernacular design, being the indigenous science of building, can serve as a valuable source of inspirations for climate adaptation solutions. Based on a literature review and comparative analysis, this paper seeks to uncover the governing principles behind the success of vernacular design in the face of a changing climate and explore climate adaptation solutions and strategies that can be transferred from the vernacular to the modern-day design, following a Vernomimicry approach. Thus, in this paper, the vernacular builders' strategies are analysed and the vernacular adaptation solutions in the face of hazards such as floods, extreme precipitation, heatwaves, storms, and drought, at both urban and building scale are investigated. On this basis, five core design principles were identified and are encouraged to be included in modern adaptation designs, namely: simplicity and adaptability in construction, transient structures and local materials, resource efficient and site specific- design, community involvement in building practices, and holistic integration with the urban context.

Chapter 6: Navigating Nearly Zero-Energy Strategies for Urban Climate Change Adaptation and Mitigation

By: Osama Omar

Climate change stands out as one of the most pressing challenges of the twenty-first century, exacerbating resource consumption, environmental degradation, human health risks, and occupant discomfort. In response, nearly zero-energy strategies emerge as pivotal solutions for curbing energy consumption and environmental impact in both existing and new constructions. The built environment significantly contributes to greenhouse gas emissions and resource depletion, with buildings consuming 32% of global energy and emitting 19% of greenhouse gas emissions, according to the International Energy Agency. This underscores the urgency of enhancing energy efficiency and reducing carbon emissions in buildings worldwide.

Zero-carbon strategies are indispensable for mitigating emissions during building operation. This chapter delves into the application of nearly zero-energy strategies in the housing sector to combat climate change, employing a literature review approach. Nearly zero-energy buildings, envisioned as structures powered by clean energy to fulfill their energy needs, offer a promising yet ambiguously defined solution. Global discrepancies in defining "clean energy" have led to confusion and hindered the integration of zero-energy building goals into building codes. Despite challenges, the concept of nearly zero-energy buildings catalyzes efforts to reduce energy demand and fossil fuel consumption worldwide, aligning with renewable energy objectives. This study proposes a range of nearly zero-energy strategies applicable to the housing sector worldwide to mitigate and adapt to climate change. Emphasizing resource efficiency and environmental responsibility throughout the building's life cycle is critical in realizing effective nearly zero-energy building designs. By adopting these strategies, the housing sector can make significant strides towards sustainability and climate resilience on a global scale.

Chapter 7: Enhancing Climate Resilience Through Urban Landscape Design: Strategies for Green Infrastructure and Adaptation

By: Reza Farhadi, Hadi Soltanifard, and Behdad Alizadeh

Climate change presents urgent global challenges that demand immediate action to mitigate its impacts. Urban landscape design has emerged as a promising approach to addressing climate change and its consequences. This chapter focuses on the relationship between urban landscape planning and green infrastructure design to promote climate resilience and adaptation. Urban Green Infrastructure (UGI) planning is a strategic approach to developing multifunctional networks of blue and green areas that can deliver various environmental, social, and economic advantages while improving cities' climate resilience. Community participation in urban planning, fostering social cohesion, and creating inclusive spaces are emphasized as key pillars that can withstand climate change impacts. Practical initiatives, such as community gardens and neighborhood networks, are explored as effective means of enhancing community resilience. It is necessary to argue how green infrastructure planning can be applied to integrate planning strategies into community-adaptive planning. This is to cope with urban climate change and promote local resilience. To address climate change effectively, it is essential to understand the variety of types of green infrastructure, their implications, and ecological design. Adopting the 'traditional' grey infrastructure or technologically-based solutions is preferable. The planning process includes a number of structured adaptive solutions derived from ecological planning and climate change adaptation in order to facilitate multiple planning responses at the same time. This will enhance planning flexibility in practice. The challenges and prospects of urban landscape design in the context of climate change adaptation. In order to fully realize the potential of urban landscape design in addressing climate change, it emphasizes the importance of interdisciplinary collaborations, policy support, and financial investment. An emphasis is placed on nature-based solutions and community involvement when designing resilient and sustainable cities. It is a valuable contribution for scholars, urban designers, professionals, and policy-makers interested in climate change adaptation and urban planners. As a final point, we discuss adaptation strategies in urban landscape design.

Chapter 8: Prioritizing Energy Efficiency and Low-Carbon Approaches: Climate Adaptation Strategies in Urban West Asia

By: Akram Ahmed Noman Alabsi

The rising demand for energy and the accompanying economic pressures, combined with the negative effects of greenhouse gas emissions and the resulting climate change, necessitate immediate action to mitigate future consequences. As a result of rapid urbanization fueled by population growth and economic development, many West Asian nations have neglected environmental considerations and sustainable strategies. As a consequence, complex environmental challenges have emerged, and the region faces significant risks from climate change, which threatens ecosystems, economies, and infrastructure dependent on natural resources. This research aims to develop standards and strategies for climate adaptation in West Asian nations that prioritize energy efficiency and a low-carbon approach. Examining the influence of region-specific climatic factors and exploring traditional local technologies used for climate effect control, as well as evaluating their compatibility with modern approaches, are essential components. The research methodology employed here focuses on learning from the past to improve energy utilization practices in modern architecture. In addition, this study provides a comprehensive review of prior research that examined the development and improvement of techniques and strategies to improve the thermal performance of buildings, thereby advancing energy efficiency. The research outcomes address the existing gaps in climate adaptation standards and guidelines for thermal comfort in West Asia, demonstrating the potential for significant energy savings in future housing compared to existing homes. The projected energy savings for the five climate zones of West Asia are approximately 80% during winter and 60% during summer. In addition, the study presents green design strategies and energy-efficient methods tailored specifically for West Asian arid regions with high temperatures. However, additional research is required to simulate the energy performance of each technology in order to fully comprehend their energy efficiency potential.

PART III: Advanced and Emerging Design Models and Tools

In the third part, we focus more on 'design methods', in order to close the gap of incorporating mitigation and adaptation solutions into planning and design practices. The development of advanced and emerging design models and tools is presented in terms of integrating historical wisdom into modern strategies, enabling transformative capacity via urban design climate workshop, exploring integrated approach to pinpointing the urban design needs and strategies, and understanding the co-benefits of urban green and blue spaces. Below is the summary of the seven chapters included in this part.

Chapter 9: Cooling Pakistan: Exploring Historical Wisdom and Modern Strategies for Passive Cooling in Pakistan

By: Shanzae Naeem, Aisha Aziz, and Baojie He

The urban population is increasing rapidly in Pakistan especially in large cities as people are migrating from Rural to Urban areas for better life opportunities. This phenomenon has far-reaching effects on the region's climate. Lahore being the second largest city in the country, has undergone a massive increase in its urban population; and the urban population of the world will increase almost 75% by 2030 due to urbanization. This increase in population is resulting in increased resource consumption within the urban context, which in turn affects the urban climate, giving rise to phenomena such as the UHI (urban heat island) effect. The climate of Lahore is semi-arid therefore, the problems mentioned above increase the consumption of energy for cooling purposes in buildings. In this chapter, we will discuss the operational energy increase with the expansion of urbanization and its negative effects on the urban climate. The increase in air conditioning due to hot weather in Lahore has caused a massive surge of almost 60% in electricity consumption. These energies release GHG Emissions which play a key role in global warming and climate change. Moreover, Carbon dioxide (CO_2) is the most significant among GHGs, contributing more than 70% of atmospheric concentrations. So, it is vital to control the production of GHG to take a step towards net zero carbon emissions as Pakistan is one of the signatories of the 2015 Paris Agreement. This problem however can be solved by adopting Passive evaporative cooling techniques. Introduced to this part of the region by the Mughals, this array of different strategies helped Mughal-era designers create comfortable indoor living conditions. It will also be beneficial for the sustainability of the urban climate of Lahore and if opted in contemporary architectural designs, can help reduce electricity consumption resulting in controlling carbon emissions.

Chapter 10: The Urban Design Climate Workshop: Integrating Mitigation and Adaptation Strategies to Confront Urban Climate Change

By: Jeffrey Raven, Mattia Leone, and Giovanni Nocerino

21st century urban design must configure densely occupied urban settlements that offset undesirable local outcomes through climate-sensitive urban design. To bridge between climate science and climate action, policymakers/stakeholders need tools and methods to identify, configure and evaluate evidence-based Urban Climate Factors at urban and local scales. The goal is to configure densely occupied urban settlements that offset undesirable local outcomes through climate-sensitive strategies. The challenge is configuring these districts to reduce the impact of increased urban heat island (UHI) and flooding due to the changing climate while enhancing an equitable, high-quality, low-carbon lifestyle. Enabling transformative climate action in cities requires expanding on the traditional influence and capabilities of urban planning and urban design. In response, we have developed a climate-driven planning process called the Urban Design Climate Workshop whose urban climate factors overlap with stakeholder-driven performance indicators.

Chapter 11: Understanding Urban Heat Islands in Tunisian City Centres: An Integrated Approach

By: Messaoudi Abir

A comprehensive study of urban climate is necessary to better understand the presence of heat islands in each part of the city centres of Sousse, Tunis, and Sfax. Temperature has a direct impact on energy consumption and the well-being of residents. However, despite the rapid and continuous urbanization of cities, few studies have specifically focused on the effects of urban heat islands (UHI). In this study, we aim to analyse the urban heat island effect in the city centres of Sousse, Tunis, and Sfax by examining the relationships between land surface temperature (LST) and spatial characteristics, composition, and configuration of impervious and natural spaces. We investigate how intensive urbanization, evapotranspiration (IDWI), urban morphology, and the radiativity of building materials influence the formation of urban heat islands in these specific contexts. To conduct this in-depth analysis, we will utilize geospatial data and various approaches, including statistics, urban landscape metrics, and spatial analysis techniques. These methods will help us understand how the characteristics of the urban landscape and urban development choices contribute to the formation of urban heat islands. Several indices will be considered, including the Normalized Difference Vegetation Index (NDVI), the Normalized Difference Built-up Index (NDBI), and land surface temperature (LST). Correlation analyses will enable us to quantify the strength and direction of the relationships between different indices and urban climate parameters. For example, we will assess whether there is a positive correlation between the density of impervious spaces (measured by the NDBI) and land surface temperature. Similarly, we will examine the correlation between the NDVI and land surface temperature to evaluate the mitigating role of green spaces in reducing urban heat islands. By employing these statistical methods, we will obtain precise and quantitative results regarding the relationships between indices and urban climate parameters. These findings will help us understand the mechanisms of urban heat islands in the city centres of Sousse, Tunis, and Sfax. Key questions to be addressed include the extent and intensity of urban heat islands in these city centres, the correlation between impervious space density and land surface temperature, and the role of green spaces in mitigating urban heat islands. This information will be essential for guiding urban planning and development decisions, aiming to create more resilient and sustainable urban environments. We will be able to identify the most vulnerable areas to urban heat islands and propose targeted urban design and greening strategies to mitigate these adverse effects. By understanding the complex mechanisms of urban heat islands in Tunisian city centers, we can create more pleasant, energy-efficient, and climate-resilient urban environments. This will contribute to improving the well-being of residents and creating sustainable cities for future generations.

Chapter 12: The Rehabilitation of Riverbanks in Beijing's Urban Area in the Light of an Ambitious Policy to Enhance Nature and Quality of Life for Residents

By: Stéphanie Boufflet, and Vassilia Pirierros

Faced with the urgency of the climate issue, awareness has grown and substantial progress has been made in recent years in the design of green and blue spaces in the megalopolis of Beijing. This trend intended to be in favour of biodiversity and the well-being of city residents continues as a further milestone in a long-term strategy for balanced development between man and the living world in the city. We contextualise our research with regard to the place of nature in Chinese ancestral culture and the green and functionalist policies implemented since the 2000s. The 2016–2035 Urban Master Plan is an extension of this, with more ambitious ecological and humanist objectives that we see as an echo of COP 15. Three field studies of recently developed riverbanks in this framework, all located in the densely populated built area of Beijing, allow us to suggest that China has moved towards a contextualization of landscape project, even if attention to the living world is still mainly ornamental. Nevertheless, these developments have enabled residents to appropriate these spaces in their own way. Three years of restricted mobility as part of the global pandemic have undoubtedly had a hand in these new forms of expression in public spaces. They reflect new links between citizens and nature in the urban area, which we believe will be decisive in the coming decades.

PART IV: Action Plans and Policy Formulation

In the fourth part, we underline the importance of adopting action plans, schemes, policies, and regulations in enhancing the implementation capacity of climate-sensitive urban planning and design. This part presents the actions in terms of urban climate change policy formulation, community engagement and participatory design, policies for public green spaces, and delivery of climate-adaptive urban planning and design in territory education. Below is the summary of the four chapters included in this part.

Chapter 13: Approaches to Urban Climate Change Policymaking: Challenges and Way Forward

By: Kirti Kusum Joshi, and Ajay Chandra Lal

The world today faces the intertwined challenges of rapid urbanization and climate change. Urban areas, vulnerable to climate impacts, also exacerbate climate change through greenhouse gas emissions. Therefore, it is crucial to address climate change as an urban issue and formulate appropriate policies. Drawing on Nepal's experience as a climate-vulnerable, rapidly urbanizing country, this chapter compares two primary approaches to urban climate change policymaking. The first approach integrates urban concerns into the broader climate change discourse, emphasizing understanding climate change's implications on urban areas and identifying urban challenges within that context. While this approach provides access to global knowledge and international climate funds, there is a risk of overlooking urban nuances amidst broader climate goals. Conversely, the second approach embeds climate concerns within urban policymaking, integrating climate change strategies into urban policies. This approach acknowledges the distinct characteristics of urban areas, allowing targeted interventions. But there is a risk of compartmentalizing climate change as a secondary concern within urban planning, undermining broader climate change efforts. This chapter emphasizes the necessity of an integrated approach that balances urban concerns with climate goals. This entails incorporating climate considerations into urban planning while aligning urban policies with broader climate change efforts.

Chapter 14: Building Climate-Resilient Cities Through Community Engagement and Participatory Design

By: Amir Tayyebi

This chapter investigates community engagement and participatory design as pivotal in forging climate-resilient urban strategies. These approaches involve local communities in the design process, fostering ownership, knowledge enhancement, and social cohesion for sustainable urban resilience. Key concepts of community engagement and participatory design are examined, emphasizing their significance in climate resilience. The chapter outlines integration methods into urban design, such as workshops and collaborative platforms, and the empowerment of communities to input into climate-resilient spaces. Worldwide case studies illustrate the benefits of these approaches, including heightened climate risk awareness, vulnerability identification, social equity, and tailored solutions. Challenges like inclusivity, interest conflicts, and technical-community balance are also discussed. The chapter concludes with guiding principles for practitioners and policymakers to enhance community engagement and participatory design effectiveness, advocating for trust, capacity building, perspective integration, and enduring community-designer-policy partnerships. Embracing these methods, cities can leverage collective wisdom for more sustainable, inclusive, and resilient urban environments against climate change.

Chapter 15: Analysing Policies for Public Green Spaces of Indian Cities

By: Piyush Pandya, Gagan Jain, and Shweta Prasad

As urbanization rapidly increases, cities around the world are recognizing the critical role of public urban green spaces (PUGS) in enhancing environmental quality and residents' well-being. Recent studies have shown that green spaces contribute to the economy by increasing individual productivity, attracting investments, mitigating urban heat islands, and making cities climate resilient—thereby making PUGS essential for urban sustainability. In India, however, urban areas face significant challenges in integrating and maintaining these green spaces due to fragmented governance and inadequate policy frameworks. As urban areas fall under the State List in the Indian Constitution's Seventh Schedule, they are primarily governed by Urban Local Bodies (ULBs). This diminishes the role of state and central governments in advisory positions, creating a lack of accountability, and city managers often lack the capacity to understand complex issues. Despite numerous state and central programs, schemes, and policies for PUGS, the absence of logic-based goals with measurable milestones remains a significant challenge. This chapter highlights these challenges within central and state-level policies and significant city-level initiatives. It critically examines Indian urban policies and governance systems at both the union and state levels concerning public urban green spaces (PUGS). The analysis includes a study of best practices from cities in India and abroad. Drawing on global best practices, the findings are supplemented with suggestions for managing urban green spaces in India, considering cultural and behavioural responses, successful financial models, and recent technologies. The study relies on literature surveys and policy reviews, which contribute directly to the development of PUGS in India.

Chapter 16: Teaching Practice: Analyzing and Transmitting the Wisdom of Climate-Adaptive Construction of Traditional Heritage Settlements

By: Xu Li, Pengcheng Liu, and Baojie He

Climate change and increasingly prominent energy demands exert significant constraints to urban sustainable development. The energy consumption of buildings is determined by both individual buildings and urban morphology. Therefore, it is imperative to improve environmental quality through urban design. To support this, this study aims to report how the education shapes theoretical and practical components in design courses by the teaching practice of the environmental performance a historic Ciqikou district in Chongqing, China. By analyzing the construction wisdom inherent in local climate adaptation, this study is expected to provide theoretical guidance for climate-responsive urban design practices. The teaching practice includes two parts: theory part and design part. The theory part consists of the interpretation of spatial characteristics and principles of multi-scale urban form adaptation to climate, and the analysis of research progress of relevant topics all over the world. In the design part, students were expected to analyse the wisdom of traditional settlements adapted to local climate at different scales, and explore the inheritance of the traditional settlements. Through environmental performance simulation and multi-objective optimization based on genetic algorithm, the students explored the spatial patterns and their combinations adapted to the regional wind and thermal environmental conditions; investigated the behavioural patterns of people adapting to the microclimate; and summarized the wisdom of building adapted to the local climate. Overall, the teaching practice is based on the combination of green design and cultural heritage, aiming to cultivate students' ability to comprehensively use the relevant knowledge of meteorology, environmental science, and planning to analyze the climate adaptive characteristics of settlements and apply them to design. After the course, students are expected to initially master the characteristics of climate-adaptive urban form, responsive environmental performance assessment software, the application of climate-adaptive urban design theories and methods, and innovative solutions from different perspectives.

References

1. Afroz M, Chen G, Anandhi A (2023) Drought-and heatwave-associated compound extremes: a review of hotspots, variables, parameters, drivers, impacts, and analysis frameworks. Front Earth Sci 10:914437

2. Cui Y, Yin M, Cheng X, Tang J, He BJ (2024) Towards cool cities and communities: Preparing for an increasingly hot future by the development of heat-resilient infrastructure and urban heat management plan. Environ Technol Innov 34:103568
3. Eliason M (2024) Building for people: designing livable, affordable, low-carbon communities. Island Press
4. Han D, An H, Cai H, Wang F, Xu X, Qiao Z, ... An Y (2023) How do 2D/3D urban landscapes impact diurnal land surface temperature: Insights from block scale and machine learning algorithms. Sustain Cities Soc 99:104933
5. He BJ, Wang J, Liu H, Ulpiani G (2021) Localized synergies between heat waves and urban heat islands: implications on human thermal comfort and urban heat management. Environ Res 193:110584
6. He BJ, Wang J, Zhu J, Qi J (2022) Beating the urban heat: situation, background, impacts and the way forward in China. Renew Sustain Energy Rev 161:112350
7. He BJ, Wang W, Sharifi A, Liu X (2023) Progress, knowledge gap and future directions of urban heat mitigation and adaptation research through a bibliometric review of history and evolution. Energy Buildings 287:112976
8. Jain AK (2023) Climate resilient, green and low carbon built environment. Springer Nature
9. Kenny GP, Tetzlaff EJ, Journeay WS, Henderson SB, O'Connor FK (2024) Indoor overheating: a review of vulnerabilities, causes, and strategies to prevent adverse human health outcomes during extreme heat events. Temperature 11(3):203–246
10. King DA (2004) Climate change science: adapt, mitigate, or ignore? Science 303(5655):176–177
11. Kumar P, Sahani J, Rawat N, Debele S, Tiwari A, Emygdio APM, ... Pfautsch S (2023) Using empirical science education in schools to improve climate change literacy. Renew Sustain Energy Rev 178:113232
12. Oke TR, Mills G, Christen A, Voogt JA (2017) Urban climates. Cambridge University Press
13. Peng Y, Gao Z, Ding W (2017) An approach on the correlation between urban morphological parameters and ventilation performance. Energy Procedia 142:2884–2891
14. Qi J, Ding L, Lim S (2022) A decision-making framework to support urban heat mitigation by local governments. Resour Conserv Recycl 184:106420
15. Raihan A (2023) A review of the global climate change impacts, adaptation strategies, and mitigation options in the socio-economic and environmental sectors. J Environ Sci Econ 2(3):36–58
16. Roth M, Lim VH (2017) Evaluation of canopy-layer air and mean radiant temperature simulations by a microclimate model over a tropical residential neighbourhood. Build Environ 112:177–189
17. Santamouris M (2020) Recent progress on urban overheating and heat island research. Integrated assessment of the energy, environmental, vulnerability and health impact. Synergies with the global climate change. Energy Build 207:109482
18. Thomson G, Newton P, Newman P, Byrne J (2019) Guide to low carbon precincts. CRC Low Carbon Living
19. Tsoulou I, He R, Senick J, Mainelis G, Andrews CJ (2023) Monitoring summertime indoor overheating and pollutant risks and natural ventilation patterns of seniors in public housing. Indoor Built Environ 32(5):992–1019
20. Yu S, Chen Z, Yu B, Wang L, Wu B, Wu J, Zhao F (2020) Exploring the relationship between 2D/3D landscape pattern and land surface temperature based on explainable eXtreme Gradient Boosting tree: a case study of Shanghai, China. Sci Total Environ 725:138229

Urban Climate Monitoring, Assessing and Forecasting

Monitoring, Assessing, and Forecasting Urban Climate Issues and Challenges

Tengku Nilam Baizura Tengku Ibrahim, Nur Azalina Suzianti Feisal, Wai Yan Cheah, Nurzawani Md Sofwan, Nur Faseeha Suhaimi, Nursalfarina Abdul Samat, and Noor Haziqah Kamaludin

Abstract The urgency of climate change demands the development of sustainable urban systems. Although advanced modeling holds significant potential for designing such systems, current tools have two primary limitations: lack of user-friendliness, which hampers accessibility for a broad range of stakeholders, and they struggle to integrate up-to-date climate data, especially regarding extreme weather events. Accurate forecasting is essential for protecting life and property through climate adaptation, mitigation actions, and disaster risk reduction strategies. Dynamic approaches that view cities as evolving systems are crucial for enabling humanity to adapt, survive, and thrive amidst volatile shocks and stresses. This chapter identifies a

T. N. B. Tengku Ibrahim · N. Md Sofwan
Faculty of Health Sciences, Universiti Teknologi MARA, Cawangan Pulau Pinang, Kampus Bertam, 13200 Kepala Batas, Pulau Pinang, Malaysia

N. A. S. Feisal
Department of Diagnostic and Allied Health Science, Faculty of Health and Life Sciences, Management and Science University, 40150 Shah Alam, Selangor, Malaysia

W. Y. Cheah
Centre of Research in Development, Social and Environment (SEEDS), Faculty of Social Sciences and Humanities, Universiti Kebangsaan Malaysia, 43600 UKM Bangi, Selangor Darul Ehsan, Malaysia

N. Md Sofwan
Environmental Health Program, Faculty of Health Sciences, Universiti Teknologi MARA, Cawangan Sarawak, Kampus Samarahan, 94300 Kota Samarahan, Sarawak, Malaysia

N. F. Suhaimi
Department of Environmental and Occupational Health, Faculty of Medicine and Health Sciences, Universiti Putra Malaysia, 43400 UPM Serdang, Selangor, Malaysia

N. Abdul Samat
Department of Environmental Health, Faculty of Health Sciences, MAHSA University, Bandar Saujana Putra, 42610 Jenjarom, Selangor, Malaysia

N. H. Kamaludin (✉)
Occupational Health and Safety Risk Management (OHSeRM) Research Initiative Group and Center of Environmental Health and Safety, Faculty of Health Sciences, Universiti Teknologi MARA, 42300 Puncak Alam, Selangor, Malaysia
e-mail: noorhaziqah@uitm.edu.my

B. He et al. (eds.), *Urban Climate and Urban Design*, Urban Sustainability,
https://doi.org/10.1007/978-981-96-1521-6_2

significant research gap in creating user-centric urban climate modeling tools that seamlessly incorporate the latest urban climate, state-of-the-art for the monitoring and assessing urban climate issues, data integration, sensing, and visualization. The chapter also discusses challenges and provides recommendations for addressing urban climate issues, aiming to formulate effective climate adaptation and mitigation strategies, ultimately benefiting human health and societal well-being.

Keywords Climate change · Urban climate · Extreme weather · Urban ecology · Urban health risks

1 Introduction

Large cities experience the urban heat island effect, where central areas are significantly warmer than surrounding suburbs. Tropical and arid cities face a brutal intensification of the urban heat island effect during summer [14]. This translates to scorching temperatures in the city center, causing significant discomfort for residents. The extreme heat poses a particular health risk for vulnerable populations, leading to heat stress, illness, and even fatalities. The ups and downs of a city's climate, from place to place and season to season, play a big role in how healthy people are and how well buildings and roads hold up. Creating a dense sensor network becomes crucial with urban climate growing in everything from energy-saving buildings to heatwave warnings. This network would allow us to continuously monitor the fine-grained variations in climate across the city, both in space and over time.

Urban areas around the globe are now facing unparalleled problems as a result of climate change, which presents substantial risks to the welfare of their inhabitants [35]. The need for adopting a proactive and all-encompassing methodology in comprehending, monitoring, and predicting urban climate concerns has reached an unprecedented level of importance because climate change is rapidly intensifying, leading to more frequent and severe extreme weather events like heatwaves, floods, and droughts [69]. These events have a significant impact on urban areas, causing damage to infrastructure, disrupting essential services, and threatening public health [42]. This review discusses the complex array of difficulties that metropolitan regions encounter within the framework of climate change and seeks to investigate effective approaches for monitoring, evaluating, and predicting these concerns. As the trend of global urbanization continues, the effects of climate change on densely populated metropolitan areas are becoming more evident [60]. The interplay between urbanization and climate change engenders a range of concerns, including elevated temperatures [84] and atmospheric contamination [107], as well as the occurrence of severe weather phenomena [23], water shortages [34], and the subsequent ramifications for public health [38]. In the present setting, it is crucial to use efficient methodologies [8] for monitoring, evaluating, and predicting urban climate concerns [104]. These techniques are essential for making well-informed decisions and fostering the creation of resilient urban landscapes [66]. The chapter aims to explore the intricacies of direct and indirect monitoring and evaluation approaches, elucidating their efficacy in capturing the intricacies of urban climate concerns. The study will further examine significant urban climate problems, such as the substantial urban heat island effect, the complex interplay between climate and air pollution, the heightened occurrence of severe weather events, and the urgent challenges surrounding water pollution and shortages.

2 Key Urban Climate Issues and Challenges

Urbanization is accelerating considerably more quickly in developing countries due to rural–urban migration, technological change, and rapid population growth. By 2050, it's anticipated that two-thirds of the world's population will reside in cities. The Sustainable Development Goals (SDGs) and the 2030 Agenda can both be hampered by the effects of the climate crisis. Urbanization has significant climatological and meteorological effects on regional synoptic features which are associated with various issues and challenges [86]. The detrimental effects of cross-cutting urban climate issues make them a key challenge in achieving sustainable cities. Recent concerns about urban climate issues and challenges have pushed for the urban viewpoint on sustainable development to make cities and human settlements safe, resilient, and sustainable. The consequences of urban warning extend beyond the urban ecology, energy consumption, social stability, and environmental, and health impacts. Table 1 provides a comprehensive overview of the key issues and challenges cities face due to climate change. It explores the negative impacts on human health and the environment, outlining methods for measuring these effects.

2.1 Urban Heat Island Effect

Urban heat island (UHI) is a phenomenon that occurs when air and surface temperatures in urban areas are significantly higher than those in nearby suburban areas [98]. UHI can be categorized into two major types: the canopy urban heat island (CUHI) and the surface urban heat island (SUHI). Numerous studies have investigated the connection between UHI and climate change [50]. The impact of climate change on the urban microclimate depends on various factors including the features of the urban canopy, land use patterns, and local climate [105]. Chen et al. [15] have developed a technique to assess the canopy UHI intensity at high resolution using remote sensing and machine learning models. The results show that the combined effects of built-up areas, heatwaves, and human activities can intensify the canopy UHI intensity, while the vegetation and water bodies give a significant cooling impact [109]. Heatwaves (HW) are prolonged episodes of high heat that can exacerbate the UHI intensity by altering sensible and latent heat fluxes and influencing wind speed [58]. Numerous studies have demonstrated a beneficial synergistic relationship between UHI and

Table 1 Overview of urban climate issues, challenges, health and environmental impacts, and measurement indicators

Urban climate issues	Challenges	Factors	Health impacts	Effects on environment	Measurement indicators	References
Urbanization and climate change	Accelerated urbanization, impact on SDGs and 2030 Agenda	Rapid population growth, rural–urban migration, technological change	Increased vulnerability to climate hazards, social instability and urban infrastructure strain	Habitat loss, increased carbon footprint, resource overexploitation	Population density, urbanization rate, infrastructure strain	Shen et al. [86], United Nations (2019)
Urban heat island effect	Higher urban temperatures, synergy with heatwaves	Urban canopy features, land use patterns, local climate variations	Increased thermal stress, higher morbidity and mortality, exacerbated heat-related illnesses	Increased energy consumption, reduced air quality, altered local weather	Temperature differentials, heatwave frequency, energy consumption	Voogt and Oke [98], Zhao et al. [109], Founda and Santamouris [26], Jiang et al. [43]
Climatic impacts on air pollution	Pollutant formation and dispersion, haze from biomass burning and fossil fuels	Temperature inversion, sea breeze, rainfall and humidity, wind patterns	Respiratory and cardiovascular diseases, Increased asthma and allergies, health impacts of transboundary haze	Decreased biodiversity, acid rain, soil and water contamination	Air quality index (AQI), pollutant levels, respiratory disease incidence	Hassan et al. [33], Hanaoka and Masui [32], Latif et al. [55], Largeron and Staquet [54], Niu et al. [77]

(continued)

Table 1 (continued)

Urban climate issues	Challenges	Factors	Health impacts	Effects on environment	Measurement indicators	References
Extreme weather events	Increased heatwaves, floods and windstorms, Impact on cyclone distribution, poor urban planning in hazard-prone areas	Greenhouse gas emissions, global temperature rise, urban development in hazard-prone areas	Mental health issues, physical injuries and fatalities, indoor overheating during heatwaves	Ecosystem disruption, coastal erosion, loss of flora and fauna	Frequency and severity of weather events, damage assessment, evacuation rates	Jongman [44], Murakami et al. [73], Wan et al. [100], Arriazu-Ramos et al. [4]
Water pollution and scarcity	Urban sprawl and agricultural runoff contribute to pollution, including nitrate contamination in groundwater	Industrial discharges, agricultural runoff, and inadequate sewage treatment	Health problems, waterborne diseases and malnutrition	Degraded water quality lead to aquatic life disruption and reduced freshwater availability	Water quality index (WQI), nitrate levels; access to clean water	Musacchio et al. [74], Ward et al. [102], Rafiei-Sardooi et al. [81]

(continued)

Table 1 (continued)

Urban climate issues	Challenges	Factors	Health impacts	Effects on environment	Measurement indicators	References
Impact on human health	Climate hazards linked to mental illnesses, Increased vector-borne diseases, impact on children's health and pregnancy	Changes in precipitation and temperature patterns, high ambient temperatures, climate-induced stress	Mental illnesses, children's health problems, pregnancy complications	Disrupted ecosystems, altered disease vector patterns, impact on food security	Incidence of climate-related illnesses, disease prevalence and health service utilization rates	Contreras et al. [17], Kim et al. [49], da Silva et al. [89], Lamy et al. [53], Anderko et al. [2], Carolan-Olah and Frankowska [12]
Extreme precipitation and flood	Warming air temperatures are associated with an increased capacity to hold moisture, leading to rise in the intensity and frequency of precipitation events, including floods	Flood vulnerability within a region is a complex phenomenon influenced by local topography, the age and condition of infrastructure, the health of the surrounding watershed and historical and projected precipitation trends	Extreme precipitation and flooding events are associated with an elevated risk of respiratory illnesses due to both direct and indirect pathways	A potential consequence of climate change is the intensification of extreme precipitation events and subsequent floods, particularly within the context of the seasonal cycle of water availability	Heavy precipitation events are quantified through frequency analysis, return period estimation (average time between events of equal or greater intensity), or direct precipitation measurement (e.g., total rainfall in a 24-h period)	Bolan et al. [9], Munyai et al. [72], Aune et al. [6]

HW [26]. For instance, the UHI intensity could be 3 times or more intense during HW in Szeged, Hungary [96]. Ao et al. [3] discovered that the HW raised the UHI intensity in Shanghai by around 1 °C. People who live in urban areas are particularly affected during heatwave outbreaks because UHI exacerbates thermal stress [24]. In contrast, the heatwave event in 2018 in the Ontario-Quebec area reveals that there were no substantial changes in the overall UHI, but there were considerable diurnal variations [43]. The investigation observed that cloud formations, precipitation, and wind speed may have a significant impact on UHI [87].

2.2 Climatic Impacts on Air Pollution

Climate-related variables affect the formation, dispersion, transportation, and deposition of pollutants [33]. The concentration of air pollutants is influenced by climatic factors such as temperature inversion [54], sea breeze [70], rainfall, relative humidity [20], and wind [33]. Haze, an atmospheric phenomenon caused by biomass burning and fossil fuel combustion, releases high levels of air pollutants [32]. Haze is formed due to high levels of particles and gas-to-particle conversion, where it occurs in stable weather conditions such as low wind speed and inverted thermodynamic structure [77]. Haze from biomass burning poses a serious threat to human health and economic impact, as well as has the potential to contribute to global warming and climate change due to the emission of large amounts of greenhouse gases [37]. Smoke from forest fires in neighboring countries was transported to Malaysia, causing transboundary haze [55]. Air pollution is the largest environmental health threat, with the greatest burden being found in Southeast Asia and the Western Pacific Regions [103]. Urbanization brings a huge demand for road infrastructure growth which leads to an increase in motor vehicle ownership to meet long-distance traveling needs (Poi et al. 2021). Heavy reliance on petroleum fuels by road transport has resulted in a significant increase in exhaust emissions, contributing to air quality degradation in cities [90].

2.3 Extreme Weather Events

An increase in the average global temperature has led to other changes throughout the atmosphere, on the land, and in the ocean. Extreme weather events such as heat occurrences, widespread flooding, or windstorms are becoming more frequent due to climate change [44]. Murakami et al. [73] detected a climatic change in the global spatial distribution of typical cyclones, which may be partially attributed to the rise in greenhouse gas emissions. A long-term study on the drought characteristics across China concluded that potential evapotranspiration (PET) is the dominant climatic factor for meteorological drought. PET is the total amount of water lost by a plant's process of transpiration and evaporation from the earth's surface, which are both influenced by temperature, humidity, sunlight, and wind [100]. In Pamplona, North Spain, an energy simulation study on the impacts of an extremely warm summer with heatwaves and typical meteorological conditions discovered that indoor overheating hours (IOH) are intensified during extremely warm summers. The influence of microclimate increases the IOH by an average of 7.5% during extremely warm summers, while there is no increase in IOH during a typical meteorological year. The highest IOH was observed in older neighborhoods with less green areas, densely packed higher structures, and urbanization [4]. Poor urban development planning near hazard-prone areas increases the risks of extreme events to the population. Hence, climate risks should be considered in the planning for sustainable urban development.

2.4 Extreme Precipitation and Flood

Communities all around the world encounter significant challenges as caused by the rising incidences of extreme precipitation and flooding over the past decade. Climate change is a major contributor to extreme weather events, with global temperatures continuing to rise and causing higher rainfall events. Increased precipitation is expected as a result of the global hydrological cycle becoming more intense due to a rising climate [21]. Extreme precipitation will become more prevalent due to global warming in the majority of the world because of a decrease in the environment's ability to hold more water, as predicted by the Clausius-Clapeyron connection [94]. Conversely, modifications in atmospheric dynamics could lead to regional reduction in precipitation extremes [79]. Particularly, flood consequences approximately 1 billion individuals globally, more than any other natural catastrophe [75]. An estimated 81 billion people who are directly at risk of a flood in 100 years [75]. Flooding could result in devastating impacts on the economic system since so many of the largest cities around the globe and agricultural regions are situated near coastal areas and floodplains [88]. The incidence and severity of flooding episodes are on increasing globally, causing tremendous costs to communities. Since 1980, flooding and tropical cyclones have been attributed to approximately fifty percent of the US's total losses of more than two trillion dollars [88].

Flooding, however, has far-reaching consequences beyond economic damage. Communities may experience long-lasting repercussions from major social disruptions brought on by flooding, such as fatalities, serious injuries, homelessness, and displacement. These effects are particularly severe in cities, where dense populations and critical infrastructure are frequently located in floodplains. Early warning systems have been developed to help communities in preparing for and responding to natural disasters, as a result of increasing risks of flooding caused by climate change and population growth in vulnerable areas. A valuable source of information to direct emergency response efforts and lessen the worst effects of flooding is provided by high-resolution impact-based early warning systems that make use of hydrological models and satellite data [75]. Communities have to make investments in comprehensive flood management plans that integrate long-term resilience planning, early warning systems, and infrastructure upgrades in order to protect vulnerable populations as extreme precipitation and flooding events become more frequent and intense.

2.5 Water Pollution and Scarcity

Urban sprawl accelerates the vulnerability of water resources due to increasing demands of water consumption and water pollution caused by human activity puts additional strain on the availability of clean water. Groundwater-fed has become an increasingly important source of irrigation because it can sustain the intense summer droughts brought on by climate change. Nitrate contamination in groundwater because of intensive fertilizer application has gained much attention. Nitrate pollution in the Po Plain of Northern Italy for example, has been increasingly impacted by extreme events due to climate change, thus escalating tensions between several water users [74]. Groundwater-fed irrigation generates higher nitrate due to basin-scale groundwater recirculation and low dilution effect. Long-term accumulation of nitrogen in groundwater may result in environmental problems and health implications, including increasing the risk of colorectal cancer, thyroid illness, and neural tube anomalies [102]. Water scarcity refers to a shortage of water resources and the inability to meet water demands for human subsistence. It has emerged as a significant global problem that could have detrimental effects on the environment and humans. Assessment of the water scarcity in an arid basin in Iran indicated that climate change has a greater impact on water yield and water supply than land use change. Such changes have significant impacts on the hydrologic balance at the regional scale, precipitation, evapotranspiration, soil moisture, and infiltration [81]. Incorporating climate adaptation into planning and practices protects people and the environment.

2.6 Impact on Human Health

The increasing frequency of climate hazard occurrences today has increased the evidence of climate change on human health [17]. Studies in Brazil and Peru presented findings on the relationship between climate change hazards, such as ambient temperature, floods, and drought with mental illnesses [49]. Common mental illnesses studied are trauma, depression, anxiety, and domestic abuse [89]. Rainfall and temperature have influenced the distributions of *Aedes albopictus*. Therefore, changes in precipitation and temperature patterns due to climate change are also responsible for increasing and spreading dengue cases globally. High ambient temperatures promote rapid vector development, increase the frequency of blood meals, and shorten the extrinsic incubation period, hence augmenting the potential for transmission [53]. A systematic review by Anderko et al. [2] found a relationship between climate change and a wide range of health problems in children, including asthma, allergies, malnutrition, low birth weight, and post-traumatic stress disorder. Exposure to high ambient temperatures also increases the risk of pregnancy and preterm births. Sudden temperature changes cause the body unable to adapt quickly [12]. It is crucial to monitor the possible relationship between climate and human health because it may lead to a wide range of environmental health risks that will regularly impact vulnerable populations and may require significant changes to the structures of the healthcare system as one strategy of climate adaptation.

3 Monitoring and Assessing Urban Climate Issues

The primary drivers of climate change are the release of greenhouse gases (GHGs) and other short-lived atmospheric pollutants resulting from the combustion of fossil fuels and alteration in land use patterns [78]. The acceleration of urbanization is progressively recognized as the predominant factor in this phenomenon. Presently, cities accommodate more than 50% of the world's population and it is projected that by 2050, over two-thirds of the urban population will be dwelling in cities and contribute to over 70% of the worldwide GHG emissions originating from the ultimate consumption of energy sources [29]. Beyond the broader patterns of global warming, urban regions encounter a distinct local phenomenon known as the urban heat island (UHI) effect [51]. This effect arises due to the dense presence of impermeable surfaces, alterations in airflow due to urban structures, and the release of heat from residential and industrial activities [95]. The advantage and disadvantage of monitoring and assessing technique for urban climate issues has been summarized in Table 2.

3.1 Direct and Indirect Monitoring and Assessing

Observing urban areas from space provides several benefits, including extensive coverage, comprehensive and repeated assessments, and the capacity to consistently measure crucial physical characteristics using validated techniques. The impact of urbanization on climate can be decomposed into direct and indirect effects [68]. To comprehend and tackle urban climate problems and difficulties effectively, it is crucial to continuously observe and analyze the temperature fluctuations within urban regions [76]. Meteorological data collected from ground-based stations is often regarded as the benchmark, but it is not without its constraints [16]. Despite the different types of measurements to analyze urban meteorology, and especially to improve knowledge about urban climate issues effect. This is particularly important to

Table 2 Advantage and disadvantage of monitoring and assessing technique for urban climate issues

Technique/method	Advantage	Disadvantage	References
Direct and indirect monitoring and assessing	• Provide real-time information on temperature variations • Applicable for assessing urban heat island (UHI) effects • Not expensive and affordable	• Limited information for large area and different industries, topology and geology • Required labour operator	Zumwald et al. [112], Azevedo et al. [7]
Sensing	• Enhanced spatial–temporal analysis of the air pollution • Time flexibility • Rapid access and near real time capabilities information • Capable in data collection during hazy or cloudy	• Limited information for large area and different industries, topology and geology • Expensive • Required labor operator • Depending on the weather • High demanding on data storage and processing capacity • Technical limitation for extreme weather event	Liu et al. [61], Manfreda et al. [63]
Data integration	• Automated weather stations reduce human errors • Provide extensive network for more accurate information with the compliment of satellites and radar • Incorporated with Artificial Intelligence to compute numerous complex algorithms instantly • Easily access data through mobile application real-time data on meteorological • Forecasting and modelling for meteorological data and air quality	• Weather station sensor required multiple approaches in providing accurate and real time information such as satellites and radar • Expensive and required skilled analysts • Interference data obtain during rainy, haze and cloudy	Ioannou et al. [39], Asha et al. [5], van der Wal et al. [113], Manfreda et al. [63]

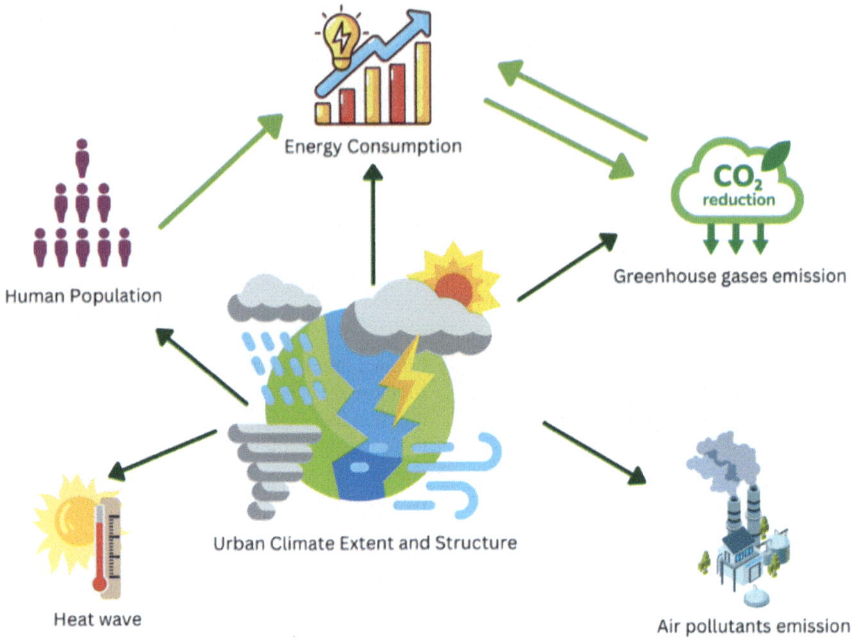

Fig. 1 Direct and indirect monitoring and assessing urban climate extent and structure

install weather stations at various locations across the city to directly measure temperature data these stations can provide real-time information on temperature variations [112]. Prior studies conducted in the Netherlands using thermometer observations have yielded valuable insights into urban temperature patterns. In the current study, the researchers analyzed two years' worth of hourly near-surface air temperature data. This approach is generally applicable for assessing urban heat island (UHI) effects, which are often not adequately captured by the official network of weather stations [7]. Mapping urban temperature distribution can be accomplished through three primary methods, in-situ measurements followed by statistical interpolation and employing statistical prediction models in conjunction with additional explanatory factors [28]. As urban areas sprawl, it plays a critical role in climate change. The expansion fuels environmental issues, but also bear the brunt of its consequences (Fig. 1).

3.2 *Sensing*

Weather stations have been built on the principle of monitoring the dynamics of the weather and climate of a particular area. Such areas usually are limited and bound by administrative districts. For a large city, few weather stations and sensors might not be able to report the weather information at any given time adequately. As such, homogeneity of weather prediction in an administrative district might not be possible, especially in a large area comprising different industries, topology, and geology. Many countries in the world have established national meteorological departments to sustainably monitor and predict the dynamics of weather and climate to provide accurate information for the various needs of the people as well as preparedness against disaster. Most of the national meteorological departments use multiple approaches and methods to provide the most accurate and real-time meteorological data, such as through satellites, radar, and sensors on the earth's surface. Common meteorological data include temperature, humidity, wind speed, and precipitation [25]. Many of the previous studies done on urban climate only highlight the air temperature, though the World Meteorological Organization has also suggested relative humidity and wind speed to be monitored by a weather station [14]. To tackle the urban climate issue, weather stations should be equipped with advanced air pollutant sensors that enable real-time monitoring and assessment of air pollution levels to facilitate targeted mitigation strategies.

Satellite images thus provided invaluable evidence of urban climate issues. A comparison of satellite images of Jaipur City, India, between the years 2000 and 2011, has indicated progressive UHI effects in the area [40]. There is evidence of significant changes in land use/land cover (LULC) demonstrated by the decline of agricultural cropland, vacant lands, and sparse and dense vegetation while showing stark increases in high and low-dense built areas. The expansion of UHI areas was further supported by the increases of more than 2 °C of Land Surface Temperature (LST) in the area. In a similar study in Penang, Malaysia, remote sensing and geospatial techniques were used to assess the data on the changes in LULC within 25 years [82]. In this study, evidence shows the decline of forest cover, dramatic expansion of urban areas, and increases of LST that further exacerbate the UHI effects in the region. A prediction algorithm, a cellular automata-based artificial neural network (CA-ANN), has been able to project the extent of UHI effects by 2030 in Dhaka, Bangladesh [1]. This facilitates the identification of critical areas vulnerable to UHI and the mitigation strategies. The addition of greenspace such as urban forests, parks, and green roofs are examples of good methods to reduce the expected impact of UHI [30].

3.3 Data Integration

The automated weather station has been used on a large scale to monitor the meteorological and climate information. As the name stands, this technology has made human intervention unnecessary during its operation, thus enabling the construction of massive numbers of automated weather stations to provide extensive networks for more accurate data, including in most remote and rural areas. The current development has made it possible for the automated weather stations to be operated using wireless technologies and in the future, incorporated with Artificial Intelligence to compute numerous complex algorithms instantly at the edge [39]. The Internet of

Things (IoT) has enabled smartphone applications, for instance, to easily access data and empower communities to contribute real-time data on meteorological and air quality parameters. Weather station equipped with sensors to monitor both meteorological and air pollutants data in real-time integrates with an HTML programming platform to provide easy access for the users [65]. The design of the user interface allows the monitoring of air quality at any time everywhere. Perhaps, early detection and prediction could provide more effective urban climate monitoring. Environmental Toxicology for Air Pollution Monitoring using Artificial Intelligence technique (ETAPM-AIT) is an innovation able to measure air pollutants and analyze the data to predict air quality in the future [5].

All monitoring and assessment methods and techniques have their advantages and disadvantages. Direct and indirect monitoring are often used because they are inexpensive and provide real-time information. However, combining sensing and data integration can significantly enhance the accuracy of spatial–temporal analysis and modeling forecasts [63].

4 Forecasting Urban Climate

Metropolises are battling growing heat-related issues resulting from the synergistic impacts of urban heat islands and global warming. Accurate depictions of urban areas and processes at the micro, regional, and global levels are required due to the persistent trends of growing urbanization, dynamic urban geography, and expanding populations, as well as to take into consideration the reliant responses. The complexity of urban climates is increased by climate change, necessitating that cities obtain comprehensive data at the decision-making level about urban climates. Current global and regional climate models and observation networks are unable to offer this level of detail. To address this, it is crucial to incorporate crucial urban processes into climate models and explore innovative approaches that can enhance our ability to capture both natural and human elements accurately. Advances in climate modeling with urban resolution are critical for generating capabilities that can capture the complex fluctuations that occur in urban settings. To comprehensively discover the impact of climate change on urban areas, it is crucial to employ (1) simulations of urban atmospheric conditions at thorough spatial scales, incorporating global climate scenarios along with projections of urban expansion and population growth, accompanied by estimates of associated uncertainties, particularly addressing coastal potential risks for cities situated along shorelines. Furthermore, (2) obtaining urban climate evaluations, particularly in urban areas in the southern region of the world, and (3) employing high-resolution spatial data on urban structure, human behavior, and energy use are critical aspects of this understanding [31]. Urban climate modeling provides insights into complex relationships and interactions within the urban domain, making it a useful research and decision-making tool for climate and pollution forecasting.

Climate change is bringing more heat waves and making cities hotter. This will lead to more people getting sick, worse air pollution, and less water available [48].

Cooling strategies like urban greening offer a double win, which can make cities cooler for residents during heat waves caused by climate change, while also reducing the greenhouse gas emissions that contribute to the problem. Running simulations of how cities grow lets communities see how it has changed in the past and predict how they will sprawl and develop for Land Use Land Cover (LULC) in the future. Leveraging advancements in computational capabilities, Urban Growth Models (UGMs) have emerged as powerful tools for forecasting potential trajectories of LULC change and development [45]. Urban climate modeling provides insights into complex relationships and interactions within the urban domain, making it a useful research and decision-making tool for climate and pollution forecasting. Because it makes complex relationships in urban settings easier to understand, it is a guiding resource for urban climate and pollution policy and practice. Recent advances have greatly improved the ability to model complex environmental and urban systems in computer and software capabilities. Recently, these technologies have become more widely available to decision-makers at all levels, from national governments to private property owners, throughout the world. The models are used to forecast urban activities' full global impact, transcending geopolitical boundaries. These models are widely used in urban research and, in different forms, are essential for tackling future challenges. These days, data about social and physical phenomena is mostly stored and accessible digitally, which provides the basis for models that not only indicate current interactions but also explore possible futures. As a result, these models can support decision-making processes that aim to improve mitigation and adaptation to potential urban impacts, including crisis and disaster management. Their usefulness resides in their ability to recognise problems, evaluate causative elements, and track the effectiveness of strategies.

5 Data Integration and Visualisation

In the fast-paced world of urban climate research, the capacity to seamlessly integrate and vividly illustrate diverse data is unparalleled. Proficient data integration and visualization excel at simplifying intricate details and connecting theoretical knowledge and tangible strategies. As cities confront escalating climate challenges, these powerful tools can significantly shape the policies formulated and their ensuring impact. This section delves deep into the meticulous techniques that convert vast data sets into clear, actionable visual narratives, becoming instrumental in advancing urban climate resilience. Meanwhile, this section begins by exploring the drawbacks of solely relying on traditional data sources like administrative records and commercial datasets. While considering alternative sources of structured and unstructured data offers potential solutions, these new approaches introduce their challenges, which will be examined later. Figure 2 provides a block diagram of the data integration and visualization in urban climate change detailing its data collection and management, spatial analysis, and mapping and visualization techniques.

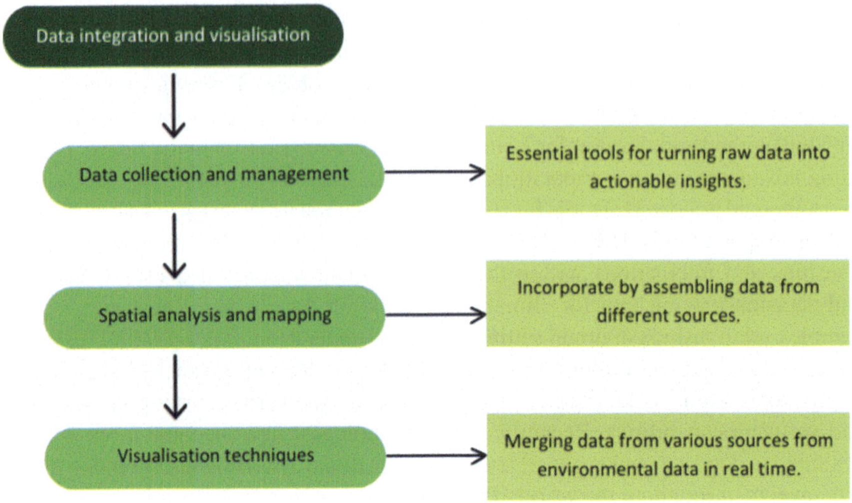

Fig. 2 Block diagram of the data integration and visualization in urban climate change

5.1 Data Collection and Management

As the complexities of climate change unfold, there's a growing need for solid data to support both research and policymaking. This critical need drives the integration of diverse data sources, each elucidating the multifaceted challenges of climate variations. At the forefront, monitoring networks epitomise technological progress in climate research. These rigorously designed networks, spread across urban areas, continually monitor, and record various climate parameters, from temperature shifts to humidity changes. Their real-time data offers timely insights crucial for immediate actions [36, 67]. Beyond urban boundaries, satellite imagery offers a panoramic view of climatic trends. Satellite-derived data covering vast expanses reveals patterns often missed at ground level, enabling a macro-level understanding of climatic changes, and aiding in proactive mitigation [68]. While technology offers quantitative data, understanding the human aspect of climate issues requires a distinct methodology. Moreover, understanding urban climate issues requires analysing the socio-economic fabric of urban settings. Socio-economic indicators provide insights into how climate challenges intersect with societal factors. Integrating data on income, employment, education, and more helps understand the multifaceted impacts of climate challenges, guiding the development of solutions that address environmental and societal aspects [14]. To summarise, addressing the multifaceted nature of climate issues demands a comprehensive approach to data collection and management. We can foster a profound understanding of urban climate challenges by merging quantitative data from monitoring networks and satellites with qualitative survey findings and framing them within socio-economic contexts.

While technology offers quantitative data, understanding the human aspect of climate issues requires a distinct methodology. Surveys, in this context, are invaluable. They capture community sentiments, individual experiences, and inherent adaptive strategies. Such qualitative data is crucial for crafting interventions that align with urban residents' realities, ensuring that systems are efficient and compassionate [10]. Moreover, understanding urban climate issues requires analysing the socio-economic fabric of urban settings. Socio-economic indicators provide insights into how climate challenges intersect with societal factors. Integrating data on income, employment, education, and more helps understand the multifaceted impacts of climate challenges, guiding the development of solutions that address environmental and societal aspects [14]. To summarise, addressing the multifaceted nature of climate issues demands a comprehensive approach to data collection and management. We can foster a profound understanding of urban climate challenges by merging quantitative data from monitoring networks and satellites with qualitative survey findings and framing them within socio-economic contexts. Such a well-rounded view forms the foundation for devising effective and inclusive urban climate strategies.

5.2 Spatial Analysis and Mapping

In the diverse field of urban climate studies, understanding spatial data is paramount for insightful analysis. As urban landscapes transform, their climatic attributes shift, revealing patterns and trends across various spatial dimensions. Capturing these intricate spatial details necessitates advanced tools, underscoring the importance of spatial analysis and mapping [56]. Central to this is Geographical Information Systems (GIS). As an epitome of technological advancement, GIS integrates multiple data layers to generate dynamic maps, vividly illustrating the spatial dynamics of climate parameters. From urban heat islands to flood-prone areas and green spaces, GIS provides a comprehensive perspective, helping stakeholders comprehend the multifaceted factors influencing urban climate [108]. However, GIS is not just for visualisation. Its analytical capabilities empower researchers to uncover essential patterns and connections [46]. For example, overlaying socio-economic data on temperature maps can reveal areas where vulnerable populations face heightened heatwave risks [93]. Such holistic analysis, melding environmental and societal aspects, aids in formulating precise mitigation measures. Complementing GIS, and other spatial tools also enhances our grasp of urban climate data [11]. Advanced statistical approaches identify nuanced trends, spotlighting gradual changes that might be overlooked. Spatial regression models, for example, elucidate how variables, from urban infrastructure to green cover, impact local climate metrics [110]. This clarity empowers policymakers to take decisive actions, from urban greening initiatives to designing resilient infrastructure [13]. A crucial use of spatial tools is pinpointing areas of vulnerability and resilience [99]. By capitalising on these tools, we can aspire to shape resilient urban spaces and promote their residents' well-being.

5.3 Visualisation Techniques

In the complex sphere of urban climate studies, the saying "a picture is worth a thousand words" rings especially true. Amidst an overwhelming influx of data, the challenge often is not just data acquisition but also effective communication. Visualisation techniques seamlessly bridge this gap, translating raw data into actionable insights. The pivotal roles of data integration and visualisation in urban climate monitoring, assessment, and forecasting cannot be downplayed. Graphs and charts have revolutionised data representation in this discipline. These visual aids distil vast data into digestible formats, offering clear insights into temporal fluctuations [111]. A line graph, for instance, can succinctly depict the trajectory of escalating urban temperatures over time, enabling stakeholders to comprehend climate patterns swiftly. Similarly, a bar chart can compare yearly rainfall patterns, spotlighting disparities that might be missed in sheer data. These visual tools empower stakeholders, aiding informed decision-making vital for urban climate challenges.

Mapping offers a profound exploration of the spatial attributes of urban climate data. Beyond traditional maps, contemporary versions, often enhanced by GIS, are gaining traction. GIS has transformed our perception and interpretation of spatial urban climate patterns [52]. For example, GIS can visualise urban heat islands, flood risks, or high-pollution areas, providing an invaluable spatial context for tailored interventions [111]. Such visual aids enable policymakers to allocate resources effectively [67]. Infographics, blending design, content, and data are potent communication tools. They often narrate a story, integrating elements like narratives, illustrations, and data visualisations [64]. Their potential to captivate and engage makes them instrumental in brand visibility [91]. In today's information-saturated age, infographics simplify intricate data, making it relatable. For instance, an urban air quality infographic could incorporate data on pollutants, their sources, health impacts, and remedial measures [101], much like the WHO Ambient Air Quality Database. By elucidating complex data, infographics significantly broaden awareness and encourage wider action. Visualising urban climate data transcends mere technicality,it is inherently communicative. In today's data-centric world, the value of adept visualisation can't be ignored. Whether it is the precision of charts, the spatial insights from maps, or the comprehensive stories of infographics, each has a distinct role in converting data to insights and insights to actions [62]. As urban areas confront diverse climate challenges, these tools will remain central to informed urban climate policies [67]. This accentuates the importance of ongoing advancements in visualisation methods. Moving ahead, creating advanced yet user-friendly tools will be vital. The trajectory of urban climate strategies hinges on our capacity to visualise and comprehend data innovatively. The visualization techniques, results and benefits in urban climate case study has been summarized in Table 3.

Table 3 Summary of visualization techniques, results and benefits in urban climate case studies

Case study	Focus area	Visualization techniques	Results	Advantages and excellence
Mosteiro-Romero et al. [71]	Microclimate data and building energy simulations	3D visualization, modeling	Improved accuracy of urban climate models and their impact on energy consumption	Enhanced model accuracy; Realistic depiction of urban climate impacts; Better energy management
MacLachlan et al. [62]	High-resolution aerial imagery and geospatial tools for temperature variations and urban planning	High-resolution imagery, GIS	Optimized urban planning for heat mitigation	High-detail imagery; Precise urban planning; Effective heat mitigation strategies
Reaver [83]	Augmented Reality (AR) for urban planning with youth participants in Oslo, Norway	Augmented reality	Enhanced understanding of urban planning and increased confidence in design proposals among youth participants	Interactive and engaging; Improved participant confidence; Real-time design visualization
Dhunnoo et al. [18]	Virtual Reality (VR) for climate change education in Surfers Paradise, Australia	Virtual reality	Demonstrated VR's effectiveness in climate change education by engaging professionals to visualize the impacts of flooding	Immersive experience; Effective educational tool; Increased awareness of climate impacts
Veisi et al. [97]	AI models (ANNs, RF, GBR) for forecasting Universal Thermal Climate Index (UTCI) in various cities	Machine learning, AI-based Visualization	Accurate forecasts of UTCI, highlighting the role of AI and ML in optimizing energy usage and improving urban planning	High predictive accuracy; Efficient energy usage optimization; Enhanced urban planning

(continued)

Table 3 (continued)

Case study	Focus area	Visualization techniques	Results	Advantages and excellence
Kazeem et al. [47]	AI and ML technologies in enhancing construction processes and developing sustainable communities	Machine learning, AI-based visualization	Showcased how AI and ML can optimize energy usage, monitor environmental conditions, and improve planning and management in urban environments	Improved construction processes; Sustainable community development; Effective resource management
Gandharum et al. [27]	Monitoring urban expansion and agricultural land loss in West Java, Indonesia using Google Earth Engine and Landsat	Drone imagery, Satellite imagery, GIS	Revealed significant urban expansion at the expense of agricultural land, emphasizing the need for sustainable urban planning	Comprehensive coverage; Frequent updates; Enhanced urban planning and monitoring
Li et al. [59]	Monitoring global vegetation status and changes using remote sensing data	Remote sensing, satellite imagery	Provided insights into vegetation health and changes, highlighting areas of improvement and degradation	Macro-level climate understanding; Detailed vegetation health insights; Proactive mitigation strategies

6 Urban Climate Change Resilience

Global climate change has brought extreme weather events like heat waves, hurricanes, rising sea levels, sandstorms, urban heat islands, and many more, which have severely to cities that consist of the largest population and those coastal cities. Urban climate change resilience refers to the city that is sustainable with its development and resource consumption, as well as able to withstand and bounce back from the disaster, specifically climate change [41]. The resilience has not just with regards to the ‘defense system’ of a city to withstand the effect of climate change, restore the city to its original condition, as well as its ability to mitigate climate change in the first place, in the city area. Urban climate change and urban transport are very closely related, thus, the priority mitigation action should be focused on rectifying this urban transport issue. In the following section, comprehensive sustainable urban transportation strategies are summarized, from a global perspective, followed by ways to reduce the impacts on the urban population via innovative infrastructure programs to mitigate the urban heat island effect and air pollution. Tremendous sustainable strategies across urban regions must be planned, managed, and implemented in the long run, to achieve urban climate change resilience.

6.1 Global Effort on Sustainable Urban Transport

Transport plays the most crucial role in the urban and country's development, by providing key access for people to education, healthcare, employment, markets, commerce, finance, recreation, and many key services. Nevertheless, urban development is always linked to the most frequent traffic congestion, excessive greenhouse gas (GHG) emissions, urban heat islands, noise pollution, and urban air pollution, as well as the health issues of the citizens and road accidents. United Nation has reported that greenhouse gas emission has more than doubled since 1970, with 80% of the increase caused by road vehicles. The CO_2 emission from transportation in the United States accounted for 38% (1.7 Billion metric tons) of all energy-related emissions, including electric power (33%), industrial (17%), residential (7%), and commercial (5%) (www.cbo.gov/publication/58419). Most of the GHG emissions are generated by the transportation sector [22]. International Energy Agency (IEA) claimed that the transportation sector emitted 8258 million tons of CO_2 globally in 2018, has accounted for 24.6% of total CO_2 emissions from fuel combustion worldwide [57]. The transition of fossil fuel-based transport to low-carbon urban transport requires appropriate policies to accelerate the transition in cities at varied phases. Efforts include the development and optimization of urban transport structures, enhancement in energy efficiency for green vehicles, improvement in urban planning and infrastructure for transit, and many more.

In China, the made-in-china 2025 strategy has aimed to reduce urban severe air quality issues and enhance economic growth using fuel cells [80]. Shenzhen City in China has targeted to increase the green energy vehicle ownership to 1 million units in 2025 to phase out fuel-dependency vehicles [57]. Similarly, Germany, in 2020, worked to have 1 million electric vehicles in operation to reduce GHG emissions [92]. Japan has also promoted hydrogen usage via fuel cells in electric vehicles [80]. California has proposed a zero-emission vehicle (ZEV) program to promote the adoption of plug-in vehicles and fuel-cell vehicles [57]. Moreover, there are as well countries that have decided to ban pure internal combustion engine vehicles, with Germany, France, United Kingdom, Canada, and Spain in 2040, India in 2030, and the Netherlands in 2050 [80, 92]. This has undeniably reflected the effort towards the transition of fossil fuel-based vehicles towards low carbon vehicles, to reaching net carbon emissions.

Sen et al. [85] claimed that half of the world's population lives in city and urban areas. Urban transport CO_2 emissions range from 20% to 60% of most cities worldwide [57]. Despite the challenges of increasing urban population growth, urban migration, economic growth, and social pressures, countries have set a strategic plan towards net zero emissions. Paris Agreement, since November 2016, has proposed the global effort to ensure global average temperature increases are kept below 2 °C. 130 countries have proposed their target to reach carbon neutrality in April 2021 [57]. Within these strategic plans, many specific efforts focus on sustainable urban transport and green city development. Table 4 summarizes the strategies of various countries towards net zero emissions, specifically to sustainable transportation approaches and green cities.

6.2 *Adaptation Strategies of Existing Motorized Transport for Urban Transportation Resilience*

Urban transportation resilience is increasingly important to rectify the unpredicted intensity and frequency of disasters brought by climate change and GHG emissions. Serdar et al. [85] claimed that the concept of urban transportation resilience should include risk management, vulnerability, reliability, robustness, flexibility, and survivability. The infrastructure of the road system should absorb sudden disturbances like floods and recover its typical traffic flow in the shortest and acceptable time frame. Urban transportation with effective infrastructure, in the first place, should be enhanced. The framework should be developed to assess resilience for specific regions. Yang et al. [106] stated that cyclone is the primary source of stress for Hong Kong, China region, sea-rising, earthquake, and rainfall are stressful for Qatar and the Arabian Peninsula region, and earthquake for Japan. Moreover, data tracking applications in transportation networks could be helpful for resilience assessment to respond to disasters. Global Positioning System (GPS) records could be applied to track via databases in taxi services to remove travel errors, normalized trips to calculate the shortest travel distance and time spent, assess the resilience of transportation networks during disaster, and reduce GHG emissions [19]. In addition, geographic information system (GIS) data could be applied to assess transportation networks towards flooding, considering the town planning with sewer system, transportation, and many more. Overall, technological innovation should be fully applied for planning, predicting, projecting, modeling, and monitoring for enhanced urban transportation resilience.

7 Conclusion

More than half of the world's people live in cities, so it's important to keep an eye on and understand how urban climate changes, especially how things like greenhouse gas emissions and changing land use trends affect the climate. Because of the Urban Heat Island impact, air pollution, extreme weather events, lack of water, and health risks related to climate change, we need to act. Ground-based weather

Table 4 International commitment and strategies towards greenhouse gas emission reduction (Ma et al. 2022)

Countries	Target	Strategies
United States of America	• 17% below 2005 levels in 2020 • 26–28% below 2005 levels in 2025 • 50–52% below 2005 levels in 2030 • Net Zero in 2050	• Electrify end uses and switch to other clean fuels
China	• CO_2 emission peak before 2030 • Carbon neutrality by 2060	• Incorporate ambitious climate and energy targets with 2030/2060 climate targets • Enhance bottom-up actions and continue to support low-carbon pilots • Accelerate renewable energy development • Embrace 5G-based infrastructure as short-term economic stimulus, accelerate emissions reduction across various sectors, and avoid carbon-intensive projects like coal-fired power plants and coal-to-chemicals facilities
India	• Achieve 500 GW non-fossil fuel energy capacity by 2030 • Meet 50% of energy needs with renewable by 2030 • Cut CO_2 emissions by 1 million tons by 2030 • Lowercarbon intensity by 45% by 2030 • Reach net zero in 2070	• Low carbon energy: – Replace legacy fossil fuel energy infrastructure with renewables – Reduce emissions from legacy fossil fuel infrastructure through enhanced efficiencies • Green mobility: – Shift in transport modal mix (from road to rail) – Green fuel and green vehicles • Green buildings, infrastructure and cities: – Efficient urban planning – Low carbon buildings and infrastructure
United Kingdom	• 78% reduction from 1900 levels by 2035 • Net Zero in 2050	• Generate electricity to meet low carbon demand across all sectors • Utilize green energy for public, business, transportation, industrial and residential buildings • Support energy supply for hard-to-reach areas, reduce oil and gas usage, and use carbon capture technology

data is complemented by tracking from space using tools like satellites, giving a full picture of important physical features over time. When IoT and AI are combined with advanced tracking technologies, they make real-time monitoring more accurate and easier to reach. It is important to include factors like temperature and wind speed to fully understand how urban climate changes over time. Also, remote sensing methods like geographic analysis and satellite images are very helpful for showing how the climate is changing in cities and finding places that are most at risk. When talking about climate adaptation, healthy urban travel stands out as a key factor. Strategies for reducing the damage, like switching to low-carbon forms of transport in cities and promoting green energy, are necessary steps. Infrastructure, policy changes, and teaching that support public transportation, car sharing, and non-motorized transportation use are all important parts of mitigating climate change. Improving resilience through risk management, vulnerability assessment, and infrastructure changes are also part of adaptation methods for motorized transportation that are already in use. Sustainable urban development, based on close tracking, cutting-edge technologies, and open participation by all members of the community, is the key to solving the tough problems that towns around the world are facing.

Acknowledgements The authors would like to convey their heartfelt appreciation to all organizations engaged in this effort. Expressions of appreciation are extended to the colleagues who participated in and supported this study.

References

1. Al-Faisal A, Al-Kafy A, Al Rakib A, Akter KS, Jahir DMAJ, Sikdar MS, Ashrafi TJ, Mallik S, Rahman MM (2021) Assessing and predicting land use/land cover, land surface temperature and urban thermal field variance index using landsat imagery for dhaka metropolitan area. Environ Chall 4:100192
2. Anderko L, Chalupka S, Du M, Hauptman M (2020) Climate changes reproductive and children's health: a review of risks, exposures, and impacts. Pediatr Res 87:414–419
3. Ao X, Wang L, Zhi X, Gu W, Yang H, Li D (2019) Observed synergies between urban heat islands and heat waves and their controlling factors in Shanghai, China. J Appl Meteorol Climatol 58(9):1955–1972
4. Arriazu-Ramos A, Ruis GR, Izquierdo JJP, Gutierrez AS, Monge-Barrio A (2023) From urban microclimate to indoor overheating: analysis of residential typologies during typical climate series and extreme warm summer. Energy Build 299:113620
5. Asha P, Natrayan L, Geetha BT, Beulah JR, Sumathy R, Varalakshmi G, Neelakandan S (2022) IoT enabled environmental toxicology for air pollution monitoring using AI techniques. Environ Res 205:112574
6. Aune KT, Davis MF, Smith GS (2022) Extreme precipitation events and infectious disease risk: a scoping review and framework for infectious respiratory viruses. Int J Environ Res Public Health 19(1):165
7. Azevedo JA, Chapman L, Muller CL (2016) Quantifying the daytime and night-time urban heat island in Birmingham, UK: a comparison of satellite derived land surface temperature and high resolution air temperature observations. Remote Sens 8(2):153–170

8. Baker RE, Mahmud AS, Miller IF, Rajeev M, Rasambainarivo F, Rice BL, Takahashi S, Tatem AJ, Wagner CE, Wang L-F, Wesolowski A, Metcalf CJE (2022) Infectious disease in an era of global change. Nat Rev Microbiol 20(4):193–205
9. Bolan S, Padhye LP, Jasemizad T et al (2024) Impacts of climate change on the fate of contaminants thorugh extreme weather events. Sci Total Environ 909:168388
10. Boulanger SOM (2023) Urban adaptation to climate change state of the art: evaluating the role of adaptation assessment frameworks through a systematic and bibliometric analysis. Sustainability 15(13)
11. Cao H, Liu J, Chen J, Gao J, Wang G, Zhang W (2019) Spatiotemporal patterns of urban land use change in typical cities in the greater Mekong subregion (GMS). Remote Sensing 11(7):801
12. Carolan-Olah M, Frankowska D (2014) High environmental temperature and preterm birth: a review of the evidence. Midwifery 30:50–59
13. Casali Y, Aydin NY, Comes T (2022) Machine learning for spatial analyses in urban areas: a scoping review. In: Sustainable cities and society, 85. Elsevier Ltd.
14. Chen X, Yang J (2022) Urban climate monitoring network design: existing issues and a cluster-based solution. Build Environ 214:108959
15. Chen S, Yang Y, Deng F, Zhang Y, Liu D, Liu C, Gao Z (2022) A high-resolution monitoring approach of canopy urban heat island using a random forest model and multi-platform observations. Atmos Meas Tech 15(3):735–756
16. Colston JM, Ahmed T, Mahopo C, Kang G, Kosek M, Sausa Junior FD, Shrestha PS, Svensen E, Turab A, Zaitchik B (2018) Evaluating meteorological data from weather stations, and from satellites and global models for a multi-site epidemiological study. Environ Res 165:91–109
17. Contreras C, Aguilar M, Eappen B (2018) Community strengthening and mental health system linking after flooding in two informal human settlements in Peru: a model for small-scale disaster response. Global Mental Health 5:e11
18. Dhunnoo Y, Carter A, O'Hare D, Birt J, Skitmore M (2023) Improving climate change awareness through immersive virtual reality communication: a case study. Sustainability 15(17):12969
19. Donovan B, Work DB (2017) Empirically quantifying city-scale transportation system resilience to extreme events. Transp Res Part C: Emerg Technol 79:333–346
20. Dotse SQ, Dagar L, Petra MI, De Silva LC (2016) Influence of Southeast Asian Haze episodes on high PM10 concentrations across Brunei Darussalam. Environ Pollut 219:337–352
21. Durack PJ, Wijffels SE, Matear RJ (2012) Ocean salinities reveal strong global water cycle intensification during 1950 to 2000. Science 336(6080):455–458
22. EPA (2020) Fast Facts on Transportation Greenhouse Gas Emissions 1990–2018. E Invent US Greenh Gas Emiss Sink 1990–2018:1–4. https://www.epa.gov/greenvehicles/fast-facts-transportation-greenhouse-gas-emissions. Accessed 16 Sept 2021
23. Ebi KL, Vanos J, Baldwin JW, Bell JE, Hondula DM, Errett NA, Hayes K, Reid CE, Saha S, Spector J, Berry P (2021) Extreme weather and climate change: population health and health system implications. Annu Rev Public Health 42(1):293–315
24. Estrada F, Botzen WJW, Tol RSJ (2017) A global economic assessment of city policies to reduce climate change impacts. Nat Clim Chang 7:403–406
25. Feng S, Hu Q, Qian W (2004) Quality control of daily meteorological data in China, 1951–2000: a new dataset. Int J Climatol 24:853–870
26. Founda D, Santamouris M (2017) Synergies between urban Heat Island and heat waves in Athens (Greece), during an extremely hot summer (2012). Sci Report 7(1):10973
27. Gandharum L, Hartono DM, Karsidi A, Ahmad M (2022) Monitoring urban expansion and loss of agriculture on the north coast of west java province, Indonesia, using Google Earth engine and intensity analysis. Sci World J 1:3123788
28. Garzon J, Molina I, Velasco J, Calabia A (2021) A remote sensing approach for surface urban heat island modeling in a tropical colombian city using regression analysis and machine learning algorithms. Remote Sens 13(21):4256

29. Gielen D, Boshell F, Saygin D, Bazilian MD, Wagner N, Gorini R (2019) The role of renewable energy in the global energy transformation. Energ Strat Rev 24:38–50
30. Gunawardena KR, Wells MJ, Kershaw T (2017) Utilising green and bluespace to mitigate urban heat island intensity. Sci Total Environ 584–585:1040–1055
31. Hamdi R, Kusaka H, Doan QV, Cai P, He H, Luo G, Kuang W, Caluwaerts S, Duchêne F, Van Schaeybroek B, Termonia P (2020) The state-of-the-art of urban climate change modeling and observations. Earth Syst Environ 4:631–646
32. Hanaoka T, Masui T (2019) Exploring effective short-lived climate pollutant mitigation scenarios by considering synergies and trade-offs of combinations of air pollutant measures and low carbon measures towards the level of the 2 °C target in Asia. Environ Pollut 261:113650
33. Hassan N, Hashim Z, Hashim J (2015) Impact of climate change on air quality and public health in urban areas. Asia Pacific J Public Health 28(2):38–48
34. He C, Liu Z, Wu J, Pan X, Fang Z, Li J, Bryan BA (2021) Future global urban water scarcity and potential solutions. Nat Commun 12(1):4667
35. Hoffmann R (2022) Contextualizing climate change impacts on human mobility in African Drylands. Earth's Future 10(6):e2021EF002591
36. Hsu A, Khoo W, Goyal N, Wainstein M (2020) Next-generation digital ecosystem for climate data mining and knowledge discovery: a review of digital data collection technologies. Front Big Data 3:525690
37. Huang K, Fu JS, Hsu NC, Gao Y, Dong X, Tsay SC, Lam YF (2013) Impact assessment of biomass burning on air quality in Southeast Asia and East Asia during BASE-ASIA. Atmos Environ 78:291–302
38. Huo Y, Yang M, Lei M, Wu P, Tian X, Yang T, Wang Z, Mao Y (2023) Advancements in water and ecological risk assessment in the Yellow River Delta, North China: Implications for legal frameworks in environmental protection and public health. Water Environ J 1–7
39. Ioannou K, Karampatzakis D, Amanatidis P, Aggelopoulos V, Karmiris I (2021) Low-cost automatic weather stations in the Internet Of Things. Information 12(4):146–167
40. Jalan S, Sharma K (2014) Spatio-temporal assessment of land use/land cover dynamics and urban heat island of Jaipur city using satellite data. The International Archives of the Photogrammetry, Remote Sensing and Spatial Information Sciences XL-8: 767–772
41. Jamali A, Robati M, Nikoomaram H, Farsad F, Aghamohammadi H (2023) Urban resilience and climate change: developing a multidimensional index to adapt against climate change in the Iranian Capital City of Tehran. Urban Science 7(1):7
42. Ji JS, Xia Y, Liu L, Zhou W, Chen R, Dong G, Hu Q, Jiang J, Kan H, Li T, Liu Q et al (2023) China's public health initiatives for climate change adaptation. Lancet Reg Health—West Pac 40:100965
43. Jiang S, Lee X, Wang J, Wang K (2019) Amplified urban Heat Islands during heat wave periods. J Geophys Res: Atmos 124(14):7797–7812
44. Jongman B (2018) Effective adaptation to flooding risk. Nat Commun 9(1):1986
45. Karimi F, Sultana S (2024) Urban expansion prediction and land use/land cover change modeling for sustainable urban development. Sustainability 16(6):2285
46. Kasniza Jumari NAS, Ahmed AN, Huang YF, Ng JL, Koo CH, Chong KL, Sherif M, Elshafie A (2023) Analysis of urban heat islands with landsat satellite images and GIS in Kuala Lumpur Metropolitan City. Heliyon 9(8)
47. Kazeem KO, Olawumi TO, Osunsanmi T (2023) Roles of artificial intelligence and machine learning in enhancing construction processes and sustainable communities. Buildings 13(8):2061
48. Kearl Z, Vogel J (2023) Urban extreme heat, climate change, and saving lives: lessons from Washington state. Urban Climate 47:101392
49. Kim Y, Kim H, Gasparrini A (2019) Suicide and ambient temperature: a multi-country multi-city study. Environ Health Perspect 127:117007
50. Koken PJ, Piver WT, Ye F, Elixhauser A, Olsen LM, Portier CJ (2003) Temperature, air pollution, and hospitalization for cardiovascular diseases among elderly people in Denveri. Environ Health Perspect 111(10):1312–1317

51. Kookana RS, Dreschsel P, Jamwal P, Vanderzalm J (2020) Urbanisation and emerging economies: issues and potential solutions for water and food security. Sci Total Environ 732:139057
52. Kraak MJ, Ormeling F (2020) Cartography: visualization of geospatial data, 4th ed. CRC Press. https://doi.org/10.1201/9780429464195
53. Lamy K, Tran A, Portafaix T, Leroux MD, Baldet T (2023) Impact of regional climate change on the mosquito vector Aedes albopictus in a tropical island environment: La Reunion. Sci Total Environ 875:162484
54. Largeron Y, Staquet C (2016) Persistent inversion dynamics and wintertime PM10 air pollution in Alpine valleys. Atmos Environ 135:92–108
55. Latif MT, Othman M, Idris N, Juneng L, Abdullah AM, Hamzah WP, Khan, Sulaiman NMN, Jewaratnam J, Aghamohammadi N, Sahani M, Xiang CJ, Ahamad F, Amil N, Darus M, Varkkey H, Tangang F, Jaafar AB (2018) Impact of regional haze towards air quality in Malaysia: a review. Atmos Environ 177:28–44
56. Lee D, Oh K, Jung S (2019) Classifying Urban Climate Zones (UCZs) based on spatial statistical analyses. Sustainability 11(7):1915
57. Li W, Bao L, Li Y, Si H, Li Y (2022) Assessing the transition to low-carbon urban transport: a global comparison. Resour Conserv Recycl 180:106179
58. Li D, Sun T, Liu M, Wang L, Gao Z (2016) Changes in wind speed under heat waves enhance urban Heat Islands in the Beijing metropolitan area. J Appl Meteorol Climatol 55(11):2369–2375
59. Li L, Xin X, Zhao J, Yang A, Wu S, Zhang H, Yu S (2023) Remote sensing monitoring and assessment of global vegetation status and changes during 2016–2020. Sensors 23(20):8452
60. Liu H, Zou L, Xia J, Chen T, Wang F (2022) Impact assessment of climate change and urbanization on the nonstationarity of extreme precipitation: a case study in an urban agglomeration in the middle reaches of the Yangtze river. Sustain Cities Soc 85:104038
61. Liu H, Huang B, Cheng X, Yin M, Shang C, Luo Y, He BJ (2023) Sensing-based park cooling performance observation and assessment: a review. Build Environ 110915
62. MacLachlan A, Biggs E, Roberts G, Boruff B (2021) Sustainable city planning: a data-driven approach for mitigating urban heat. Front Built Environ 6:519599
63. Manfreda S, McCabe MF, Miller PE, Lucas R, Pajuelo Madrigal V, Mallinis G, Ben Dor E, Helman D, Estes L, Ciraolo G (2018) On the use of unmanned aerial systems for environmental monitoring. Remote Sens 10(4):641
64. Mc Sween-Cadieux E, Chabot C, Fillol A, Saha T, Dagenais C (2021) Use of infographics as a health-related knowledge translation tool: protocol for a scoping review. BMJ Open 11:46117
65. Megantoro P, Pramudita B, Vigneshwaran P, Yurianta A, Winarno H (2021) Real-time monitoring system for weather and air pollutant measurement with HTML-Based UI application. Bull Electr Eng Inform 10(3):1669–1677
66. Mehryar S, Sasson I, Surminski S (2022) Supporting urban adaptation to climate change: what role can resilience measurement tools play? Urban Climate 41:101047
67. Middel A, Nazarian N, Demuzere M, Bechtel B (2022) Urban climate informatics: an emerging research field. Front Environ Sci 10.867434
68. Milesi C, Churkina G (2020) Measuring and monitoring urban impacts on climate change from space. Remote Sens 12(21):3494
69. Mishra BK, Kumar P, Saraswat C, Chakraborty S, Gautam A (2021) Water security in a changing environment: concept, challenges and solutions. Water 13(4)
70. Monteiro A, Gama C, Cândido M, Ribeiro I, Carvalho D, Lopes M (2016) Investigating ozone high levels and the role of sea breeze on its transport. Atmos Pollut Res 7(2):339–347
71. Mosteiro-Romaro M, Maiullari D, Esch MPV, Schlueter A (2020) An integrated microclimate-energy demand simulation method for the assessment of urban districts. Front Built Environ 6:553946
72. Munyai RB, Chikoore H, Musyoki A, Chakwizira J, Muofhe TP, Xulu NG, Manyanya TC (2021) Vulnerability and adaptation to flood hazards in rural settlements of Limpopo Province, South Africa. Water 13(24):3490

73. Murakami H, Delworth TL, Cooke WF, Hsu PC (2020) Detected climatic change in global distribution of tropical cyclones. Earth, Atmos Planet Sci 117(20):10706–10714
74. Musacchio A, Mas-Pla J, Soana E, Re V, Sacchi E (2021) Governance and groundwater modelling: Hints to boost the implementation of the EU Nitrate Directive. The Lombardy Plain Case, N Italy. Sci Total Environ 782:146800
75. Najafi H, Shrestha PK, Rakovec O, Apel H, Vorogushyn S, Kumar R, Thober S, Merz B, Samaniego L (2024) High-resolution impact-based early warning system for riverine flooding. Nat Commun 15(1):3726
76. Nazarian N, Lipson M, Norford LK (2023) Chapter 4—Multiscale modeling techniques to document urban climate change. Urban climate change and heat islands. In: Characterization, impacts, and mitigation, pp 123–164
77. Niu H, Hu W, Zhang D, Wu Z, Guo S, Pian W, Cheng W, Hu M (2016) Variations of fine particle physiochemical properties during a heavy haze episode in the winter of Beijing. Sci Total Environ 77:1005–2010
78. Perera F (2018) Pollution from fossil-fuel combustion is the leading environmental threat to global pediatric health and equity: solutions exist. Int J Environ Res Public Health 15(1):16
79. Pfahl S, O'Gorman PA, Fischer EM (2017) Understanding the regional pattern of projected future changes in extreme precipitation. Nat Clim Chang 7(6):423–427
80. Pramuanjaroenkij A, Kakaç S (2023) The fuel cell electric vehicles: The highlight review. Int J Hydrogen Energy 48(25):9401–9425
81. Rafiei-Sardooi E, Azareh A, Shooshrari SJ, Parteli EJR (2022) Long-term assessment of land-use and climate change on water scarcity in an arid basin in Iran. Ecol Model 467:109934
82. Rahaman ZA, Al Kafy A, Saha M, Rahim AA, Almulhim AI, Rahaman SN, Fattah MA, Rahman MT, Kalaivani S, Faisal AA, Al Rakib A (2022) Assessing the impacts of vegetation cover loss on surface temperature, urban heat island and carbon emission in Penang City, Malaysia. Build Environt 222:109335
83. Reaver K (2023) Augmented reality as a participation tool for youth in urban planning processes: case study in Oslo, Norway. Front Virtual Real 4:1055930
84. Santos MM, Lanzinha JCG, Ferreira AV (2021) Review on urbanism and climate change. Cities 114:103176
85. Serdar MZ, Koç M, Al-Ghamdi SG (2022) Urban transportation networks resilience: indicators, disturbances, and assessment methods. Sustain Cities Soc 76:103452
86. Shen P, Wang M, Liu J, Ji Y (2023) Hourly air temperature projection in future urban area by coupling climate change and urban heat island effect. Energy Build 279:112676
87. Shu C, Gaur A, Leon LW, Lacasse MA (2023) Evolution of the local climate in Montreal and Ottawa before, during and after a heatwave and the effects on urban heat islands. Sci Total Environ 890(10):164497
88. Shu EG, Porter JR, Hauer ME, Sandoval Olascoaga S, Gourevitch J, Wilson B, Pope M, Melecio-Vazquez D, Kearns E (2023) Integrating climate change induced flood risk into future population projections. Nat Commun 14(1):7870
89. Da Silva I, de Almeida DS, Hashimoto EM, Martins LD (2020) Risk assessment of temperature and air pollutants of temperature and air pollutants on hospitalizations for mental and behavioral disorders in Curitiba, Brazil. Environ Health 19:79
90. Sofwan NM, Latif MT (2021) Characteristics of the real-driving emissions from gasoline passenger vehicles in the Kuala Lumpur urban environment. Atmos Pollut Res 12:306–315
91. Spicer JO, Coleman CG (2022) Creating effective infographics and visual abstracts to disseminate research and facilitate medical education on social media. Clin Infect Dis 74(3):e14–e22
92. Sun X, Li Z, Wang X, Li C (2019) Technology development of electric vehicles: a review. Energies 13(1):90
93. Teo YH, Makani MABH, Wang W, Liu L, Yap JH, Cheong KH (2022) Urban heat island mitigation: GIS-based analysis for a tropical city Singapore. Int J Environ Res Public Health 19(19):11917

94. Trenberth KE, Dai A, Rasmussen RM, Parsons DB (2003) The changing character of precipitation. Bull Am Meteor Soc 84(9):1205–1218
95. Tzavali A, Paravantis JP, Mihalakakou G, Fotiadi A, Stigka E (2015) Urban heat island intensity: a literature review. Fresenius Environ Bull 24:12b
96. Unger J, Skarbit N, Kovacs A, Gal T (2020) Comparison of regional and urban outdoor thermal stress conditions in heatwave and normal summer periods: a case study. Urban Climate 32
97. Veisi O, Tehrani AA, Gharaei B, Shakibamanesh A (2024) Using artificial intelligence for predicting universal thermal climate index based on different urban conditions: a comparative study of machine learning models. SSRN
98. Voogt JA, Oke TR (2003) Thermal remote sensing of urban climates. Remote Sens Environ 86(3):370384
99. Waly NM, Ayad HM, Saadallah DM (2021) Assessment of spatiotemporal patterns of social vulnerability: a tool to resilient urban development Alexandria, Egypt. Ain Shams Eng J 12(1):1059–1072
100. Wan L, Bento VA, Qu Y, Qiu J, Song H, Zhang R, Wu X, Xu F, Lu J, Wang Q (2023) Drought characteristics and dominant factors across China: insights from high resolution daily SPEI dataset between 1979 and 2018. Sci Total Environ 901:166362
101. Wang Y, Cai G, Yang L, Zhang N, Du M (2022) Monitoring of urban ecological environment including air quality using satellite imagery. PLoS ONE 17(8):e0266759
102. Ward MH, Jones RR, Brender JD, De Kok TM, Weyer PJ, Nolan BT, Villanueva CM, Van Breda SG (2018) Drinking water nitrate and human health: an updated review. Int J Environ Res Public Health 15(7):1557
103. World Health Organization (2022) Ambient (Outdoor) Air Pollution. https://www.who.int/en/news-room/fact-sheets/detail/ambient-(outdoor)-air-quality-and-health. Accessed 19 Dec 2022
104. Xiang X, Li Q, Khan S, Khalaf OI (2021) Urban water resource management for sustainable environment planning using artificial intelligence techniques. Environ Impact Assess Rev 86:106515
105. Xu Z, Liu X, Luo X, Zhu S, Zhang N, Guo Q (2022) Urban heat island under the background of urbanization: a Case Study in Nan Jing City, China, pp 391–410
106. Yang Z, Mahendran R, Yu P, Xu R, Yu W, Godellawattage S, Li S, Guo Y (2022) Health effects of long-term exposure to ambient PM2.5 in Asia-Pacific: a systematic review of cohort studies. Curr Environ Health Report 9:130–151
107. Ye B, Jiang J, Liu J, Zheng Y, Zhou N (2021) Research on quantitative assessment of climate change risk at an urban scale: review of recent progress and outlook of future direction. Renew Sustain Energy Rev 135:110415
108. Yu D, Fang C (2023) Urban remote sensing with spatial big data: a review and renewed perspective of urban studies in recent decades. Remote Sens 15(5):1307
109. Zhao C, Jensen JLR, Weng Q, Currit N, Weaver R (2020) Use of local climate zones to investigate surface urban heat islands in Texas. GIScience Remote Sens 57(8):1083–1101
110. Zhu J, Wright G, Wang J, Wang X (2018) A critical review of the integration of geographic information system and building information modelling at the data level. ISPRS Int J Geo Inf 7(2):66
111. Zhu W (2021) A study of big-data-driven data visualization and visual communication design patterns. Sci Program 6704937
112. Zumwald M, Knusel B, Bresch DN, Knutti R (2021) Mapping urban temperature using crow-sensing data and machine learning. Urban Clim 35:100739
113. van der Wal T, Abma B, Viguria A, Prévinaire E, Zarco-Tejada PJ, Serruys P, van Valkengoed E, van der Voet P (2013) Fieldcopter: unmanned aerial systems for crop monitoring services. In: Precision agriculture'13. Wageningen Academic Publishers, pp 169–175

Heatwave Resilience: What Do We Know About Its Monitoring, Assessing and Forecasting?

Chenyu Yang, Junqing Tang, Pengjun Zhao, and Jing Wang

Abstract In the studies of climate change, the intensity, frequency, and duration of heatwave situations have escalated and are projected to continuously worsen in coming years, leading to detrimental effects on ecosystems, infrastructure, public health, and social life. To effectively mitigate the adverse effects of heatwaves and foster sustainable development, it is essential for urban areas to prioritize the consideration of the concept of "heatwave resilience". However, the current approaches to addressing heatwave resilience, including its assessment, monitoring, and forecast, are still in the early phase and need further development. In this chapter, we structured the content into six sections. We begin by introducing the background and significance of heatwave resilience and elaborating on related concepts and their significance. Next, we illustrate the impacts of heatwaves on modern cities. We then delve into the assessment framework and methods of heatwave resilience based on its key determinants, followed by a discussion on the monitoring methods, tools, and platforms available. We also survey the forecasting methods, highlighting various predictive model frameworks that enhance preparedness for heatwave impacts. Lastly, we discuss future directions and possibilities for building heatwave resilience. We have put forth a set of suggestions and implications for policy-makers to address heatwave resilience, which serves as a novel paradigm for appraising prospective policy shifts, government grants, and community education engagements. This holds significance for maintaining sustainable urban development as well as public health and social justice.

Keywords Heatwave · Climate change · Heatwave resilience assessment · Heatwave resilience monitoring · Heatwave resilience forecasting

C. Yang · J. Tang (✉) · P. Zhao · J. Wang
School of Urban Planning and Design, Shenzhen Graduate School, Peking University, Shenzhen 518055, China
e-mail: junqingtang@pku.edu.cn

B. He et al. (eds.), *Urban Climate and Urban Design*, Urban Sustainability,
https://doi.org/10.1007/978-981-96-1521-6_3

1 Introduction

In the context of global warming, the frequency of extreme heatwaves is projected to significantly increase on a global scale [9]. However, while heatwave frequency, duration, and cumulative heat have exhibited predominantly significant increasing patterns since 1950, it is important to note that these trends have not followed a consistent, unvarying trajectory [35]. According to data from the global observational dataset from 1950 to 2017, there is clear evidence of significant increasing trends in heatwave frequency, duration, and cumulative heat since 1950, with a notable increasing trend in recent decades. Furthermore, on a global scale, heatwave events encompass vast regions across the globe, exhibiting varying degrees of intensity amplification in different geographical areas [35]. For example, in the year 2003, a severe heatwave event occurred in Western Europe, marked by temperatures that reached their highest levels since the beginning of the sixteenth century [26]. Subsequently, in 2009, a heatwave in southeastern Australia resulted in the loss of 374 lives, which was twice the fatality count of the subsequent bushfire event [49]. Additionally, the 2010 Russian heatwave, lasting over a month, claimed about 54,000 lives [30].

In response to the challenges posed by such climate change, the Paris Agreement was officially adopted during the 2015 Paris Climate Conference. Its overarching objective is to limit the rise in the global mean temperature to a level substantially below 2 °C above pre-industrial levels, with a dedicated commitment to keeping the temperature increase within the narrower margin of 1.5 °C [9]. The critical need to restrict the temperature increase to 2 °C or less is underscored by the profound consequences that surpassing this threshold could entail for both human society and natural ecosystems. This would subject them to exacerbated impacts of climate change. As articulated by climate scientist Daniela Jacob from the Climate Service Center Germany, the ongoing trend of global warming is unprecedented, signifying an increased occurrence of extreme weather events, characterized by heightened intensity and prolonged durations.

In response to the prescribed temperature thresholds of 1.5 °C and 2 °C set by the Paris Agreement, Dosio et al. [9] conducted pertinent research. Their findings indicate that in a world experiencing a 1.5 °C temperature increase, approximately 13.8% of the global population will encounter severe heatwaves at least once every five years. This proportion significantly escalates to nearly three times higher, reaching 36.9%, under a 2 °C warming scenario, representing a difference of approximately 1.7 billion people.

Currently, the Earth's temperature continues to rise and is progressively nearing the 2 °C temperature threshold. According to the "2022 Global Climate Status" report released by the World Meteorological Organization (WMO), the global average temperature exceeded pre-industrial levels by approximately 1.15 °C in 2022, leaving only a limited margin of 0.85 °C before reaching the 2 °C limit. Furthermore, in July 2023, the world witnessed the hottest month ever recorded, with the global average temperature surging to approximately 1.5 °C above pre-industrial levels. This worldwide temperature increase has been accompanied by the occurrence of severe regional heatwaves. For example, China achieved a new national record with temperatures reaching 52.2 °C in Xinjiang in 2023 [5]. Concurrently, vast areas of the United States have been enduring extreme heat, causing ocean temperatures off the Florida coast to resemble those of a hot tub, which raises concerns about marine ecosystems. Additionally, long-lasting high temperatures in southern Europe resulted in a significant increase in hospitalizations due to heat-related issues (2023). The extreme heatwave events of 2023 are no longer infrequent occurrences and should be regarded as the prevailing standard (2023). In this context, addressing the impact of heatwaves has become a significant scientific and societal challenge in contemporary human society.

Before embarking on an investigation into strategies for mitigating the impact of heatwaves, it is crucial to establish a precise definition of high-temperature heatwaves. These heatwaves are characterized by elevated temperatures, increased humidity levels, and long durations, resulting in discomfort among individuals and potential threats to public health and safety. Furthermore, they lead to a rise in energy consumption and disrupt various societal activities and productivity due to adverse weather conditions [13].

The absence of a universal definition of heatwaves is primarily attributed to the influence of local acclimatization and adaptation on the effects of extreme heat. Even within a specific region, there can be multiple interpretations of what constitutes a "heatwave" event, often dependent on different temperature thresholds or timeframes [47]. In general, distinctions in heatwave definitions revolve around three fundamental aspects: (1) the selection of a temperature metric, (2) the level of intensity, usually represented by a specific temperature threshold; and (3) the duration over which the heatwave persists [53]. Within the realm of heatwave definitions, a common contemporary research approach is to utilize a criterion that considers a daily maximum temperature of 35 °C (approximately within the top 1% of temperature values) enduring for three or more consecutive days [47].

It is imperative to recognize that heatwaves impose impacts on a diverse array of systems, influencing not only human health but also human-made infrastructure and natural ecosystems [34]. Given the intricate nature of cities as complex and vast systems, heatwaves can have notably harsh consequences within urban environments, particularly when urban heat island effects amplify extreme temperatures. In this context, urban systems exhibit increasing susceptibility [16]. Therefore, in response to the susceptibility of urban systems, the concept of resilience is proposed. According to Luthar et al. [27], the definition of resilience is a dynamic process of effectively coping with adversity and maintaining positive adaptation. Based on the concepts of heatwave and resilience, we define heatwave resilience as follows: heatwave resilience is defined as the ability of various segments in the ecosystem, infrastructure, human health, and social life to effectively cope with challenges from heatwave and maintain positive adaptation. Therefore, the development of heatwave resilience is an essential undertaking for modern cities, especially megacities in the world, as they grapple with the growing reality of global warming and the increasing frequency of heatwaves. The incorporation of measures to bolster resilience against extreme heat in urban planning is of paramount importance on a global scale (2023).

Currently, most of the research related to heatwaves primarily focuses on the heatwave itself, while the approaches to addressing heatwave resilience, including heatwave resilience assessment, monitoring, and forecasting, are still in their initial phases and require further development. Due to the fact that there is still a lack of a comprehensive review of these approaches to heatwave resilience, in this chapter, our focus is on the resilience side, including how we can use different measures to assess, monitor, and forecast heatwave resilience in infrastructure systems, ecosystems, and public health systems, though this series of processes may also encompass the heatwave itself.

2 Impacts of Heatwave on Modern Cities

Because heatwave resilience falls under the umbrella of the broader concept of urban resilience, we kick off our review journey by first looking at how heatwaves might pose significant impacts on overall urban resilience in the most significant domains. In this vein, to date, the influence of climate change-induced extreme high-temperature weather on urban resilience mainly encompasses three key domains, namely urban ecosystems, urban infrastructure systems, and urban public health systems, respectively.

- *Impact on urban ecosystem*

Heatwaves have a primary impact on urban ecosystems, affecting both the urban green environment and the communities of urban lives. Urban green areas play a crucial role in mitigating the urban heat island effect, reducing energy consumption, and enhancing air quality. However, prolonged exposure to extreme heat and drought can subject plants to stress, thereby compromising their capability to provide essential ecosystem services. Changes in urban vegetation can trigger a ripple effect across the entire ecosystem. Reduced plant diversity disrupts insect populations, impacting processes like pollination and ecological interactions. Altered food and shelter availability affect urban wildlife, changing their distributions and behaviors. Heatwaves strain urban forests, aggravating water shortages as trees and plants demand more water. This competition for water resources highlights the need for proactive urban planning. In summary, urban heatwaves challenge ecosystems by stressing plant life, reducing biodiversity, and affecting wildlife. To enhance urban resilience, cities should integrate human and physical geography, focusing on green infrastructure, urban planning, and climate adaptation measures [52].

- *Impact on infrastructure system*

Urban infrastructures, including transportation, electrical systems, medical services, industrial installations, agricultural facilities, and livestock-related infrastructure, are all subject to varying impacts caused by the detrimental consequences of heatwaves.

For instance, certain regions in the United States are predicted to experience an energy deficit of nearly 20%, as projected temperature extremes outpace the capacity of existing infrastructure [31]. This will place added pressure on medical authorities, potentially resulting in a surge in ambulance callouts and hospital admissions. Moreover, heatwaves in Australia have led to the warping of railway tracks [29] and placed significant stress on power supply systems [6]. The insufficient power supply leads to reduced access to air conditioning and other efficient cooling measures for individuals, resulting in the consequent adverse effect on public health. The consequences of extreme heat also reverberate through agricultural industries. The Russian grain harvest, for instance, experienced a substantial 30% reduction following the 2010 heatwave [2], primarily due to the sensitivity of crops to high temperatures, which can adversely affect grain-filling and reproductive processes [1]. Other crops like rice are similarly vulnerable to extreme temperature fluctuations [23]. Furthermore, bovine livestock and their milk production also suffer adverse effects under such high-temperature conditions [11].

- *Impact on public health system*

The effects of severe heatwaves on human well-being typically do not occur immediately [34]; however, they can give rise to heat-related illnesses, particularly among socially and economically vulnerable populations [17]. Heatstroke tends to worsen underlying health conditions, disproportionately affecting the elderly, young individuals, and those engaged in outdoor labor and frequently leading to fatalities days, or even months, later. Furthermore, when contrasting nocturnal heatwaves with their daytime counterparts, the former poses a significantly graver threat to human well-being. This is attributed to the fact that humans require cooler nighttime temperatures for recovery, enabling them to better cope with extreme heat conditions during the subsequent day [34]. Furthermore, some side effects that combine heatwaves and other environmental impacts could also be harmful to public health. For instance, poor air quality stemming from large-scale wildfires might be caused by heatwaves, and increased air pollution can have detrimental effects on public health, making a city's inhabitants more vulnerable to climate-related challenges. The convergence of these elements underscores the essential requirement for flexible and resilient healthcare systems within urban areas to enhance their capacity to withstand and respond to these challenges [54].

Acknowledging the increased susceptibility in the above three domains under the urban climate change context, it becomes possible to undertake specific initiatives aimed at bolstering urban resilience against heatwave impacts and adopting strategies to mitigate these consequences. These strategies may encompass investments in climate-adaptive infrastructure, the development of green spaces, and enhanced disaster preparedness. Such actions can enhance their ability to confront the challenges posed by urban climate and, at the end of the day, safeguard the welfare and prosperity of urban populations. In this vein, it is crucial to have a better understanding of the concept of "*heatwave resilience*", indicating how cities, all evidence people living within, could effectively cope with heatwaves, and build more resilience into the systems. By doing so, we need a better understanding of how to assess, monitor, and forecast heatwave resilience. In the following sections, we will frame our discussion around these three aspects.

3 Heatwave Resilience Assessment

This section focuses on the assessment approaches of heatwave resilience. It first outlines the framework for evaluating heatwave resilience and the criteria for selecting indicators. Subsequently, through an extensive review of previous research, the assessment techniques and tools related to heatwave resilience are synthesized, encompassing a variety of systems, platforms, and data. Finally, the existing challenges in this field are summarized, and directions for future research are put forth.

As part of urban resilience, heatwave resilience plays a crucial role in influencing the overall urban systems in terms of building overall resilience. But first, it is essential to note that assessing heatwave resilience, in fact, is to evaluate resilience that relates to the impact of heatwaves, which one can learn the critical indicators of assessing it from a background understanding of how urban resilience can be effectively evaluated first. Creating urban resilience assessment frameworks is an effective approach to simplify the intricacies of urban resilience and elucidate the connections among its various facets. Ideally, an effective urban resilience assessment framework should encompass all dimensions of urban systems [40]. Table 1 provides a brief summary of resilience assessment methods, including those specific to heatwave resilience. The specific details of these methods will be introduced in the following text.

To offer a conceptual framework for developing resilience assessment tools, Sharifi and Yoshiki [40] analyzed 29 existing resilience assessment frameworks and identified five major dimensions of urban resilience: materials and environmental resources, society and well-being, economy, the built environment and infrastructure, and governance and institutions. Each of these dimensions encompasses a comprehensive list of related criteria. According to this framework, a series of indicators could be chosen to build the heatwave resilience assessment framework. This includes indicators related to socioeconomic characteristics, safety and well-being, equity, and diversity within the dimensions of society; indicators related to robustness and redundancy of critical infrastructure, infrastructure efficiency, and land use and urban design within the dimensions of the built environment and infrastructure dimension; indicators related to contingency, emergency, and recovery planning within the dimensions of governance and institutions dimension. Similarly, Li and Zhai [24] developed an assessment indicator system for urban disaster resilience in the country, utilizing the Resilience Baseline Model (BRIC) and expert consultation methods in the context of China's comprehensive system covers six essential dimensions: economic resilience, social resilience, environmental resilience, community resilience, infrastructure resilience, and organizational resilience. By selecting the proper indicators in this framework, it could also be adaptable to assess heatwave resilience.

Table 1 Summary of resilience assessment methods

Type	Literature	Key information	Object	Source
Assessment framework	Urban resilience assessment: Multiple dimensions, criteria, and indicators	It analyzed 29 resilience assessment frameworks, identifying five major dimensions of urban resilience, with indicators to build a heatwave resilience assessment framework	Urban resilience	Sharifi and Yoshiki [40]
	Research on urban disaster resilience assessment and enhancement strategies in China	It developed an assessment indicator system for urban disaster resilience, utilizing the Resilience Baseline Model (BRIC) and expert consultation methods	Urban disaster resilience	Li and Zhai [24]
Assessment indicators	Indicators for monitoring urban climate change resilience and adaptation. Sustainability	It provided indicators set to measure urban climate resilience and monitor adaptation activities, constructed through literature reviews, online surveys, and expert ratings	Urban climate resilience	Feldmeyer et al. [12]
	Indicators and monitoring systems for urban climate resiliency	It constructed the urban climate change and resiliency indicator and monitoring system	Urban climate resilience	Solecki and Cynthia [41]
	Heat stress risk and resilience in the urban environment	It identified three critical dimensions of heatwave resilience assessment	Heatwave resilience	Hatvani-Kovacs et al. [15]

(continued)

Table 1 (continued)

Type	Literature	Key information	Object	Source
	A heat vulnerability index: spatial patterns of exposure, sensitivity and adaptive capacity for Santiago de Chile	It developed the Heat Vulnerability Index (HVI), which measures vulnerability through adaptive capacity, exposure, and sensitivity	Heatwave vulnerability	Inostroza et al. [18]
	Design, monitoring and evaluation of resilience interventions: conceptual and empirical considerations	It determined heatwave resilience indicators based on resilience causal frameworks	Heatwave resilience	Béné et al. [3]

Nevertheless, one of the foremost challenges in establishing a heatwave resilience assessment framework is the selection of suitable indicators. Due to the lack of a unified definition for resilience and the complexity of interactions between various systems involved, there is no clear standard for selecting heatwave resilience indicators. To address this issue, some scholars have made efforts to classify and select indicators for assessing resilience against heatwaves. For example, Feldmeyer et al. [12] have provided an indicator set to measure urban climate resilience and monitor adaptation activities. These indicators were constructed through literature reviews, online surveys, and expert ratings. Using a similar method, The New York City Panel on Climate Change (NPCC) has also constructed the Urban Climate Change and Resiliency Indicator and Monitoring System [41], where the assessment of heatwave resilience is an important component of it.

Based on the characteristics of resilient cities, some scholars have also constructed indicators to assess heatwave resilience. For instance, Hatvani-Kovacs et al. [15] identified three critical dimensions within heatwave resilience assessment: heatwave intensity, heatwave impact, and population heat stress resilience. Among these three critical dimensions," heatwave intensity" refers to the severity of a heatwave, typically determined by the extent and duration of high temperatures. It measures how extreme the heatwave conditions are and their potential impact on various aspects of society. "Heatwave impact" covers the consequences of heatwaves on urban infrastructure and public health. "Population heat stress resilience" depends on the heat stress resistance of the built environment, adaptation capacity, and vulnerability. Another example is the Heat Vulnerability Index (HVI), which was developed using spatial analysis in a Geographic Information System (GIS) [18]. In this approach, the HVI is computed as a function of adaptive capacity (A), exposure (E), and sensitivity (S) to identify variations in vulnerability levels through combinations of individual indices. Land Surface Temperature is employed to gauge exposure, while sensitivity is evaluated using factors such as the elderly population, very young population, disabled population, family structure, education level, and unemployment. Adaptive capacity is assessed using indicators including access to communication technologies, access to water supply, material index, access to medical services, roads, and the Normalized Difference Vegetation Index (NDVI).

In addition, other scholars have determined heatwave resilience indicators based on resilience causal frameworks [3]. In this framework, the indicators are categorized into the following three components, namely the ex-ante component, which includes resilience capacity and the initial vulnerability level; the disturbance component, referring to heatwave intensity; and the ex-post component, including resilience capacity and vulnerability level.

Another primary concern in constructing a heatwave resilience assessment framework is the selection of assessment methods. This process aligns with the identification of heatwave resilience indicators, depending on the specific scope of the research. Typically, there are three primary categories of methods for evaluating urban resilience: quantitative approaches, qualitative approaches, and a combination of both [46]. Scorecard Toolkit and Resilience Matrix, Resilience Workbooks, and Multilateral [44] are predominantly used in cases where a resilience assessment framework encompasses multiple systems and stakeholders for evaluating heatwave resilience. A typical example is the Urban Climate Change and Resiliency Indicator and Monitoring System constructed by The New York City Panel on Climate Change Collaboration [41]. Quantitative methods encompass Low-Dimensional Metrics, System State-Based Models, Network-Based Approaches, Ontology-Based Interdependence Approaches, Bayesian Network (BN) Tools, Agent-Based Simulation Tools [44], which are often utilized for heatwave resilience assessment together with simulation and prediction.

From the review, unfortunately, it is indicative that studies focusing on directly assessing heatwave resilience are extremely lacking, which shows the criticality of considering this issue in future studies. In conclusion, we have identified the following challenges that persist in the field of heatwave resilience assessment.

- The characteristics of heatwave events induced by climate change, such as their uncertainty, non-linear patterns, and long timeframes, present difficulties in designing a resilience assessment framework and formulating adaptation programs [8].
- While aggregate monitoring and evaluation systems for assessing heatwave resilience are in development, only a limited number are fully operational.
- Individuals striving to assess resilience encounter significant obstacles related to data accessibility, comprehensiveness, and quality [43].
- We believe that the development of composite indicators is an essential auxiliary tool for urban resilience assessment, while high-level indicators must be complemented with local knowledge and contextual information to enable meaningful decision-making [43].

In light of the challenges we have outlined, the assessment of urban heatwave resilience offers numerous prospects in the meantime. Firstly, the development of a multidimensional heatwave resilience assessment framework promotes collaborative efforts and broad engagement. Secondly, addressing the scale of resilience and identifying the primary system for resilience construction helps to establish a comprehensive and holistic assessment framework, which takes the complexities and conflicts inherent in resilience assessment into account. Lastly, regarding the challenge of data accessibility, establishing data integration and visualization methods, such as the Data Integration and Visualization En Masse (DIVE) web-based data catalog could be of particular help, which provides a partial solution to the resilience data challenges observed in the creation of the national index [43].

4 Heatwave Resilience Monitoring

Monitoring heatwave resilience is crucial for urban sustainability and general well-being in cities. It facilitates the issuance of early warnings and the implementation of measures like providing emergency medical services and establishing temporary shelters to minimize damage. Additionally, monitoring social vulnerability assists municipalities in aiding marginalized groups, such as the elderly, low-income households, and homeless individuals. Heatwave resilience monitoring is an integral component of urban climate adaptation, aiding urban planners in effectively addressing climate change by implementing sustainable measures such as increasing urban green spaces, revamping building designs, and curbing carbon emissions to mitigate the detrimental consequences of climate change on urban areas and their inhabitants.

This section primarily elucidates the procedures for monitoring heatwave resilience. However, prior to delving into the specifics of heatwave resilience monitoring, it is imperative to acknowledge that the primary concerns in this process revolve around the careful selection of assessment indicators and the formulation of a working definition and scope of heatwave resilience. Moving forward with the monitoring process, attention is directed to two principal areas: (1) monitoring the heatwave itself and (2) observing dynamic changes within the system. Drawing from an extensive review of previous studies, a comprehensive overview of the tools, models, and methods employed in this endeavor is provided. In the realm of data collection for monitoring heatwave resilience, three main data collection tools are currently in use (summarized in Table 2): sensor technology, satellite remote sensing, and internet platforms that support data collection and mining. These tools have a vital role in the field of heatwave resilience monitoring, offering a range of data collection and analytical options, contributing to its diversity and comprehensiveness.

The utilization of sensors is extensive in the domain of heatwave resilience monitoring. For example, in Japan, the DoCoMo environmental sensor network has been deployed to observe weather conditions at around 4,000 mobile phone stations [33]. Additionally, sensors could be integrated with other devices, providing

Table 2 The summary of heatwave resilience monitoring methods

Type	Literature	Methods	Description	Source
Sensor technology	Participatory sensing data tweets for micro-urban real-time resiliency monitoring and risk management	DoCoMo environmental sensor network	It observed weather conditions at around 4,000 mobile phone stations	Murakami et al. [33]
	Field-based Heatwave Risk Assessment of Outdoor Workers Measured by Wearable Sensors	A blend of wearable sensors and surveys	It assessed the heat stress experienced by integrating sensors with a daily questionnaire	Kim and Yoo [22]
Satellite remote sensing	Policy-relevant indicators for mapping the vulnerability of urban populations to extreme heat events: a case study of Philadelphia	Land Surface Temperature (LST) data integrating ground-based data	It combined ground-based data with satellite surface temperature data to identify heat vulnerability areas	Weber et al. [50]
	Satellite monitoring of summer heat waves in the Paris metropolitan area	NOAA-AVHRR images and HRV image	It analyzed the spatial fluctuations in land surface temperature (LST)	Dousset et al. [10]
	Profile and concentric zonal analysis of relationships between land use/land cover and land surface temperature: Case study of Shenyang, China	Landsat 8 image	It investigated relationships between land use/land cover types and land surface temperature (LST) patterns	Zhao et al. [56]
Internet platform	Participatory sensing data tweets for micro-urban real-time resiliency monitoring and risk management	Social media platforms	It used geo-tagged tweets related to heat, along with sparse weather sensor data, to analyze urban temperatures	Murakami et al. [32]

(continued)

Table 2 (continued)

Type	Literature	Methods	Description	Source
	Google search patterns monitoring the daily health impact of heatwaves in England: How do the findings compare to established syndromic surveillance systems from 2013 to 2017?	Internet search data	It compared daily trends in syndromic surveillance indicators of heat-related morbidity with symptom-based Google search terms associated with heatwaves	Green et al. [14]
	Global & geographical mapping and visualization method for personal/ collective health data with 5D world map system	5D World Map System	It analyzed natural hazards and environmental anomalies, including ambient temperature	Tropmann-Frick ([48]
	Mapping Urban Resilience for Spatial Planning—A First Attempt to Measure the Vulnerability of the System. Sustainability	GIS tools	It visualized the influence of various shocks and linear pressures on sensitivities within Moncalieri	Brunetta and Salata [4]
	Agile IoT for critical infrastructure resilience: Cross-modal sensing as part of a situational awareness approach	The Internet of Things (IoT)	It gathered temperature fluctuations and generated actionable insights by linking digital and physical objects	Russell et al. [38]
	Advances in computer vision-based civil infrastructure inspection and monitoring	Artificial Intelligence (AI)	It created a platform that utilized Unmanned Aerial Vehicles (UAVs) for the automated collection of remote data, including videos and photographs	Spencer et al. [42]

researchers with a wider range of monitoring techniques to obtain data that match their research objectives. Another approach involves tower monitoring, which allows the surveillance of surface temperatures and the collection of data pertinent to heat-wave resilience. Tower monitoring typically entails the placement of sensors, instruments, or data collection equipment on tall towers or structures to measure a wide range of environmental and atmospheric variables [33]. Moreover, integrating sensors with other data collection techniques, such as questionnaires, allows for the capture of qualitative data that might be difficult to monitor. This, in turn, offers an alternative means of monitoring urban heatwave resilience. For instance, some studies concentrate on narrower scales and relatively limited sample sizes, undertaking more extensive public health surveillance through a blend of sensors and surveys [22]. By making use of wearable sensors, they assess the heat stress experienced by outdoor workers on the worksite, simultaneously administering a daily questionnaire to gauge their psychological comfort levels during their job tasks [22].

Remote sensing technology is also widely employed in various applications, including monitoring surface temperatures, and it serves as a vital tool in heatwave resilience monitoring. By consistently recording thermal infrared radiation, one can determine the radiative temperature of the Earth's surface. Remote sensing technology provides extensive spatial coverage and high resolution, enabling the monitoring of land surface temperature variations across diverse geographic regions and temporal scales. Here, we have outlined several applications of remote sensing in the context of heatwave resilience monitoring. An important characteristic of remote sensing data, which is supplied by organizations like the National Oceanic and Atmospheric Administration (NOAA) and the US Geological Survey (USGS), is its ability to provide extensive coverage over various scales and at regular intervals [7, 50]. For instance, Weber et al. [50] emphasize the potential for effectively combining ground-based observations with satellite data on surface temperature to delineate heat vulnerability areas in Philadelphia. Another example is that Dousset et al. [10] harnessed satellite remote sensing imagery in their study of summer heatwaves. In this research, a series of 61 NOAA-AVHRR (Advanced Very High-Resolution Radiometer) images and one high-resolution visible (HRV) image from SPOT were employed to analyze the spatial fluctuations in land surface temperature (LST) throughout the diurnal cycle during the heatwave. Similarly, remote sensing devices such as MODIS (with a resolution of 500 m), Aster (with a resolution of 90 m), and Landsat 8 (with a resolution of 30 m) are employed for higher spatial resolution land surface temperature monitoring [33], also, Zhao et al. [56] used Landsat 8 image to investigate relationships between land use/land cover (LULC) types and land surface temperature (LST) patterns in Shenyang, China. Additionally, airborne monitoring is also a remote sensing method. To observe micro-scale heatwave conditions, an airborne monitoring operation is carried out in partnership with SkyMap Co. Ltd. The images captured during each monitoring session are merged using mosaic processing to create a surface temperature map. This map effectively reflects the elevated temperatures on rooftops and the lower temperatures near rivers. However, this approach has several drawbacks, including its high cost, limited daily monitoring feasibility, and restricted coverage within the monitoring area [33].

Internet platforms, providing chances to collect crowd-sourcing data and monitoring, are a recent addition to the monitoring toolbox in the context of modern technological developments. It is based on a bottom-up, multi-agent participation approach, utilizing the vast network-generated data from individuals. This data is extracted and processed to obtain the relevant information required for assessing heatwave resilience, ultimately fulfilling the objective of heatwave resilience monitoring. Social media platforms, such as Twitter, provide real-time and high-resolution spatial information that could be valuable for estimating climate conditions. Murakami et al. [32] employed geo-tagged tweets, which are considered participatory sensing data, for the analysis of urban temperatures. They began by identifying tweets related to heat, which they referred to as "*heat-tweets*", using a statistical model framework that utilized copula modeling techniques. In their study, they combined these heat-tweets with sparsely distributed weather monitoring sensor data to inform the spatiotemporal Gaussian process interpolation of temperatures, as demonstrated in their application example. Furthermore, internet search data can be used to monitor the resilience of the public health system to heatwaves, which represents one facet of monitoring urban heatwave resilience A retrospective observational study was conducted in England from June 2013 to September 2017. This study aimed to compare the daily trends in validated syndromic surveillance indicators related to heat-related morbidity with symptom-based Google search terms associated with heatwaves [14].

Regarding heatwave monitoring platforms, there are specialized platforms designed for monitoring both heatwaves and their impact on urban systems. These platforms collect operational data associated with urban systems affected by heatwave disasters [33]. For example, Tropmann-Frick [48] introduced the application of the 5D World Map System as an "Environmental AI System", which is an integrated platform, not just natural hazards, but also environmental anomalies, including ambient temperature. This system facilitates the implementation of the United Nations' Sustainable Development Goals (SDG) 9 and 11 across various levels, including the global, regional, national, sub-regional, and city levels. By utilizing the Sensing-Processing-Actuation (SPA) functions of the 5D World Map System, comprehensive climate-induced hazard risk analyses can be conducted in diverse contexts. Additionally, the system is capable of assessing future risks by integrating global climate scenarios downscaled to specific regions and countries.

Generally speaking, it is clear that there is a lack of monitoring platforms for heatwave resilience. Furthermore, the existing studies predominantly emphasize the monitoring of heatwave resilience in specific systems and lack an all-encompassing heatwave resilience monitoring platform.

Following the previous discussion, we turn our attention to platforms developed for monitoring heatwave resilience and the integration of data, including GIS, Using GIS tools for spatial analysis is instrumental in gaining insights into urban vulnerabilities and resilience, facilitating the evaluation of disaster risks and resilience, including those posed by heatwaves. This assists urban planners in pinpointing regions that are at a higher risk of being affected by such disasters. The GIS-based measurement of resilience with indices or scorecards is commonly employed in previous research [36, 37, 45, 51]. For example, Brunetta and Salata [4] developed a GIS-based method for visualizing the influence of various shocks and linear pressures on sensitivities within Moncalieri, Italy, as the initial phase in gauging resilience.

The Internet of Things (IoT) can also serve as a platform for monitoring heatwave resilience. As described by Russell et al. [38], IoT can gather temperature fluctuations.

It involves linking digital and physical objects, such as smartphones, transportation infrastructure, and energy assets, through effective information exchange over the Internet. This connectivity allows for data collection, communication, processing, and the generation of actionable insights for various applications, services, and decision-making processes.

We can also leverage Artificial Intelligence (AI) and Machine Learning (ML) as described by Spencer et al. [42] to create a platform that utilizes Unmanned Aerial Vehicles (UAVs) for the automated collection of remote data, including videos and photographs. This data can then be processed and inspected using engineering algorithms to interpret the condition of infrastructure such as roads, railways, and pipelines, allowing for automation and speed, especially in response to the challenges posed by the heat wave impact.

Additionally, the prevailing trend in resilience research is the development of digital twin technology. An interactive urban digital twin has the potential to serve as a valuable tool for facilitating collaboration among diverse stakeholders to formulate and execute adaptive strategies aimed at reducing the risks associated with disasters resulting from extreme natural events. This tool should provide support throughout the entire iterative process, encompassing assessment, implementation, and monitoring. Currently, urban digital twin technology, a contemporary digital technology, is predominantly applied to enhance flooding resilience in urban areas, while its application in heatwave resilience remains significantly underexplored. Consequently, there is a growing need for more scholars to conduct comprehensive research in this particular domain in the future [20].

From the above, the current difficulties and challenges in heatwave resilience monitoring include the following and we also elucidated the future development opportunities in this field.

- Lack of a unified heatwave resilience measurement framework: Currently, there is still no consistent framework for measuring resilience.
- Improvement in the temporal and spatial accuracy of monitoring data: Due to the real-time and dynamic nature of resilience at the urban level, it is necessary to use time-continuous and spatially detailed data in related research. Taking temperature monitoring as an example, a challenge is obtaining open-source weather data at the micro-level, which poses a difficulty for research.
- Emphasis on single-system resilience monitoring: Current heatwave resilience monitoring research primarily focuses on the monitoring of resilience in specific urban systems, while comprehensive heatwave resilience monitoring research has not been sufficiently explored.
- "Hybridized" natural and engineered conditions: Urban areas and cities are characterized by their high complexity, consisting of numerous integrated and interrelated systems [41]. This complexity presents difficulties for monitoring systems and for discerning clear cause-and-effect relationships.

In turn, for future opportunities, the primary task is to establish a common resilience measurement framework to standardize the assessment of resilience across different urban systems. This could help decision-makers gain a more accurate understanding of a city's resilience under different conditions, enhancing the scientific rigor and accuracy of decision-making.

Secondly, improving the temporal and spatial accuracy of monitoring data is an urgent challenge. With ongoing technological advancements, in addition to using more precise satellite remote sensing data, meteorological observations, and sensor networks to capture micro-scale changes in urban environments, we also advocate for the establishment of a bottom-up data-sharing system. Through the collective efforts of individuals in data collection, an open data platform can be constructed, enabling more accurate acquisition of micro-level data, such as temperature, humidity, wind speed, and more. This will facilitate a finer assessment of urban resilience.

Another important opportunity lies in conducting more research on more comprehensive heatwave resilience monitoring. While current research primarily focuses on monitoring the resilience of individual urban systems, we can encourage more research to explore holistic heatwave resilience at the interurban, regional, and global scales. This will enhance our understanding of interactions between urban systems and provide comprehensive information to urban and regional decision-makers for better preparedness and response to future heatwave events.

Furthermore, it's worth noting that the continuous development of machine learning and artificial intelligence technologies offers new opportunities for heatwave resilience monitoring. These technologies can be utilized for data analysis, the construction of predictive models, and the development of real-time decision support systems. International cooperation is also a significant opportunity. Heatwave issues transcend city, regional, and national boundaries, so cross-border collaboration can facilitate the sharing of best practices, data, and experiences. Through international cooperation, we can better address global heatwave challenges and work together to build more resilient and sustainable urban environments.

5 Predicting Heatwave Resilience

Heatwave resilience prediction is based on effective assessment and accurate monitoring. Studies focusing on forecasting heatwave resilience (summarized in Table 3) are also relatively underdeveloped in the field.

We found that most of the relevant previous studies focused on forecasting heatwave per se. For instance, Zhu and Li [55] employed a spatiotemporal projection model (STPM) to extend the forecast window (ranging from 5 to 30 days in advance) for surface air temperature (SAT) and heatwaves in China. However, due to the evolving relationship between predictors and predictands caused by climate change, these kinds of efficient predictive models are experiencing difficulties like accuracy reduction. To overcome this challenge, Khan et al. [21] endeavored to create a heatwave prediction model resilient to climate change using machine learning (ML) algorithms, including Support Vector Machines (SVM), random forest, and artificial neural networks. Besides, Jacques-Dumas et al. [19] proposed that physics-based

Table 3 Summary of heatwave resilience predicting methods

Type	Literature	Method/model used	Key information	Source
Heatwave prediction	Urban resilience assessment: Multiple dimensions, criteria, and indicators	Spatiotemporal Projection Model (STPM)	It forecasted surface air temperature (SAT) and heatwaves 5 to 30 days in advance	Zhu and Li [55]
	Research on Urban Disaster Resilience Assessment and Enhancement Strategies in China	Machine Learning Algorithms (SVM, Random Forest, Artificial Neural Networks)	It developed a heatwave prediction model resilient to climate change, improving prediction accuracy	Khan et al. [21]
	Indicators for Monitoring Urban Climate Change Resilience and Adaptation. Sustainability	Deep learning algorithms	It used climate model outputs for training to predict prolonged extreme heatwaves	Jacques-Dumas et al. [19]
Resilience prediction	Indicators and monitoring systems for urban climate resiliency	Resilience calculation framework	It assessed the current performance and future resilience of systems using performance and capacity indicators and validated the framework with ABM	Schwind et al. [39]
Warning systems	Heat stress risk and resilience in the urban environment	Heatwave Early Warning Systems (HEWS)	It minimizes avoidable health issues during heatwaves by providing timely notifications of preventive measures	Lowe et al. [25]
	A heat vulnerability index: spatial patterns of exposure, sensitivity and adaptive capacity for Santiago de Chile	National Severe Weather Warning Service (NSWCWS)	It established a direct link between temperatures and mortality risk to create a more cohesive health forecasting system	Masato et al. [28]

weather forecast systems and climate models could be harnessed for predicting heatwave occurrences. Their research explores the utilization of deep learning algorithms trained with outputs from a climate model as an alternative method for forecasting prolonged, extreme heatwaves. Such machine learning prediction algorithms are anticipated to become increasingly valuable in the future, addressing the three principal scientific challenges in climate extreme studies: the scarcity of historical data, issues related to model sampling, and addressing model biases.

Additionally, some studies focus on the methods and platforms to predict resilience itself as a subject of inquiry. For instance, Schwind et al. [39] introduced a computational framework for predicting resilience, known as the "Resilience Calculation Framework." This framework incorporates two key metrics: "performance indicators" and "capacity indicators." Performance indicators are utilized to evaluate the current performance of a system, while capacity indicators assess the system's future resilience, taking into account its current state, standards, environmental assumptions, and potential decisions. Expanding on this prediction framework, they presented an innovative method based on performance indicators to gauge the future resilience of the system, considering uncontrollable environmental factors and government actions. For each possible future scenario, the same set of performance indicators is assessed, and the results are consolidated into a single value, defining the performance-based capacity indicator. This value represents the predicted resilience of the system [39]. Furthermore, the researchers used agent-based modeling (ABM) to validate this prediction framework [39]. This framework offers a robust tool for assessing the resilience of systems in diverse contexts and is applicable to resilience forecasting in various urban systems, including heatwave resilience.

As previously mentioned, there is a notable lack of research in the field of heatwave resilience predicting, and few existing studies nearly emphasize a single urban subsystem resilience predicting in heatwave conditions. These studies tend to have a narrow scope, with limited attention given to predicting the overall heatwave resilience of the entire urban system. Taking the public health system as an example, there is currently a widespread focus on predicting and issuing warnings regarding the impacts of heatwaves, and numerous nations have instituted heatwave early warning systems (HEWS). These systems are designed to minimize avoidable health problems during heatwaves by delivering timely notifications of preventive measures to vulnerable populations. HEWS utilizes a variety of indicators to trigger heatwave warnings, which include factors like the maximum temperature, the heat index (which takes into account both air temperature and relative humidity), synoptic conditions (airmass), or various combinations of temperature and duration [25]. Similarly, aiming to enhance the information accessible to responders in the health and social care sector, Masato et al. [28] devised a prototype for a novel health forecasting alert system, aligning it with the approach employed by the Met Office's (MO) National Severe Weather Warning Service (NSWWS). The system establishes a more direct connection between temperatures and the risk of mortality, aiming to create a health forecasting system that is more cohesive with other weather alerts.

We believe that such scarcity in forecasting heatwave resilience is largely due to the fact that how to forecast heatwave resilience is still a controversial issue. The reasons might be as follows:

- The complex interdependencies among multiple urban systems make it challenging to accurately forecast the performance of the urban mega-system under heatwave impacts because it is a multifaceted issue that is difficult to grasp comprehensively as a whole.
- The dynamic nature of resilience makes it a challenging, unpredictable property and this characteristic makes it difficult to accurately predict, which results in complex feedback loops and multilateral coupling relationships, revealing its inherent complexity. In light of this inherent complexity, grasping the dynamic and intricate adaptive features naturally presents numerous challenges for researchers. Considering that the challenges encountered in assessment and monitoring are foundational issues in forecasting, it is clear that the difficulties in predicting heatwave resilience outweigh the available methods and solutions when effective solutions for the former two are lacking.

These two challenges also represent opportunities in our future research. This is crucial since our ultimate objective is to realize effective heatwave resilience forecasting through the means of monitoring and assessment, such an accurate and effective forecasting system could enable early warning and preparedness, allowing cities to implement heatwave mitigation measures, thereby safeguarding public health and reducing heat-related illnesses and fatalities. Moreover, accurate predictions facilitate the optimization of energy usage, reducing strain on power grids during peak heat periods. This, in turn, lowers the risk of heat-induced blackouts and ensures uninterrupted essential services. Additionally, proactive heatwave forecasting can aid urban planning and infrastructure development, ensuring that cities are better equipped to handle rising temperatures and maintain overall resilience in the face of climate change.

6 Implications and Future Directions

It is evident that, in general, there are some fundamental questions that need to be addressed in heatwave resilience research, roughly categorized into three dimensions, which could form the focal points for future studies.

1. The precise definition of heatwave resilience remains a topic of ongoing ambiguity within the field.
2. There is still a lack of clarity regarding the availability of accurate measurement methods for heatwave resilience. The process of quantifying heatwave resilience continues to lack a unified standard.
3. Challenges persist in monitoring and forecasting heatwave events, with difficulties encompassing methods, theoretical frameworks, and technological pathways.

From the above review and the synthesis of the previous studies, we propose the following aspects that researchers could pay extra attention to while conducting studies related to heatwave resilience.

Firstly, Heatwave resilience and urban resilience are closely interconnected. Therefore, when building heatwave resilience, it is essential to consider its alignment with the broader urban resilience framework. This alignment should be reflected in the selection of measurement baselines, data collection, model building, and platform development to maintain consistency with the overall urban resilience construction process.

Secondly, the understanding of heatwave resilience hinges on a thorough comprehension of the heatwave itself. Thus, for effective research on heatwave resilience, we need to deepen our understanding of heatwaves, including investigating the causes of heatwaves, the interaction mechanism between heatwaves and urban heat island effects [16], and the application of monitoring, assessment, and prediction methods and tools for heatwave.

Moreover, constructing heatwave resilience requires a multi-faceted approach, primarily focusing on three major systems closely associated with heatwaves. It also entails engagement with a multitude of stakeholders, necessitating the resolution of conflicts among them. Harmonizing these conflicts is pivotal in establishing a holistic and systematic knowledge framework for heatwave resilience.

Last but not least, the assessment, simulation, and forecasting of heatwave resilience are closely intertwined with the development of Information and Communication Technologies (ICTs). We can observe that technological advancements are of significant importance in helping us overcome the challenges mentioned above and making further progress in our future research directions. Take ESPON European and Macro-regional Territorial Monitoring Tool—MRS, which was established by the European Spatial Planning Observation Network (ESPON) as an example, it aims to provide a practical and operational online platform to continuously observe the territorial development trends and patterns taking place in Europe and its macro-regions. It is a highly integrated system that encompasses information from different data sources, including climate-related information. If heatwave and its resilience module can be embedded scholars and stakeholders can comprehensively monitor the macro-level trends of heatwave evolution, and in turn, its dynamic resilience. This would indeed lead to advanced heatwave resilience studies. Even so, towards the end of this chapter, we would like to convey to the readers that investigating heatwave resilience is a collaborative effort, necessitating interdisciplinary research, and thus, the essence of this chapter is to welcome more future studies on this topic.

References

1. Barlow K et al (2013) Simulating the impact of extreme heat and frost events on wheat production: the first steps. In: 20th international congress on modelling and simulation. modeling and simulation Society of Australia and New Zealand, Adelaide, Australia

2. Barriopedro D et al (2011) The hot summer of 2010: redrawing the temperature record map of Europe. Science 332(6026):220–224
3. Béné C, Frankenberger T, Nelson S (2015) Design, monitoring and evaluation of resilience interventions: conceptual and empirical considerations. Institute of Development Studies Working Paper 459
4. Brunetta G, Salata S (2019) Mapping urban resilience for spatial planning—A first attempt to measure the vulnerability of the system. Sustainability 11(8):2331
5. China logs 52.2 Celsius as extreme weather rewrites records (2013) Reuters. https://go.nature.com/3qQGwnQ
6. Colombo AF, Etkin D, Karney BW (1999) Climate variability and the frequency of extreme temperature events for nine sites across Canada: implications for power usage. J Clim 12(8):2490–2502
7. De Sherbinin A et al (2014) Using satellite data to develop environmental indicators. Environ Res Lett 9(8):084013
8. Dinshaw A et al (2014) Monitoring and evaluation of climate change adaptation: methodological approaches
9. Dosio A et al (2018) Extreme heat waves under 1.5 °C and 2 °C global warming. Environ Res Lett 13(5):054006
10. Dousset B et al (2011) Satellite monitoring of summer heat waves in the Paris metropolitan area. Int J Climatol 31(2):313–323
11. Dunn RJH et al (2014) Analysis of heat stress in UK dairy cattle and impact on milk yields. Environ Res Lett 9(6):064006
12. Feldmeyer D et al (2019) Indicators for monitoring urban climate change resilience and adaptation. Sustainability 11(10):2931
13. General Administration of Quality Supervision, Inspection and Quarantine of the People's Republic of China, Standardization Administration of China, High Temperature and Heat Wave rating, GB/T 29457-2012
14. Green HK et al (2018) Google search patterns monitoring the daily health impact of heatwaves in England: how do the findings compare to established syndromic surveillance systems from 2013 to 2017? Environ Res 166:707–712
15. Hatvani-Kovacs G et al (2016) Heat stress risk and resilience in the urban environment. Sustain Cities Soc 26:278–288
16. He B-J et al (2021) A framework for addressing urban heat challenges and associated adaptive behavior by the public and the issue of willingness to pay for heat resilient infrastructure in Chongqing, China. Sustain Cities Soc 75:103361
17. He B-J et al (2022) Will individuals visit hospitals when suffering heat-related illnesses? Yes, but…. Build Environ 208:108587
18. Inostroza L, Palme M, De La Barrera F (2016) A heat vulnerability index: spatial patterns of exposure, sensitivity and adaptive capacity for Santiago de Chile. PLoS ONE 11(9):e0162464
19. Jacques-Dumas V et al (2022) Deep learning-based extreme heatwave forecast. Front Climate 4
20. Johannsen C, Salten J, Franke W (2022) A whole-of-government approach to climate change and the process of making cities resilient: constructing a digital twin network for urban adaptation. Environmental Informatics. Springer International Publishing, Cham, pp 165–181
21. Khan N et al (2021) Prediction of heat waves over Pakistan using support vector machine algorithm in the context of climate change. Stochastic Environ Res Risk Assess 35:1335–1353
22. Kim A, Yoo G (2023) Field-based heatwave risk assessment of outdoor workers measured by wearable sensors. Asia-Pac J Atmos Sci 59(4):447–458
23. Lanning SB et al (2011) Extreme nighttime air temperatures in 2010 impact rice chalkiness and milling quality. Field Crops Res 124(1):132–136
24. Li Y, Guofang Z (2017) Research on urban disaster resilience assessment and enhancement strategies in China. Urban Planners 33(8):5–11
25. Lowe D, Ebi KL, Forsberg B (2011) Heatwave early warning systems and adaptation advice to reduce human health consequences of heatwaves. Int J Environ Res Public Health 8(12):4623–4648

26. Luterbacher J et al (2004) European seasonal and annual temperature variability, trends, and extremes since 1500. Science 303(5663):1499–1503
27. Luthar SS, Cicchetti D, Becker B (2000) The construct of resilience: a critical evaluation and guidelines for future work. Child Dev 71(3):543–562
28. Masato G et al (2015) Improving the health forecasting alert system for cold weather and heat-waves in England: a proof-of-concept using temperature-mortality relationships. PLoS One 10(10): e0137804
29. McEvoy D, Ahmed I, Mullett J (2012) The impact of the 2009 heat wave on Melbourne's critical infrastructure. Local Environ 17(8):783–796
30. McMichael AJ, Lindgren E (2011) Climate change: present and future risks to health, and necessary responses. J Intern Med 270(5):401–413
31. Miller S, Muir-Wood R, Boissonnade A (2008) An exploration of trends in normalized weather-related catastrophe losses. Climate Extremes Soc 12:225–247
32. Murakami D et al (2016) Participatory sensing data tweets for micro-urban real-time resiliency monitoring and risk management. IEEE Access 4:347–372
33. Murakami D et al (2019) Spatiotemporal heatwave risk modeling combining multiple observations. In: IGARSS 2019–2019 IEEE international geoscience and remote sensing symposium. IEEE
34. Perkins SE (2015) A review on the scientific understanding of heatwaves—Their measurement, driving mechanisms, and changes at the global scale. Atmos Res 164:242–267
35. Perkins-Kirkpatrick SE, Lewis SC (2020) Increasing trends in regional heatwaves. Nat Commun 11(1):3357
36. Renschler CS et al (2010) A framework for defining and measuring resilience at the community scale: the PEOPLES resilience framework. Buffalo: MCEER
37. Rockefeller Foundation. 100 resilient cities. http://www.100resilientcities.org/. Accessed 21 June 2019
38. Russell L et al (2018) Agile IoT for critical infrastructure resilience: Cross-modal sensing as part of a situational awareness approach. IEEE Internet Things J 5(6):4454–4465
39. Schwind N et al (2016) Computational framework of resilience. In: Urban resilience: a transformative approach, pp 239–257
40. Sharifi A, Yamagata Y (2016) Urban resilience assessment: multiple dimensions, criteria, and indicators. In: Urban resilience: a transformative approach, pp 259–276
41. Solecki W, Rosenzweig C (2020) Indicators and monitoring systems for urban climate resiliency. Clim Change 163(4):1815–1837
42. Spencer Jr, BF, Hoskere V, Narazaki Y (2019) Advances in computer vision-based civil infrastructure inspection and monitoring. Engineering 5(2):199–222
43. Stevenson JR et al (2018) The data challenges of monitoring resilience. In: Proceedings of the information systems for crisis response and management Asia Pacific 2018 conference innovating for resilience
44. Tang J (2021) Assessment of resilience in complex urban systems. In: Industry, innovation and infrastructure. Springer International Publishing, Cham, pp 84–93
45. The Nature Conservancy. TNC Coastal Resilience Tool. http://coastalresilience.org/naturasolutions/toolkit/. Accessed 26 Oct 2018
46. Tong P (2021) Characteristics, dimensions and methods of current assessment for urban resilience to climate-related disasters: a systematic review of the literature. Int J Disaster Risk Reduct 60:102276
47. Tong S, Wang XY, Barnett AG (2010) Assessment of heat-related health impacts in Brisbane, Australia: comparison of different heatwave definitions. PloS one 5(8):e12155
48. Tropmann-Frick M (2021) Global & geographical mapping and visualization method for personal/collective health data with 5D world map system. Inf Modell Knowl Bases XXXII 333:134
49. Victorian Department of Health (2009, January) Heatwave in Victoria: an assessment of health impacts. The Victorian Government Department of Human Services Melbourne, Victoria

50. Weber S et al (2015) Policy-relevant indicators for mapping the vulnerability of urban populations to extreme heat events: a case study of Philadelphia. Appl Geogr 63:231–243
51. Winderl T (2014) Disaster resilience measurements: stocktaking of ongoing efforts in developing systems for measuring resilience. https://www.preventionweb.net/files/37916_disasterresiliencemeasurementsundpt.pdf. Accessed 26 Oct 2018
52. Xiong L, He BJ (2022) Analytical framework for the analysis of co-benefits, conflicts and trade-offs of urban heat mitigation strategies. IOP Conf Series: Earth Environ Sci 1078(1). IOP Publishing
53. Xu Z et al (2016) Impact of heatwave on mortality under different heatwave definitions: a systematic review and meta-analysis. Environ Int 89:193–203
54. Yuan W et al (2022) Heat-induced health impacts and the drivers: implications on accurate heat-health plans and guidelines. Environ Sci Pollut Res 29(58):88193–88212
55. Zhu Z, Li T (2018) Extended-range forecasting of Chinese summer surface air temperature and heat waves. Clim Dyn 50(5):2007–2021
56. Zhao Z-Q et al (2017) Profile and concentric zonal analysis of relationships between land use/land cover and land surface temperature: case study of Shenyang, China. Energy Build 155:282–295

Modeling Natural Ventilation in Urban Environments: A Review of Existing Techniques and Tools

Humera Mughal, Anber Rana, and Salma Sherbaz

Abstract The urban climate faces numerous challenges stemming from climate change and escalating populations. The built environment and lifestyle patterns exacerbate these challenges, resulting in an intensified urban heating effect. These challenges include poor air quality, heat island effects, and increase risks of floods/droughts with change in land use cover. Addressing many of these challenges hinges on effective urban ventilation strategies. Urban ventilation can be evaluated and improved using different tools and techniques. This study's primary objective is to thoroughly review these tools and techniques used in modeling urban ventilation, with a specific focus on identifying common challenges in modeling and charting future directions for improvement. The study commences by providing an overview of the need to evaluate urban ventilation. A brief overview of these tools and techniques is followed by the most prevalent issues and essential research avenues for improving urban ventilation assessment. The outcomes of this study hold considerable value in fostering sustainable urban environments, as they furnish guidance for enhancing existing techniques and tools. Specifically, this work serves as a valuable resource for the modeling community and newcomers to the field of research.

Keywords Urban breathability · Wind tunnel · Machine learning · LiDAR

H. Mughal (✉)
College of Architecture and Design, Prince Sultan University, Riyadh, Saudi Arabia
e-mail: humughal@psu.edu.sa

A. Rana
Department of Architecture, University of Washington, Seattle, USA

S. Sherbaz
Research Center for Modeling and Simulation, National University of Sciences and Technology, Islamabad, Pakistan

B. He et al. (eds.), *Urban Climate and Urban Design*, Urban Sustainability,
https://doi.org/10.1007/978-981-96-1521-6_4

1 Introduction

Globally, cities are adopting high-density and high-rise construction to accommodate the demands of the rising urban population [48]. High urbanization rates result in more complex urban microclimates, including changing population densities, urban canyon sizes, street geometries, building types and forms, land covers, and energy systems [16]. As a result, urban climate issues, such as urban heat islands (UHIs), poor air quality, and heat waves, exacerbate in intensity [27]. Hence, cities must address and adapt to these multiple issues promptly. Most of these issues are associated with poor air conditions that can be mitigated through adequate urban ventilation [52].

Urban ventilation directly impacts local air quality, thermal comfort of pedestrians, and energy use of buildings and can play a pivotal role in improving heat island effects [12, 26]. Urban ventilation is controlled by urban geometry (urban shape, area density, and aspect ratio of urban canyons) and metrological parameters (wind speed, wind direction, temperature, and humidity) [2]. To date, several indices have developed based on airflow patterns and interaction between urban morphology [33]. Urban practitioners consider these parameters and indices to design urban spaces with optimal airflows to ensure pollutants are not buildup, and good air quality and comfortable urban environment is ensured for urban residents. Urban ventilation has been evaluated by studying a list of phenomena that include turbulence, building wakes, plumes, heat transfer and thermal flows [14, 42, 45]. The phenomenon investigated, and the temporal-spatial scales of interest influence the method used for investigation [41]. Spatial classification can vary from simplified urban area models such as 2-D street canyon to real urban geometry [31]. While temporal scales can vary from less than minute to several years. Typical methods employed to study urban ventilation phenomena are in-situ measurements, laboratory measurements (such as wind tunnel tests) and, numerical simulations [17].

Despite considerable efforts dedicated to developing better urban ventilation assessment methods, tools, and techniques, there is a pressing need for a review of possible errors in urban ventilation evaluation encountered during their application. This review aims to guide future research on urban ventilation using these tools and techniques. This short review also provides a snapshot of emerging tools and techniques that can be considered for improving urban ventilation assessments. A comprehensive overview of specific urban ventilation parameters, such as wind direction, wind energy potential, air quality assessment, and trends in turbulence models, was beyond the scope of this chapter. Due to space constraints, only the most common problems and emerging trends are addressed.

2 Tools and Techniques

2.1 Experimentation Techniques

Experimental techniques may involve in situ measurement, or full scale or scaled down physical models that are designed to replicate urban environment [28]. These techniques are usually applied for lab-scale studies or short duration measurements. Common experimental apparatus includes wind or water tunnels, flow tables, and other devices that simulate air flows using genuine fluids and scaled-down copies of real buildings or topographies [34]. The wind tunnel is the most widely used experiment that usually predicts the wind flow behavior (i.e., velocity around, and forces or pressures upon structures), with applications spanning at vertical (from single-story to skyscraper design), and horizontal scales (from neighborhood to metropolis) [9, 38]. Wind tunnels are most commonly used during design stage for determining pressure coefficients for large buildings. Scaled down wind models are more common but when possible full-scale are preferred. Both type of wind tunnel models suffers from limitations. Below are some specific limitations and gaps that need to be addressed in future research:

Hybrid Models: In full-scale wind model control of boundary conditions is an issue [3, 56, 25]. Though boundary conditions can be managed in scaled-down models, but there's a threshold beyond which scaling down compromises' accuracy. Furthermore, scaled-down models often exhibit disproportionately thick walls, posing challenges in replicating realistic opening sizes [13]. Both types of wind tunnels—scaled and full-scale—are costly and time-intensive [21]. However, these experimental techniques are indispensable and required to validating CFD simulations. Development of more hybrid modelling approaches may be solution to address the problem of accuracy in scale down models. Hybrid models can give combination of numerical and experimental techniques at different scales [54]. These models would enable address issues of both CFD models and scaling down to remain within financial resources associated with project.

Measurement Instruments: Wind-tunnel measurements are generally only performed at a few selected points in the model area, and do not provide a whole image of the flow field. In addition, the type of sensor and measurement method selected for evaluating urban ventilation need to compromise between the accuracy and cost and time. Blocken et al. [5] specifically focused on pedestrian level wind performance using different measurement devices and concluded the low cost and relatively less time expensive devices are sufficient. Figure 1 provides a qualitative view of how different instruments compare with respect to time, accuracy and cost. However, in recent years advancements have emerged in the data collection methods for both wind tunnel experiments and actual field sites. For instance, work by Tong et al. [47] uses hydrological simulation to study urban environment. Therefore, a more current review that compares performance of existing instruments with newer ones in terms of their effectiveness in terms of time, accuracy, and cost.

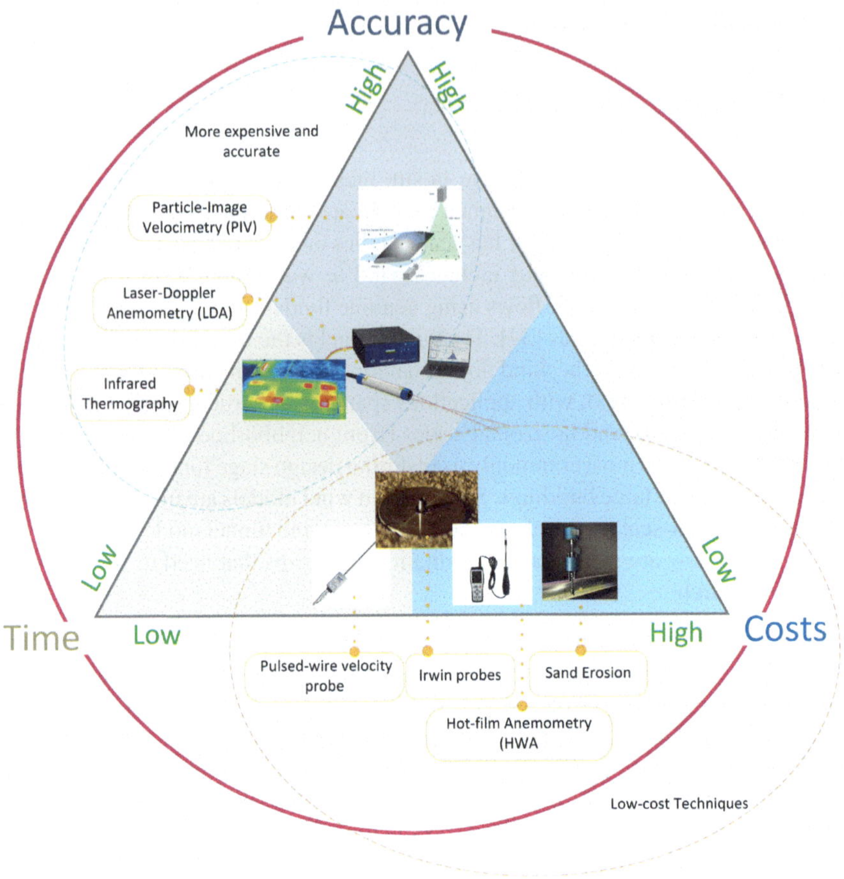

Fig. 1 Comparison of measurement instruments with respect to time, costs and accuracy (*adapted from* [5, 29])

Optimization of sampling: Normally, one year of wind direction and speed data is required for conducting urban ventilation studies [8]. This data is collected either through local meteorological observation stations or by installing sensors over masts to collect long-term data. Both these data collection techniques have limitations. The local meteorological data from public sources is usually of low resolution and cannot be used to capture wind characteristics needed for robust urban ventilation studies. Installing sensors, on the other hand, is an expensive feature that requires maintenance and repair over time. Likewise, remote sensing satellite data only provides information on horizontal spatial directions [6]. This challenge can be addressed by making use of advance data sampling techniques, such as Latin hypercube [1]. Use of these techniques would help minimize the time and quantity of samples needed, thereby cutting costs and bolstering confidence in the collected data. Table 1 shows most common advanced sampling techniques.

Table 1 Advanced sampling methods. *Source* [22, 46]

1. Latin Hypercube Sampling (LHS)	2. High-Dimensional Data Sampling
3. AI and Machine Learning-Enhanced Sampling	4. Adaptive Sampling
5. Multistage and Multiphase Sampling	6. Rank Set Sampling
7. Time-Series and Panel Sampling	8. Cluster Sampling

2.2 Numerical Techniques

Numerical techniques can be scaled to desired time and geometrical scales. Due to advancements in computer power, computational precision, and the widespread availability of software applications, computational fluid dynamics (CFD) has emerged as a primary numerical method for studying wind fields. CFD simulations are based on partial differential equations to represent near ground atmospheric conditions and provide information on important variables [7]. The accuracy of CFD models is validated using results of either actual on-site measurements or wind tunnel experiments [30, 50, 20]. CFD is extensively utilized in disciplines, such as engineering fluid analysis, building design, urban wind fields, and air pollutant dispersion [53]. The tool is handy for comparing alternate urban design options for early-stage design [18]. Compared to real-world observations and wind tunnel studies, the CFD method is cost-effective in practice, and its simulation results may accurately represent the whole wind field in the focus area [24]. CFD method enables simultaneous comparative analysis across several scenarios, an unattainable capability through actual measurements and wind tunnel operations. Some future research directions to be considered for numerical modelling are provided below (Fig. 2):

Urban stratification: The type of stratification used for analysis significantly impacts the accuracy of CFD simulations. Most work has concentrated on stable or unstable

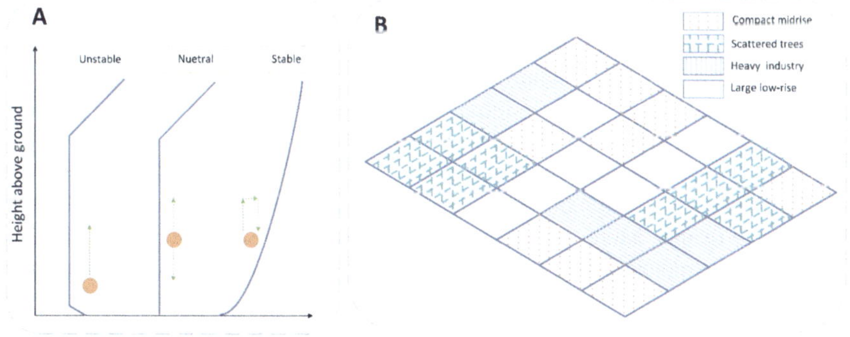

Fig. 2 Future research directions in numerical modelling need to consider (**a**) type of urban stratification and (**b**) Heterogeneity related to land use

stratification to study urban ventilation. For instance, considering only neutral stratification would result in overestimating mean ventilation. On the contrary, considering neutral stratification in areas with unstable stratification can result in the underestimation of ventilation up to 40% [11]. Similarly, at the city scale, the impact of building geometry becomes less critical, but under unstable stratification, the ventilation patterns are highly influenced by surrounding building height and density [32]. Hence, for a comprehensive estimation of ventilation potential, all types of stratification scenarios need to be considered [11].

Heterogeneity: Simplifying urban topography is a common strategy to reduce the computation times of CFD. These urban models often concentrate on a few design elements while ignoring the effect of heterogeneities present in the natural urban environment related to surface materials, sources of local heat, including radiation, and moisture in the atmosphere [7, 12, 55]. Neglecting these variations would result in an error in ventilation potential since the resulting secondary circulations are not considered. Identifying recirculation regions under air exchange heterogeneity is essential for designing building openings and improving ventilation [43]. Likewise, water bodies influence urban ventilation [39, 51]. There is a need to consider the impact of non-aligned building walls on roughness. Gronemeier [11] proposed that conducting multiple simulations of a single cube-shaped building at different orientations concerning the simulation grid can address this problem. Using perfectly aligned building arrays for models generates strong air channels. This overestimation would be especially critical when the pollutant dispersion is evaluated using urban ventilation data, which results in an underestimation of the health risks for the local population. Advanced surface models or integration with the databases used for other applications, such as Visicom [4], can help consider the main heterogeneous features, resulting in more realistic and valuable simulations. Similarly, GIS-based classification, Local climate zone (LCZ), can capture the urban diversity in CFD models [44].

Seasonal variations and extreme weather events: More work needs to be conducted on wind variations at different seasons and the impact of extreme weather events, such as heat waves or cold snaps, to conduct proper planning and implement mitigation measures. Detailed planning of specific area microscale simulations would be needed for idealized wind conditions [19]. Designing urban areas with seasonal variation and extreme weather events in mind may also require a transdisciplinary research approach that will require acquiring data from several sources and getting expert opinions from different stakeholders while generating, simulating, and interpreting CFD results. More work on urban ventilation is also needed in developing countries with exponentially expanding urban areas, such as Lahore city in Pakistan [36], so that timely and economic solutions under seasonal and extreme weather events can be determined. Another direction in which CFD techniques need to be developed are occurrence of major shift in urban residents' patterns. For instance, COVID-19 pandemic gave rise to drastic changes in building occupancy patterns and, consequentially, the energy use in buildings [35], which essentially increased the heat flows from residential areas.

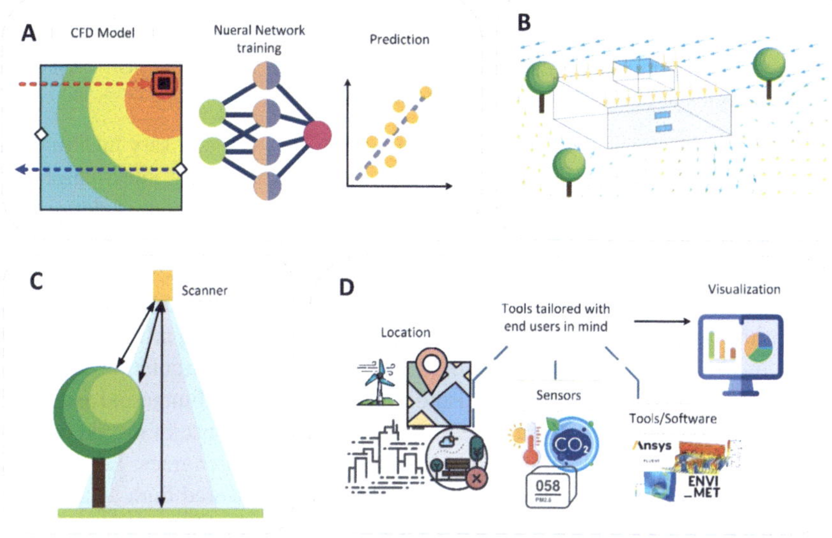

Fig. 3 Emerging techniques for urban ventilation design include **a** Machine Learning Techniques, **b** Digital twin technology, **c** Light Detection and Ranging (LiDAR) and (D) tailored CFD tools

2.3 Other Techniques

This section briefly defines some emerging techniques and tools (Fig. 3) that do not fall in the conventional experimental and numerical techniques defined in previous two sections.

Machine Learning Techniques: Machine learning techniques has been successfully used in numerous engineering and business applications. The results of machine learning applications show low prediction errors at significantly higher speeds than traditional CFD simulations. Machine learning algorithms are especially useful for reducing computation times for complex applications. Waibel et al. [49] combined machine learning with fast fluid dynamics (FFD) to generate detailed results in short time. This approach also ensured the data generated was more accurate than FFD models which are known to generate fast performance data but at lower accuracy than CFD models. Researchers have employed machine learning algorithms, such as gaussian processes and neural networks, to generate wind speeds and flows. Machine Learning algorithms, such as U-NET, have been used for extracting information from satellite images and has shown high performance [23]. Each algorithm used to study urban ventilation has its limitations. For instance, Lau et al. [23] used heuristic search algorithm to search urban corridor. The algorithm was only capable of considering one direction for wind and could not be used for areas where multiple wind directions are present. Hence more advancements in algorithms are needed to increase their capabilities to address effect of multiple wind directions as well as small and large urban obstacles. In addition, these techniques can be used to combine data of existing research related to urban geometries, local metrological conditions (isothermal and no isothermal) pollutant sources, types and concentrations to generate low-cost high benefit strategies to address urban challenges.

Digital twin technology: Digital twin technology can provide necessary information to decision makers, urban planners and policy makers, necessary information for assessing the effects of local greenery systems [10]. "A digital twin is defined as a virtual representation of a physical asset enabled through data and simulators for real-time prediction, optimization, monitoring, controlling and improved decision making" [37]. Digital twin can be useful urban planning tool to control urban air pollution and heat island effect. It was used successfully in planning urban ventilation channel in China using combination of CFD, GIS and local circulation structure research [15]. Likewise, digital twin is useful for planning microclimate and assessing outdoor thermal comfort. A team of researchers in Italy made use of digital twin to evaluate the urban thermal comfort by using three individual engines, EnergyPlus, Rhino and OpenFoam, integrated through a Python code ensured high accuracy of the modelling [10]. More work is needed to apply digital twin technologies in urban environment especially for extreme weather events for different urban morphologies. A tool combining data of digital twin libraries with machine learning tools for urban ventilation could help efficiently address urban environment problems.

Light Detection and Ranging (LiDAR): Conventional experiment methods have primarily employed either scaled down models used in wind tunnel tests or limited duration field data collected through anemometers attached to static or mobile (such as balloons, drones, helicopters) to capture urban climate data for evaluating urban ventilation [12, 40] Integration of advanced data collection techniques such as Light Detection and Ranging (LiDAR) helps in ensuring robustness of ventilation experiments. He et al. [12] used LiDAR to compare performance of conventional experimental and empirical models for dense urban area wind profiles. The results indicated high variations in the wind velocity and thermal comfort for conventional techniques. Since majority of the conventional methods have been developed based on simplified models often with set of assumptions absent in real urban climate it is essential that newer technologies with better data collection ability are integrated in urban ventilation studies.

Tailored CFD tools: The end-users for urban ventilation assessment must be kept in mind when developing or updating existing CFD tools. Urban climate CFD model users may come from numerous disciplines with minimal basic expertise; therefore, the community must provide more guidelines, recorded information, and best practices. In the same vein, there is need for developing better integration urban climate CFD models with GIS, LiDAR, city CADs and other spatial tools.

3 Conclusion

The current study aimed to explore various tools and methodologies utilized in assessing urban ventilation with a particular emphasis on identifying potential future research directions extrapolated from the existing literature. Conventional experimental approaches predominantly involve scaled-down physical models of urban landscapes in wind tunnel experiments. Meanwhile, numerical methods heavily rely on computational fluid dynamics (CFD) simulations, offering the capability to gauge urban ventilation across diverse temporal and spatial scales. In addition, experimental and CFD techniques are complimentary and both are still required to study urban ventilation. Emerging techniques, such as Machine Learning Algorithms, coupling GIS tools, remote sensing technologies, can help make the results of experimental and numerical techniques more robust, replicable, and reliable at low computation costs. However, more work must be conducted on accurate urban geometries to improve existing urban design guidelines and move towards more sustainable and healthier cities.

Acknowledgements The author would like to thank Prince Sultan University for their support.

References

1. Afzal A, Kim KY, Seo JW (2017) Effects of Latin hypercube sampling on surrogate modeling and optimization. Int J Fluid Mach Syst 10(3):240–253. https://doi.org/10.5293/IJFMS.2017.10.3.240
2. Antoniou N, Montazeri H, Wigo H, Neophytou MKA, Blocken B, Sandberg M (2017) CFD and wind-tunnel analysis of outdoor ventilation in a real compact heterogeneous urban area: Evaluation using "air delay." Build Environ 126:355–372. https://doi.org/10.1016/j.buildenv.2017.10.013
3. Belleri A, Lollini L, Dutton SM (2014) Natural ventilation design: an analysis of predicted and measured performance. Build Environ 81:123–138. https://doi.org/10.1016/j.buildenv.2014.06.009
4. Bibikom (2024) Digital maps, radio planning map (RF), RF planning. https://visicomdata.com/uk/
5. Blocken B, Stathopoulos T, Van Beeck JPAJ (2016) Pedestrian-level wind conditions around buildings: review of wind-tunnel and CFD techniques and their accuracy for wind comfort assessment. Build Environ 100:50–81. https://doi.org/10.1016/j.buildenv.2016.02.004
6. Cao H, Tao P, Li H, Shi J (2019) Bundle adjustment of satellite images based on an equivalent geometric sensor model with digital elevation model. ISPRS J Photogramm Remote Sens 156:169–183
7. Cao J, Chen Z, Kong S, Liu L, Wang R (2024) Towards urban wind utilization: the spatial characteristics of wind energy in urban areas. J Clean Prod 141981. https://doi.org/10.1016/j.jclepro.2024.141981
8. Dong S, Gong Y, Wang Z, Incecik A (2019) Wind and wave energy resources assessment around the Yangtze River Delta. Ocean Eng 182:75–89. https://doi.org/10.1016/j.oceaneng.2019.04.030
9. Duthinh D, Simiu E (2011) The use of wind tunnel measurements in building design. Wind Tunnels and Experimental Fluid Dynamics Research

10. Gholami M, Torreggiani D, Barbaresi A, Tassinari P (2023) Smart green planning for urban environments: the city digital twin of Imola. In: Smart cities: social and environmental challenges and opportunities for local authorities, pp 133–150. https://doi.org/10.1007/978-3-031-35664-3_10
11. Gronemeier T (2021) Advances in urban ventilation assessments using large-eddy simulation. https://doi.org/10.15488/11316
12. He Y, Yuan C, Ren C, Ng E (2022) Urban ventilation assessment with improved vertical wind profile in high-density cities—Comparisons between LiDAR and conventional methods. J Wind Eng Ind Aerodyn 228:105116. https://doi.org/10.1016/j.buildenv.2022.109018
13. Heiselberg P, Bjørn E, Nielsen PV (2002) Impact of open windows on room air flow and thermal comfort. Int J Vent 1:91–100. https://doi.org/10.1080/14733315.2002.11683625
14. Hertwig D, Grimmond S, Kotthaus S, Vanderwel C, Gough H, Haeffelin M, Robins A (2021) Variability of physical meteorology in urban areas at different scales: implications for air quality. Faraday Discuss 226:149–172. https://doi.org/10.1039/D0FD00098A
15. Huang Y, Peng H, Sofi M, Zhou Z, Xing T, Ma G, Zhong A (2022) The city management based on smart information system using digital technologies in China. IET Smart Cities 4(3):160–174. https://doi.org/10.1049/smc2.12035
16. Javanroodi K, Nik VM (2020) Interactions between extreme climate and urban morphology: investigating the evolution of extreme wind speeds from mesoscale to microscale. Urban Clim 31:100544
17. Javanroodi K, Nik VM, Adl-Zarrabi B (2020, November) A multi-objective optimization framework for designing climate-resilient building forms in urban areas. In: IOP conference series: earth and environmental science, vol 588, no 3. IOP Publishing, p 032036
18. Juan YH, Li Z, Lee YT, Wen CY, Yang AS (2023) Effect of wind-based climate-responsive design on city breathability of a compact high-rise city. J Build Eng 78:107773. https://doi.org/10.1016/j.jobe.2023.107773
19. Kwok YT, de Munck C, Lau KKL, Ng E (2022) To what extent can urban ventilation features cool a compact built-up environment during a prolonged heatwave? A mesoscale numerical modelling study for Hong Kong. Sustain Cities Soc 77:103541. https://doi.org/10.1016/j.scs.2021.103541
20. Ku CA, Tsai HK (2020) Evaluating the influence of urban morphology on urban wind environment based on computational fluid dynamics simulation. ISPRS Int J Geo-Inf 9(6):399
21. Labat M, Woloszyn M, Garnier G, Roux JJ (2013) Assessment of the air change rate of airtight buildings under natural conditions using the tracer gas technique. Comparison with numerical modelling. Build Environ 60:37–44. https://doi.org/10.1016/j.buildenv.2012.10.010
22. Latpate R, Kshirsagar J, Gupta VK, Chandra G (2021) Advanced sampling methods. Springer, Singapore. https://doi.org/10.1007/978-981-16-0622-9
23. Lau TK, Tsai PC, Ou HY, Lin TP (2024) Efficient and cost-effective method for identifying urban ventilation corridors using a heuristic search algorithm. Sustain Cities Soc 101:105144. https://doi.org/10.1016/j.scs.2023.105144
24. Li J, Delmas A, Donn M, Willis R (2018) Validation and comparison of different CFD simulation software predictions of urban wind environment based on AIJ wind tunnel benchmarks. In: SIMAUD '18: Proceedings of the symposium on simulation for architecture and urban design. https://doi.org/10.5555/3289750.3289777
25. Lo LJ, Banks D, Novoselac A (2013) Combined wind tunnel and CFD analysis for indoor airflow prediction of wind-driven cross ventilation. Build Environ 60:12–23
26. Luo Y, He J, Ni Y (2017) Analysis of urban ventilation potential using rule-based modeling. Comput Environ Urban Syst 66:13–22. https://doi.org/10.1016/j.compenvurbsys.2017.07.005
27. Masson V, Lemonsu A, Hidalgo J, Voogt J (2020) Urban climates and climate change. Annu Rev Environ Resour 45:411–444. https://doi.org/10.1146/annurev-environ-012320-083623
28. Martins NR, da Graça GC (2016, February) Validation of numerical simulation tools for wind-driven natural ventilation design. In: Building simulation, vol 9, no 1. Springer Berlin Heidelberg, Berlin/Heidelberg, pp 75–87

29. Mittal H, Sharma A, Gairola A (2018) A review on the study of urban wind at the pedestrian level around buildings. J Build Eng 18:154–163. https://doi.org/10.1016/j.jobe.2018.03.006
30. Montazeri H (2011) Experimental and numerical study on natural ventilation performance of various multi-opening wind catchers. Build Environ 46(2):370–378. https://doi.org/10.1016/j.buildenv.2010.07.031
31. Panagiotou I, Neophytou MKA, Hamlyn D, Britter RE (2013) City breathability as quantified by the exchange velocity and its spatial variation in real inhomogeneous urban geometries: an example from central London urban area. Sci Total Environ 442:466–477. https://doi.org/10.1016/j.scitotenv.2012.09.001
32. Park SB, Baik JJ, Ryu YH (2013) A large-eddy simulation study of bottom-heating effects on scalar dispersion in and above a cubical building array. J Appl Meteorol Climatol 52(8):1738–1752. https://doi.org/10.1175/JAMC-D-12-0255.1
33. Peng Y, Buccolieri R, Gao Z, Ding W (2020) Indices employed for the assessment of "urban outdoor ventilation"—A review. Atmos Environ 223:117211. https://doi.org/10.1016/j.atmosenv.2019.117211
34. Plate EJ (1999) Methods of investigating urban wind fields—physical models. Atmos Environ 33(24–25):3981–3989. https://doi.org/10.1016/S1352-2310(99)00140-5
35. Rana A, Kamali M, Riyadh MM, Sultana SR, Kamal MR, Alam MS, ... Sadiq R (2022) Energy efficiency in residential buildings amid COVID-19: a holistic comparative analysis between old and new normal occupancies. Energy Build 277:112551. https://doi.org/10.1016/j.enbuild.2022.112551
36. Rana IA, Bhatti SS (2018) Lahore, Pakistan-Urbanization challenges and opportunities. Cities 72:348–355. https://doi.org/10.1016/j.cities.2017.09.014
37. Rasheed A, San O, Kvamsdal T (2020) Digital twin: values, challenges and enablers from a modeling perspective. IEEE Access 8:21980–22012. https://doi.org/10.1109/ACCESS.2020.2970143
38. Ricci A, Burlando M, Freda A, Repetto MP (2017) Wind tunnel measurements of the urban boundary layer development over a historical district in Italy. Build Environ 111:192–206. https://doi.org/10.1016/J.BUILDENV.2016.10.016
39. Ronda RJ, Steeneveld GJ, Heusinkveld BG, Attema JJ, Holtslag AAM (2017) Urban finescale forecasting reveals weather conditions with unprecedented detail. Bull Am Meteor Soc 98(12):2675–2688. https://doi.org/10.1175/BAMS-D-16-0297.1
40. Roth M (2000) Review of atmospheric turbulence over cities. Q J R Meteorol Soc 126(564):941–990. https://doi.org/10.1002/qj.49712656409
41. Schlünzen KH, Grawe D, Bohnenstengel SI, Schlüter I, Koppmann R (2011) Joint modelling of obstacle induced and mesoscale changes—Current limits and challenges. J Wind Eng Ind Aerodyn 99(4):217–225. https://doi.org/10.1016/j.jweia.2011.01.009
42. Shirzadi M, Mirzaei PA, Tominaga Y (2021) LES analysis of turbulent fluctuation in cross-ventilation flow in highly-dense urban areas. J Wind Eng Ind Aerodyn 209:104494. https://doi.org/10.1016/j.jweia.2020.104494
43. Stein-Montalvo L, Ding L, Hultmark M, Adriaenssens S, Bou-Zeid E (2024) Kirigami-inspired wind steering for natural ventilation. J Wind Eng Ind Aerodyn 246:105667. https://doi.org/10.1016/j.jweia.2024.105667
44. Stewart ID, Oke TR (2012) Local climate zones for urban temperature studies. Bull Am Meteor Soc 93(12):1879–1900. https://doi.org/10.1175/BAMS-D-11-00019.1
45. Tian G, Ma Y, Chen Y, Wan M, Chen S (2024) Impact of urban canopy characteristics on turbulence dynamics: insights from large eddy simulation. Build Environ 250:111183. https://doi.org/10.1016/j.buildenv.2024.111183
46. Tufail A (2023) Chapter 6: Advanced sampling techniques—Pioneering methods for modern data challenges
47. Tong Z, Luo Y, Zhou J (2021) Mapping the urban natural ventilation potential by hydrological simulation. Build Simul 14:351–364. https://doi.org/10.1007/s12273-020-0755-6
48. United Nations (2018) 68% of the world population projected to live in urban areas by 2050, says un | UN Desa Department of Economic and Social Affairs. United Nations

49. Waibel C, Zhang R, Wortmann T (2021) Physics meets machine learning: coupling FFD with regression models for wind pressure prediction on high-rise facades. Association for Computing Machinery: New York, NY, USA, 1
50. Wang H, Wang Q, Yang X, Chen T, Lam CKC, Zhang M, Hang J (2021) Steady and unsteady turbulent flows and pollutant dispersion in 2D street canyons with novel boundary conditions and various Re numbers. Urban Climate 39:100973. https://doi.org/10.1016/j.uclim.2021.100973
51. Wolf-Grosse T, Esau I, Reuder J (2017) Sensitivity of local air quality to the interplay between small-and large-scale circulations: a large-eddy simulation study. Atmos Chem Phys 17(11):7261–7276. https://doi.org/10.5194/acp-17-7261-2017
52. Yim SHL, Fung JCH., Ng EYY (2014) An assessment indicator for air ventilation and pollutant dispersion potential in an urban canopy with complex natural terrain and significant wind variations. Atmosp Environ 94:297–306. https://doi.org/10.1016/j.atmosenv.2014.05.044
53. Zhang S, Kwok KCS, Liu H, Jiang Y, Dong K, Wang B (2021) A CFD study of wind assessment in urban topology with complex wind flow. Sustain Cities Soc 71:103006. https://doi.org/10.1016/J.SCS.2021.103006
54. Zhao Y, Chew LW, Fan Y, Gromke C, Hang J, Yu Y, Ricci A, Zhang Y, Xue Y, Fellini S, Mirzaei PA (2023) Fluid tunnel research for challenges of urban climate. Urban Climate 51:101659. https://doi.org/10.1016/j.uclim.2023.101659
55. Zhen M, Zhou D, Bian G, Yang Y, Liu Y (2019) Wind environment of urban residential blocks: a research review. Archit Sci Rev 62(1):66–73. https://doi.org/10.1080/00038628.2018.1528967
56. Zhou C, Wang Z, Chen Q, Jiang Y, Pei J (2014) Design optimization and field demonstration of natural ventilation for high-rise residential buildings. Energy Build 82:457–465. https://doi.org/10.1016/j.enbuild.2014.06.036

Mitigation and Adaptation Strategies

From Vernacular to Vernomimicry: Vernacular Design Principles for Resilient Communities

Ahmed Khoja

Abstract Climate change poses increasing threats to our built environment and the debate is gradually shifting toward its magnitude and not its occurrence. As we embrace for a new climatic era, designers and planners are forced to produce architectural design that can address the dual challenge of mitigation and adaptation. This requires incorporating a design paradigm that is "with" nature and not "despite" it. Vernacular design, being the indigenous science of building, can serve as a valuable source of inspirations for climate adaptation solutions. Based on a literature review and comparative analysis, this paper seeks to uncover the governing principles behind the success of vernacular design in the face of a changing climate and explore climate adaptation solutions and strategies that can be transferred from the vernacular to the modern-day design, following a Vernomimicry approach. Thus, in this paper, the vernacular builders' strategies are analysed and the vernacular adaptation solutions in the face of hazards such as floods, extreme precipitation, heatwaves, storms, and drought, at both urban and building scale are investigated. On this basis, five core design principles were identified and are encouraged to be included in modern adaptation designs, namely: simplicity and adaptability in construction, transient structures and local materials, resource efficient and site specific- design, community involvement in building practices, and holistic integration with the urban context.

Keywords Vernacular architecture · Climate change adaptation · Vernomimicy · Design principles · Resilient urban design · Nature design principles · Biomimicry

1 Introduction

From the Stockholm Declaration of 1972 to the Paris agreement in 2015, the global community has made progressive efforts towards reducing greenhouse gas emissions. Despite these endeavours, recent estimates by the IPCC indicate that these efforts might well fall short of keeping global warming below the critical 2 °C [33]. These

A. Khoja (✉)
Department of Architecture, Munich University of Applied Sciences, Munich, Germany
e-mail: Khoja@hm.edu

B. He et al. (eds.), *Urban Climate and Urban Design*, Urban Sustainability,
https://doi.org/10.1007/978-981-96-1521-6_5

alarming findings facilitate shifting the debate from questioning the occurrence of climate change to arguing its severity. Thus, it becomes imperative that both existing and new structures must adequately prepare to adapt to future climatic conditions without compromising the buildings' ability to meet the essential mitigation targets. Addressing this duality of climate adaptation and mitigation in the built environment is a tall order.

Since the late industrial revolution there has been a prevailing belief in the power of technology to tame and control nature, which has dictated our co-existence with the surrounding environment [8, 21] and led to extensive use of active and energy intensive building systems. Eventually, such systems have become an integral and standard component of any building since the mid-21th century. Even though these systems were initially implemented to solve a climatic problem, the current overreliance on such systems is now recognized as a part of the climate challenges they tried to fix [45]. Therefore, to be able to meet the climate mitigation and adaptation expectations we need a paradigm shift in the existing architectural practices, moving away from a design philosophy of 'despite' nature to one that works 'with' nature. This transition advocates for an architecture that is not just environmentally responsive but integrally connected with the nature's principles.

The quest to find design principles that can guide this transformation to design "with" nature has gained momentum by the turn of the century. An increasing number of scholars advocate the power of biomimetic, nature-inspired design in creating a built environment that harmonizes with nature [7: 2, [42]. In the realm of climate adaptation, biomimicry is sometimes confused with another closely related term, namely nature-based solutions (NbS). Though an overlap between both of terms exists, the work of Bianciardi and Cascini [9] has clearly showed that NbS is more concerned with the utilization of nature in order to increase resilience of a system. This places NbS within the broader field of bio-utilization rather than biomimicry. Nevertheless, Bianciardi and Cascini [9] highlighted the complementary nature of both bio-inspired and bio-based solutions in delivering inspiring answers towards creating an adapted and sustainable environment.

Applying biomimicry in architecture is a rather complex task. The inherent differences between biology and architecture present substantial challenges in translating natural efficiencies into built forms [35, 55]. Additionally, the biomimicry process is often a problem-driven and offers unique solutions to specific challenges [26], which can sometimes stand in conflict with the iterative, user-focused nature of architectural design, where each solution is ultimately a function of many societal, technological, cultural and economic compromises. Furthermore, as Gruber [24] highlighted, architects often adopt a superficial approach to biomimicry, not fully grasping the underlying natural phenomena. An example of this is the Eastgate Centre in Harare by Mick Pearce, often cited as an icon of biomimetic architecture. While its solar chimney-based atria, used for temperature regulation shares some inspiration from termite mounds, the atria functions differently from termite mounds it is claimed to be model after [53].

Given the limitations of the biomimicry approach in architecture, this paper opted to investigate the design principles found in the vernacular architecture following the vernomimicry concept [31]. The vernomimicry seeks to abstract and synthesize the design principles found in the vernacular architecture to draw up lessons that are applicable to the modern context. Vernomimicry offers an accessible approach for architects, bypassing the complexities of biological systems and abstract ecological concepts, and, instead, drawing from the practical, time-tested architectural practices. The vernacular architecture, as described by Oliver [40] as a "native science of building", demonstrates a remarkable symbiotic relationship between humans and the environment and is filled with examples of nature-like and nature aligned adaptations. Although biomimicry recognizes the vernacular architecture as a part of nature, however, the use of it in biomimetic architecture is still not fully appreciated [24].

2 Methodology

Given the unique evolution of vernacular architecture across diverse climatic and geographic regions, it offers an unequalled repository of human solutions to wide range of challenges [44]. However, this research aims not merely to catalogue unique solutions found in the vernacular architecture but rather to reveal the underlying design principles that can guide modern design so that it becomes a design "with" and not "despite" nature. Therefore, our research question is formulated as: *What are the key design principles of climate adaptation that can be learned from vernacular architecture and applied to modern architectural design practices?*

Previous research has often examined the correlation between vernacular architecture and resiliency through a broader sustainability lens [18] or has remained site/region specific or hazard/impact specific [2, 36, 54]. However, as per Pardo [41] a notable gap exists in investigating how the principles of vernacular architecture can be integrated into modern design.

Addressing this gap, our research aims to bundle the wealth of existing knowledge into a comprehensive overview of general climate adaptation design principles shared among vernacular architecture, thereby contributing to the body of knowledge on possible integration of these principles into contemporary design.

To start with, the design process that the vernacular builders used to adapt to the changes in their climate consists of two continuously evolving processes—a proactive and a reactive one [15]. The proactive processes are based on previous experiences and prevailing weather and climate conditions. Hence, the proactive processes of climatic responses are embedded in the building design tradition. The proactive solutions are site and culturally specific design responses that gave the architecture of a specific region its distinct shape, or as in the terms of Alexander [4] are part of the forces that give the resulting building and urban form its unique look. These responses impact a wide array of building parameters such as its room arrangements, openings, building material, roofing shape, envelope colure, etc.

Conversely, the reactive process in vernacular buildings occurs in response to shifts in climate or after extreme weather events, such as floods or storms, which exceeds the capacity of proactive measures [15]. Such extreme events disrupt indigenous knowledge and result in either the invention of a new proactive adaptation, which changes the perceived image of the older style, or in upgrading of the existing proactive measures to accommodate for such events in the future.

The constant development of the traditional Chinese roof curvature, sloping, pitch and the relation between the roof span and height that happened between 925 and 1504 AD presents a perfect example of how a reactive response turns into a proactive one to adapt to changing climate conditions. As shown in Fig. 5.1, the roof span to height ratio (HSR) of the Longmen temple had consistently evolved in response to climate changes, becoming steep in cold periods and then again softer and flatter in hot periods, without radically violating the harmony and coordination principles of ancient Chinese architecture.

Recognizing that reactive design processes inevitably evolve into proactive ones, this study chose to take a deeper look into proactive adaptations to answer the research question. By examining the proactive design principles, we seek to understand how vernacular architecture has been shaped to effectively serve the social, cultural, and functional expectations of its users, while successfully meeting the specific climatic requirements of its location.

Respectively, six universally (i.e., shared by many geographical locations) applicable climatic hazards or impacts had been selected, namely:

- Floods, flash floods and increase in ground water levels.
- Extreme Precipitation.
- Drought and water scarcity.
- Wind and storm hazard.
- Heat and warming trend.

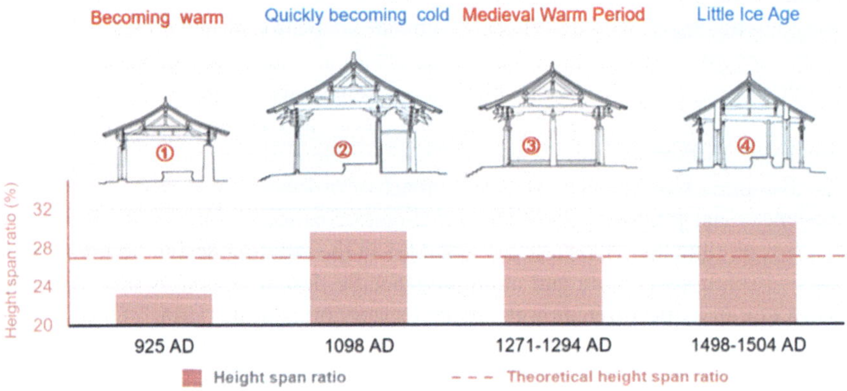

Fig. 5.1 Illustration showing the constant change of high to span ratio of Chinese roof in response to changes in local climate [34]

- Compound hazard (combinations of events that are not themselves extremes but lead to an extreme event [50]).

To each of these hazards, a search inquiry in Google Scholar (last updated Nov. 10th, 2023) has been made using the name of the hazard or climate impact, such as "flood" and "Vernacular Architecture". As the general search would return answers well in the thousands for each hazard, the results were only filtered to review articles only. For the "Compound hazard" the term was replaced with the general "resilience" term. Moreover, the search was time limited between 2010 and 2023 (Fig. 5.2).

The generated results are summarized in Table 5.1.

As depicted in Fig. 5.3, the yearly distribution of review papers published between 2010 and 2023, addressing the interaction between vernacular architecture and climatic hazards, has witnessed consistent growth over the years. This escalating

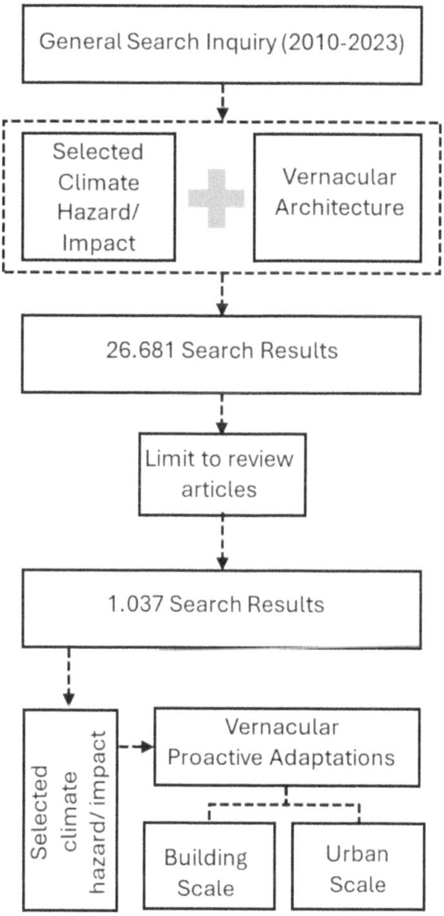

Fig. 5.2 A schematic showing the research methodology

Table 5.1 A summary of the return of the key word research made in Google scholar

Search term	Number of results	Number of results with the review articles filter
"Flood" and "Vernacular Architecture"	3.721	108
"Precipitation" and "Vernacular Architecture"	3.111	113
"Drought" and "Vernacular Architecture"	2.530	91
"Water scarcity" and "Vernacular Architecture"	310	13
"Storm" and "Vernacular Architecture"	2.190	64
"Wind hazard" and "Vernacular Architecture"	6	0
"Heat" and "Vernacular Architecture"	9.920	376
"Resilience" and "Vernacular Architecture"	4.830	272

trend reflects an increasing interest in climate adaptation solutions inherent in vernacular architecture. Notably, the growth is most pronounced in the number of papers discussing the design principles of vernacular architecture in relation to heatwaves, mirroring the mirroring a traditional topic of interest in the context of immediate environmental challenges. Post-2016, the topic of resilience has seen a surge in the number of papers published, showcasing a growing appreciation for the value of integrating indigenous knowledge and practices with contemporary architectural and urban planning strategies. However, the amount of research investigating the relationship between vernacular architecture and the climatic hazards of storms, wind, and drought remains an exception, showing a small, yet constant volume of publications over the past 13 year.

3 Summary of the Findings

Based on the selected methodology, the research progressed further by limiting the examination to the review of publications related to each hazard. This rule didn't apply to the case of wind and storm hazard as it returned only 6 results. During the review, it was observed that the building scale often overlaps with greater urban solutions. Therefore, in this research for each climate hazard, the design commonalities found in the reviews are summarized and differentiated between urban scale adaptation solution and building scale. The list found in Table 5.2 represents an example and is not an exhausting list of solutions.

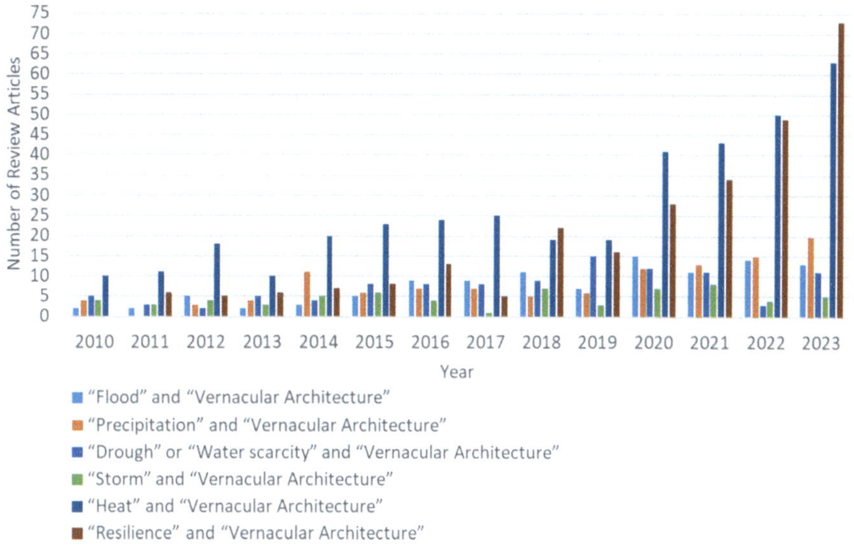

Fig. 5.3 Yearly number of review papers on vernacular architecture and selected climatic hazards (2010–2023)

Table 5.2 A Summary of vernacular design adaptation solutions applied in at the urban and building scale for a selected list of climate impacts and hazards

Hazard/ impact	Building scale		Urban scale	
	Examples of design solutions	References	Examples of design solutions	References
Floods, flash floods and increase in ground water levels	Axile location of openings to allow for flood water to exit; raised attic to store food and use a safety hideout; building is raised above the flood threshold either on stiles or a retaining wall; or the use of floating structures, seasonal homes; use water resisting ground floor materials	[14, 32]	Dual use of public space for excess water detention, recreation, and storage; increase space between pavement to enhance permeability; creating water ways and ditches; lower land as flood catchment areas; cascading irrigation systems	[14, 17]

(continued)

Table 5.2 (continued)

Hazard/impact	Building scale		Urban scale	
	Examples of design solutions	References	Examples of design solutions	References
Extreme precipitation	Gentle pitched roof, extended eaves, gutters; use of water absorbing or resistant roofing materials; water resistant and water repellent façade materials; repellent extended elements on the envelope; green façade: increased ventilation; turf, living/green roof	[1, 11, 22, 39, 48, 51]	The use of arcades, eaves, (Engawa); wet edge, raised porch around the building used to drain water under and away from building; narrow streets to reduce water intake; slopes	[10, 27, 37]
Drought and water scarcity	Indoor water collection in ponds; rainwater harvesting from the roof using a gutter system that led to cistern; sloping guiding the falling rainwater to the cistern	[20, 29]	Rainwater harvesting in underground cistern or overground ponds; building reservoirs, aqueducts, wells, controlled water channels; collective water management; use of tank-houses for water collections; water storage in ice houses	[16, 43, 49]
Wind and storm hazard	Narrow and aerodynamic shape facing the prevailing wind; gentle pitched roof with minimal eaves; courtyards as wind breaker; low rise building	[5, 10, 13]	Use wind breakers; embedding the buildings in the terrain, elongated façade parallel to the wind; compact settlement design	[5, 10, 13]

(continued)

Table 5.2 (continued)

Hazard/ impact	Building scale		Urban scale	
	Examples of design solutions	References	Examples of design solutions	References
Over heating	Light coloured building materials, seasonal rooms; reduced direct sun exposure via narrow openings and sunshades, the use of building materials with the high thermal mass, use of high ceilings to increased air exchange through cross ventilation; taking advantage, the ground's thermal capacity	[11, 12, 19, 28, 46]	Use of narrow paths and dense urban patterns to reduce solar exposure and to provide mutual shading; covered streets and passages; orientation of building in the direction of prevailing wind and desired sun path	[3, 12, 19, 47]
Compound hazard	Design with escape routes; creating emergency safe gathering space; deliberate failure points in the building structure; flexible design and ease of assembly and disassembly	[18, 23, 25, 31, 32]	Homogeneous communities with shared responsibilities and ability to repair damage; avoiding construction in risk prone areas; passing of local adaptation knowledge	[18, 23, 25, 30, 32]

4 From Vernacular to Vernomimicry: A Recipe for Climate Adapted Design

The review of climate adaptation strategies in vernacular architecture has uncovered a wealth of solutions addressing a diverse range of climatic challenges. Through careful examination of literature and practical examples, five core design principles that are commonly applied in vernacular buildings that are transferable to design modern have been distilled, which are:

Simplicity: The vernacular buildings are rather simple constructions that can be assembled and disassembled with simple tools and manpower only. The construction is usually a collaborative work that can be accomplished in a relatively short time and allows for a swift reconstruction and adaptation of the building [18, 23, 31].

Transient Structures and Local Materials: Unlike modern constructions designed for permanence, vernacular buildings often have shorter lifespans [57]. This characteristic stems from the use of locally sourced natural materials and simple techniques, necessitating regular maintenance and allowing for agile adaptation to environmental changes [14, 23]. For example, the ladders' shaped extrusions were used for constructing the building are also used for maintaining the hut envelope and are a distinct feature of the Cameroonian Musgum mud huts.

Resource, Location and Form-Efficient Design: In vernacular architecture, form is treated as inexpensive, while materials are considered expensive. This approach results in creating efficient and effective shapes, maximizing material utilization, and utilization of site specific advantages [31].

Community-Centric Building Practices: Building in the vernacular is a communal act, fostering the transfer of accumulated knowledge and traditions across generations. This communal approach ensures that constructions are harmonious, inline with the neighboring structures and align with broader community interests [41].

Integration with Urban Fabric: The vernacular architecture is embedded in the greater urban fabric to the degree that the boundaries of the urban scale and building scales are blurred and both planning scales are interwoven. This meant that both scales acted together as a single unit complementing each other [38, 56]. The effectiveness of this principle in providing appropriate climate adaptation was documented in Al-Lyaly [3] and Taleb et al. [52] studies.

5 Conclusions

This research set out to identify climate design lessons from vernacular architecture that could be applied to contemporary architectural practices. As revealed in Sect. 3, vernacular architecture is a rich source of climate adaptation solutions. However, as Pardo [41] noted, in-depth studies on the effectiveness of these solutions in a modern context are scarce, which limits their broader application and understanding its practicality and limitations in modern contexts. The diminishing number of true vernacular structures, coupled with often inadequate documentation, restricts taking full utilization of the lessons learned of the remaining examples [6]. This highlights the importance of focusing on the core concepts and principles of vernacular architecture and testing their applicability, as we risk a misinterpretation of the examples when the studies are limited historical and aesthetic appreciation of the vernacular examples. Studying the vernacular using a vernomimicry lens can help to contribute to a shift in modern architectural design philosophy from 'despite' nature to 'with' nature. To this end, this research has conducted a comprehensive literature review and comparative analysis, with the aim to discover the resilience design principles of vernacular design that can be ultimately transferred to moder-day architecture practices. The study synthesized five core design principles that the vernacular architecture can offer to the modern architectural practice to enhance the urban resiliency without sacrificing the mitigation objectives. Most notably, the vernacular architecture sublime merger between building and urban scale, private and community, nature and manmade is pivotal to its success. In vernacular architecture, these aspects interact synergistically, much like a single organism. By highlighting these core principles and illustrating how they can be integrated into today's designs through vernomimicry, the study introduces new insights into how vernacular methods can address the dual challenges of sustainability and resilience. The Incorporating these five core vernacular strategies through vernomimicry can make modern architecture more adaptable and resilient to climate challenges, while also fostering a deeper connection between the built environment and the community it serves. The findings of this research demonstrate the continued relevance of vernacular architecture as a valid source of inspiration, offering a wealth of adaptation solutions that can inspire modern architecture in addressing contemporary urban climate challenges. However, it is important not to overlook the practical limitations of vernacular design. The study and interaction with the vernacular principles, particularly through vernomimicry, advocates for a design philosophy in harmony with nature, underlining the significance of simplicity, use of local materials, site specific designs, community involvement, and integration with the broader urban context. These timeless principles from vernacular architecture offer a roadmap for sustainable and resilient building practices in the modern era.

References

1. Abass F, Ismail L, Wahab I, Elgadi A (2020) A review of green roof: definition, history, evolution and functions. In: IOP conference series: materials science and engineering
2. Aktürk G, Fluck H (2022) Vernacular heritage as a response to climate: lessons for future climate resilience from Rize, Turkey. Land 11(2):276
3. Al-Lyaly SMZ (1990) The traditional house of Jeddah: a study of the interaction between climate, form and living patterns University of Edinburgh
4. Alexander C (1964) Notes on the synthesis of form, vol 5. Harvard University Press
5. Almssad A, Almusaed A (2015) Environmental reply to vernacular habitat conformation from a vast areas of Scandinavia. Renew Sustain Energy Rev 48:825–834
6. Asquith L, Vellinga M (2005) Vernacular architecture in the twenty-first century. Taylor & Francis
7. Benyus JM (2002) Biomimicry: innovation inspired by nature. Perennial. http://books.google.de/books?id=mDHKVQyJ94gC
8. Berkebile B, McLennan J (2004) The living building: biomimicry in architecture, integrating technology with nature. http://jasonmclennan.com/draft/articles/The_Living_Building.pdf
9. Bianciardi A, Cascini G (2023) How would nature design and implement nature-based solutions? Nat Based Solut 3:100047

10. Bodach S, Lang W, Hamhaber J (2014) Climate responsive building design strategies of vernacular architecture in Nepal. Energy Build 81:227–242
11. Cañas I, Martín S (2004) Recovery of Spanish vernacular construction as a model of bioclimatic architecture. Build Environ 39(12):1477–1495
12. Chandel S, Sharma V, Marwah BM (2016) Review of energy efficient features in vernacular architecture for improving indoor thermal comfort conditions. Renew Sustain Energy Rev 65:459–477
13. Coch H (1998) Chapter 4–Bioclimatism in vernacular architecture in architecture, comfort and energy. Gallo C, Sala M (eds) Sayigh AMM Elsevier. Amsterdam
14. Cruz-Ramírez LC (2022) Identification of adaptive strategies in vernacular architecture facing flood risk. Sustain Dev Plan XII 258:393
15. Daoudi NS, Mestoul D, Lamraoui S, Boussoualim A, Adolphe L, Bensalem R (2019) Vernacular architecture in arid climates: adaptation to climate change. In: Bioclimatic architecture in warm climates. Springer, pp 119–154
16. Dehghani-sanij A, Sayigh A (2016) Cisterns: sustainable development, architecture and energy, vol 1. River Publishers
17. Dilhani K, Jayaweera N (2016) A study of flood risk mitigation strategies in vernacular dwellings of Rathnapura, Sri Lanka
18. Dipasquale L, Kisa Ovali P, Mecca S, Özel B (2014) Resilience of vernacular architecture. In VerSus: heritage for the future. Vernacular knowledge fo sustainable architecture. FUP Firenze University Press, pp 64–73
19. Fardeheb F (2008) Examination and review of passive solar cooling strategies in middle eastern and north african vernacular architecture. In: Proceedings of ISES world congress 2007, vols I–V. Solar Energy and Human Settlement
20. Fernandes JEP, Mateus R, Gervásio HMS, Silva SM, Bragança L (2016) The importance of vernacular strategies for a climate responsive building design
21. Foltz R (2003) Worldviews, religion, and the environment: a global anthology. Thomson/ Wadsworth. http://books.google.de/books?id=BqhkQgAACAAJ
22. Gansach A, Meir IA (2004) Energy-culture across altitude. PLEA2004 [21st]
23. Glassie H (2000) Vernacular architecture. Indiana University Press
24. Gruber P (2011) Biomimetics in architecture: architecture of life and buildings. Springer Wien. http://books.google.de/books?id=_6VSwKITN3EC
25. Gupta TD, Chakraborty A, Danda AA (2021) Confronting cascading disasters, building resilience: lessons from the Indian Sundarbans. ORF Occasional Paper (297)
26. Helms M, Vattam SS, Goel AK (2009) Biologically inspired design: process and products. Des Stud 30(5):606–622
27. Izumida H. Kaki Lima as contesting space between European and Asian values
28. Johnson W (1995) Keeping cool. Aramco World 46(3):10–17
29. Kelly S, Paget B, Phillips K, Smith B. Rainwater harvesting, vernacular architecture and the culture of water resource management in Bermuda
30. Khoja A, Danylenko O, Polo Lopez CS, Davis A, Essig N (2021) Socioeconomic reflections on historic buildings renovations: a portrait of rural alpine municipalities. IOP Conf Ser Earth Environ Sci 863(1):012001. https://doi.org/10.1088/1755-1315/863/1/012001
31. Khoja A, Waheeb S (2020) Vernomimicry: bridging the gap between nature and sustainable architecture. J Sustain Dev 13(1):33
32. Lari Y (2011) Build back safer with vernacular methodologies-DRR-driven post-flood rehabilitation in Sindh. Karachi. https://www.academia.edu/3618923/Build_Back_Safer_with_Vernacular_Methodologies
33. Legg S (2021) IPCC, 2021: climate change 2021-the physical science basis. Interaction 49(4):44–45
34. Li S, Ding K, Ding A, He L, Huang X, Ge Q, Fu C (2021) Change of extreme snow events shaped the roof of traditional Chinese architecture in the past millennium. Sci Adv 7(37). https://doi.org/10.1126/sciadv.abh2601

35. López M, Rubio R, Martín S, Croxford B (2017) How plants inspire façades. From plants to architecture: biomimetic principles for the development of adaptive architectural envelopes. Renew Sustain Energy Rev 67:692–703
36. Manzano-Agugliaro F, Montoya FG, Sabio-Ortega A, García-Cruz A (2015) Review of bioclimatic architecture strategies for achieving thermal comfort. Renew Sustain Energy Rev 49:736–755
37. Marra C (2011) The climate-adaptive vernacular architecture of Asia-Pacific. In: Environment design guide, pp 1–12
38. Mauree D, Naboni E, Coccolo S, Perera ATD, Nik VM, Scartezzini J-L (2019) A review of assessment methods for the urban environment and its energy sustainability to guarantee climate adaptation of future cities. Renew Sustain Energy Rev 112:733–746
39. Nguyen A-T, Tran Q-B, Tran D-Q, Reiter S (2011) An investigation on climate responsive design strategies of vernacular housing in Vietnam. Build Environ 46(10):2088–2106
40. Oliver P (2006) Built to meet needs: cultural issues in vernacular architecture. Architectural. http://books.google.de/books?id=aVRgwaT8FU4C
41. Pardo JMF (2023) Challenges and current research trends for vernacular architecture in a global world: a literature review. Buildings 13(1):162
42. Pawlyn M (2011) Biomimicry in architecture. RIBA Publications http://books.google.com/books?id=P3D7tgAACAAJ
43. Pitman LS (1992) The domestic tankhouse as vernacular architecture in rural California. Mater Cult 24(1):13–32
44. Rapoport A (2006) Vernacular design as a model system. In: Asquith L, Vellinga M (eds) Vernacular architecture in the twenty-first century: theory, education and practice. Taylor & Francis, pp 179–198. http://books.google.de/books?id=NxTPpLQWGnMC
45. Reid WV, Mooney HA, Cropper A, Capistrano D, Carpenter SR, Chopra K, Dasgupta P, Dietz T, Duraiappah AK, Hassan R (2005) Ecosystems and human well-being-synthesis: a report of the Millennium ecosystem assessment. Island Press
46. Sahebzadeh S, Dalvand Z, Sadeghfar M, Heidari A (2018) Vernacular architecture of Iran's hot regions; elements and strategies for a comfortable living environment. Smart Sustain Built Environ
47. Saleh MAE (2001) The evolution of planning & urban theory from the perspective of vernacular design: MOMRA initiatives in improving Saudi Arabian neighbourhoods. Land Use Policy 18(2):179–190
48. Salgın B, Bayram ÖF, Akgün A, Agyekum K (2017) Sustainable features of vernacular architecture: housing of Eastern Black Sea Region as a case study. Arts
49. Scarborough VL, Dunning NP, Tankersley KB, Carr C, Weaver E, Grazioso L, Lane B, Jones JG, Buttles P, Valdez F (2012) Water and sustainable land use at the ancient tropical city of Tikal, Guatemala. Proc Natl Acad Sci 109(31):12408–12413
50. Seneviratne SI, Nicholls N, Easterling D, Goodess CM, Kanae S, Kossin J, Luo Y, Marengo J, McInnes K, Rahimi M (2012) Managing the risks of extreme events and disasters to advance climate change adaptation: changes in climate extremes and their impacts on the natural physical environment
51. Sun F (2013) Chinese climate and vernacular dwellings. Buildings 3(1):143–172
52. Taleb D, Abu-Hijleh B (2013) Urban heat islands: potential effect of organic and structured urban configurations on temperature variations in Dubai, UAE. Renew Energy 50:747–762
53. Turner JS, Soar RC (2008) Beyond biomimicry: what termites can tell us about realizing the living building. In. Loughborough University: first international conference on industrialized, intelligent construction

54. Vegas F, Mileto C, Escobar AH, de Miguel ML (2022) Sustainability, risks and resilience of vernacular architecture, vol 16. Taylor & Francis, pp 817–819
55. Vincent JFV (2001) Stealing ideas from nature. In: Pellegrino S (ed) Deployable structures. Springer, pp 51–58. http://books.google.de/books?id=A8iIK5oUb-UC
56. Weber W, Yannas S (2014) Lessons from vernacular architecture, vol 2. Routledge, London
57. Wrathmell S (1984) The vernacular threshold of northern peasant houses. Vernac Archit 15(1):29–33

Navigating Nearly Zero-Energy Strategies for Urban Climate Change Adaptation and Mitigation

Osama Omar

Abstract Climate change stands out as one of the most pressing challenges of the 21st century, exacerbating resource consumption, environmental degradation, human health risks, and occupant discomfort. In response, nearly zero-energy strategies emerge as pivotal solutions for curbing energy consumption and environmental impact in both existing and new constructions. The built environment significantly contributes to greenhouse gas emissions and resource depletion, with buildings consuming 32% of global energy and emitting 19% of greenhouse gas emissions, according to the International Energy Agency. This underscores the urgency of enhancing energy efficiency and reducing carbon emissions in buildings worldwide. Zero-carbon strategies are indispensable for mitigating emissions during building operation. This chapter delves into the application of nearly zero-energy strategies in the housing sector to combat climate change, employing a literature review approach. Nearly zero-energy buildings, envisioned as structures powered by clean energy to fulfill their energy needs, offer a promising yet ambiguously defined solution. Global discrepancies in defining "clean energy" have led to confusion and hindered the integration of zero-energy building goals into building codes. Despite challenges, the concept of nearly zero-energy buildings catalyzes efforts to reduce energy demand and fossil fuel consumption worldwide, aligning with renewable energy objectives. This study proposes a range of nearly zero-energy strategies applicable to the housing sector worldwide to mitigate and adapt to climate change. Emphasizing resource efficiency and environmental responsibility throughout the building's life cycle is critical in realizing effective nearly zero-energy building designs. By adopting these strategies, the housing sector can make significant strides towards sustainability and climate resilience on a global scale.

Keywords Climate change adaptation · Climate change mitigation · Nearly zero-energy strategies · Zero carbon · Sustainable development goals

---

O. Omar (✉)
Associate Professor of Architecture, Department of Architecture and Interior Design, College of Engineering, University of Bahrain, P.O. Box 32038, Manama, Kingdom of Bahrain
e-mail: oomar@uob.edu.bh

B. He et al. (eds.), *Urban Climate and Urban Design*, Urban Sustainability,
https://doi.org/10.1007/978-981-96-1521-6_6

1 Introduction

Observations in urban areas have shown a mutual interaction between climate change and the built environment. An increase in population size, particularly due to urbanization, requires an increase in the built environment as well. Climate change is strongly influenced by these increases in land use and subsequent large-scale changes [7]. The 21st century will be marked by climate change as one of humanity's biggest challenges. Various negative environmental effects are associated with increasing anthropogenic emissions, including global temperature rise, shrinking ice sheets, warming oceans, rising sea levels, extreme weather events, and ocean acidification [26]. The same source reported 19 of the hottest 20 years as being between 2001 and 2019, and the average surface temperature has risen by nearly 1 °C since 1880. Emissions of greenhouse gases from the built environment are significant.

As well as greenhouse gas emissions being directly influenced by the built environment and land use, studies have also found that regional climate change can be directly attributed to factors such as changes in evaporation levels, precipitation regimes, increasing temperatures, changing temperature flows, and increased surface flows. In light of the projected growth of urban populations in the future, it becomes apparent that cities are likely to have a stronger triggering effect on climate change, leaving urban areas more vulnerable to floods, high water levels, heat waves, droughts, and storms [32] it is essential to focus on integrating energy-efficient and sustainable practices within urban development to combat climate change impacts. By implementing nearly zero-energy strategies in urban areas, such as enhancing building insulation, utilizing renewable energy sources, and promoting energy-efficient technologies, cities can reduce their carbon footprint and contribute to climate change mitigation efforts. Additionally, incorporating green infrastructure, sustainable transportation systems, and resilient building designs can help cities adapt to the changing climate and minimize environmental risks associated with extreme weather events. This holistic approach not only addresses the environmental challenges posed by climate change but also enhances the overall quality of life in urban areas while promoting sustainability and resilience [5, 7, 38].

Studies identified a research gap that needs to be addressed in order to understand the effects of sustainable building adaptation on climate change comprehensively. While there are existing studies that highlight the importance of sustainable building adaptation for reducing environmental impact and enhancing resource efficiency, there is a call for more research to investigate the broader implications of these strategies across social, economic, and environmental dimensions. The basis to link mitigation and adaptation with the sustainable built environment lies in the necessity to address both aspects to effectively combat climate change. Mitigation involves reducing greenhouse gas emissions to lessen the impact of climate change, while adaptation focuses on implementing strategies to reduce vulnerability to the effects of climate change. By integrating sustainable building adaptation practices, cities can mitigate their environmental footprint, enhance resilience to climate change impacts, and promote sustainable development in the built environment [1, 8, 33].

Energy-related GHG emissions from buildings account for 19% of the total global energy use and 32% of the total global energy consumption [18]. It is also expected that climate change will have a significant impact on buildings, resulting in mismatching of capacity, overloads, inefficiency and malfunctions [37, 10]. Furthermore, as temperatures increase, cooling demand will also increase and air conditioning needs will rise dramatically. In addition to thermal discomfort, illness and mortality, climate change has a detrimental effect on building occupants. A major cause of several problems in buildings is overheating, which can range from thermal discomfort to productivity loss to illness and death. As global warming continues, buildings are expected to become more overheated. It is possible to reduce the risk of overheating in both existing and new buildings if policy makers make quick decisions about adaptation interventions [15]. As a result, buildings contribute to anthropogenic climate change via a feedback loop [2].

Because of the urgency of climate change, it is critical for the built environment to improve energy efficiency, reduce energy demand and expand the use of decentralized renewable energy sources. In addition, building energy efficiency is a critical component in lowering carbon emissions during the operating phase of a building's life cycle, and in improving interior environmental quality [31].

In order to attain carbon neutrality in built environment, the concept of zero-energy buildings focuses on the ability of buildings to cover as much energy as they consume annually through on-site renewables. The EU partially adapted this notion as part of a binding regulation for virtually zero-energy buildings [14]. It has become a strong locomotive of building energy efficiency, as well as on-site solar integration. This concept is gradually expanding to the urban scale, i.e. zero-energy districts, which consider the built environments and their interactions with the surroundings in order to achieve high-energy performative goals that go beyond the buildings' scale [21, 30].

Using a literature review approach, this chapter will first examine the relationship between built environment and climate change by reviewing relevant scientific literature related to the interaction between built environment and climate change. As a result of the previous review, the problem was identified and the challenges and knowledge gap were clarified in the introduction. The second part of the research will present a range of zero-energy building strategies in order to propose a framework of zero-energy building design that aims to adapt to and mitigate climate change. Thirdly, the work will investigate how this framework influences climate-resilient strategies at the city scale and with respect to local climates. Finally, zero-energy techniques will be used to demonstrate the factors that contribute to the adaptation and mitigation of urban climate change.

2 Integrated Nearly Zero-Energy Strategies as Part of Mitigation and Adaptation Strategies in Face of Climate Change

In the face of climate change, the built environment can implement various mitigation and adaptation strategies to address the challenges posed by a climate change. Mitigation strategies primarily focus on reducing greenhouse gas emissions and limiting the extent of climate change. Adaptation strategies, on the other hand, focus on building resilience to the impacts of climate change. As adaptation strategies are often specific to specific systems or locations, they are much more varied than mitigation strategies. Adaptation and mitigation strategies designed to address climate change may also affect ecosystem functions, ecosystem services, and human well-being. In addition to reducing GHG emissions, renewable energy sources also reduce air pollution, acid rain, smog, asthma, and respiratory illnesses among humans. As a result of sea-level rise and storm surges, coastal armoring, such as the construction of seawalls, breakwaters, and jetties, may reduce damage to structures, but can negatively affect coastal ecosystems as well [24].

Architects must convert sustainability principles into practical design principles in order to create architecture with a positive impact [6]. The concept of sustainable architecture emphasizes the importance of using renewable resources, reducing waste, and creating buildings that adapt to their climatic conditions. It involves designing structures that establish a reciprocal relationship with the environment, focusing on energy efficiency, durability, and the use of renewable energy sources (Fig. 1).

In the realm of sustainable design, principles such as form, function and usability, cost-effective solutions, renewable energy, materials and recycling, durable design solutions, and constant improvement and sharing of knowledge play a vital role. Designers are urged to consider how the shape, size, and function of a product impact energy consumption, transportation costs, and usability. Additionally, the use of renewable energy sources, recyclable materials, and durable design solutions are essential for achieving sustainability goals in product design.

The previous practical design principles can be considered as best practice in nearly zero-energy design by applying them under the design steps towards achieving a nearly zero-energy design, as shown in Fig. 2 [13].

For example, the use of daylight strategies can improve the performance of buildings with a nearly zero-energy design. From an energy point of view, they will help in decreasing the energy consumption inside this type of building. Additionally, from the point of view of health, increasing the daylight factor in the indoor environment is likely to promote increased productivity within the workforce. Moreover, using energy efficiency, water conservation, and the use of sustainable building materials like clay bricks, concrete blocks, and glass blocks will not only benefit the environment but also lead to lower operating costs, reduced energy bills, and less pollution design principles for architects involve life cycle thinking, circular economy approaches, and social responsibility.

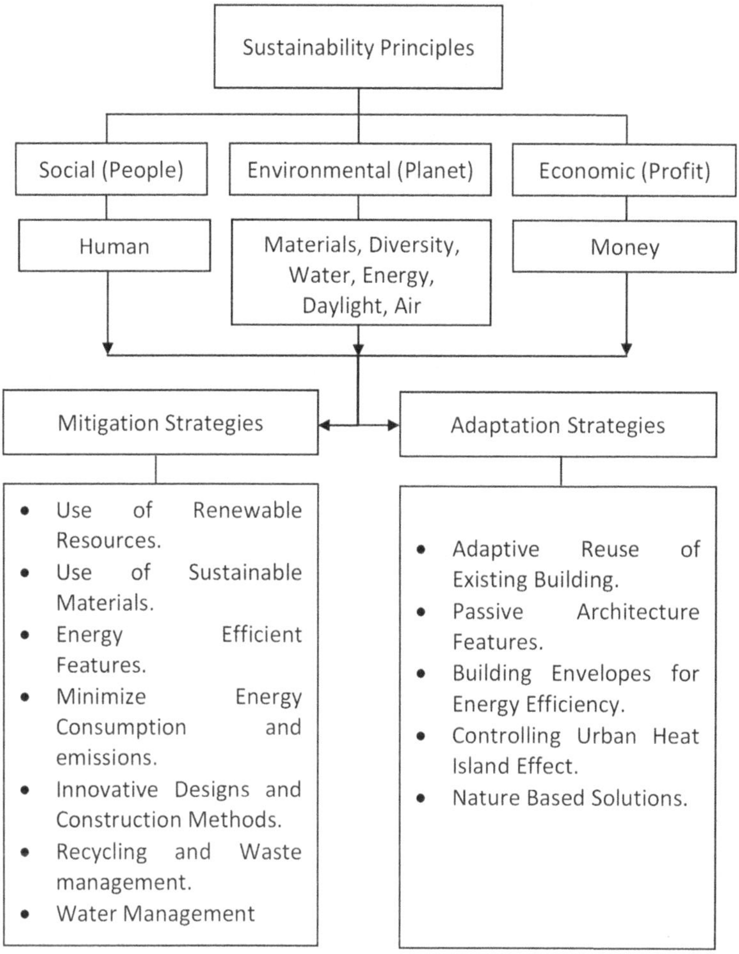

Fig. 1 Converting sustainability principles into practical design principles

The framework for the nearly zero-energy design illustrates how several factors or parameters can be easily integrated into 'research, reduce, reuse and produce' steps by ensuring the level of achievement for each factor within an indoor environment, as shown in Fig. 2. This can be achieved by integrating the four previous steps with a post-occupancy evaluation of different strategies, as shown as in Fig. 3 [13].

It is important to clarify that mitigation strategies focus on addressing the root causes of climate change by reducing emissions, transitioning to cleaner energy sources, and achieving global climate goals. On the other hand, adaptation strategies aim to build resilience, protect communities, enhance infrastructure, and address vulnerability to the impacts of climate change, ensuring a more sustainable and prepared future in the face of a changing climate. Several mitigation and adaptation

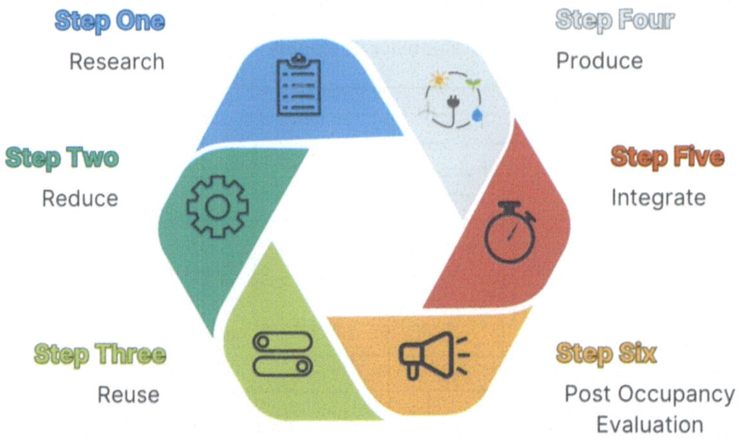

Fig. 2 Design steps towards nearly zero energy

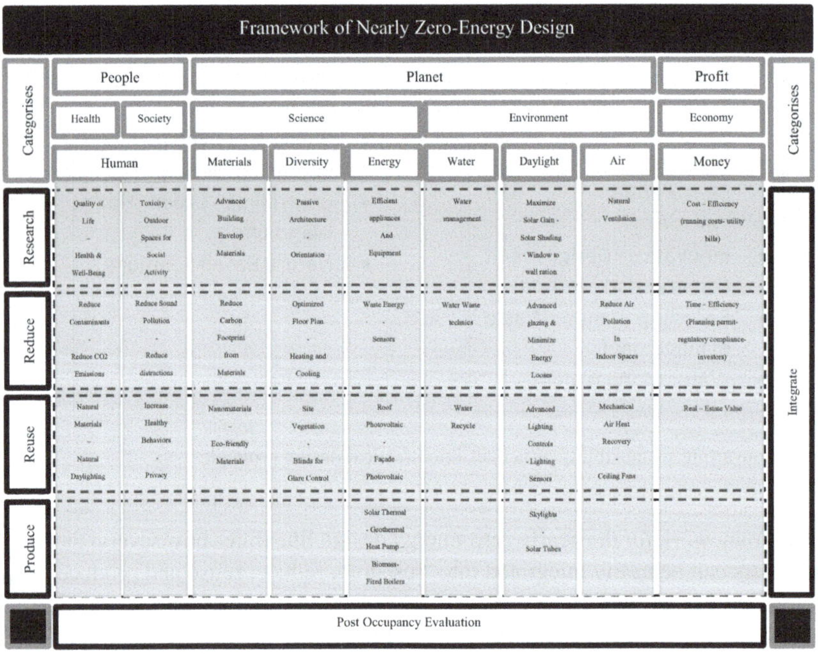

Fig. 3 Framework of nearly zero-energy design

strategies can be integrated into the design of nearly zero energy buildings to enhance their sustainability and resilience in the face of climate change. However, there are many parameters or factors that can be added to this framework to adapt it to the local climate and identity of the buildings at that location. Extending nearly zero-energy strategies across countries, especially in the Middle East region, is a major challenge, because this strategy must be integrated with the architecture style in these countries without affecting their culture.

In the pursuit of nearly zero-energy building designs, designers must prioritize three integrated pillars: Reuse, Reduce, and Produce. These pillars serve as foundational principles guiding the design process towards optimal energy efficiency and sustainability as shown as in Fig. 4. By integrating these three pillars into the design process, designers can create nearly zero-energy buildings that prioritize sustainability, energy efficiency, and environmental stewardship, thereby contributing to a more resilient and sustainable built environment:

- First, produce clean energy, through using renewable resources (on site or off site).
- Second, reduce the current consumption, through controlling the building management system and building automation system.
- Third, the reuse concept supports the design with intelligent and passive architecture features, in order to decrease consumption and enhance the health and wellbeing of the indoor space, to ensure comfortable conditions that satisfy the user. On the other hand, the designer should not forget to use advanced building materials, especially insulation materials, which play the main role in decreasing the energy consumption related to heating and cooling load.

Passive architecture, intelligent architecture, advanced building materials, renewable resources and smart appliances are the five main criteria of selection towards nearly zero-energy building designs, as shown in Fig. 5. It's important to note that passive architecture features do not contradict intelligent architecture techniques; in actual fact, they complete each other. Passive architecture features can play the role

Fig. 4 Three integrated pillars of nearly zero-energy building design concept

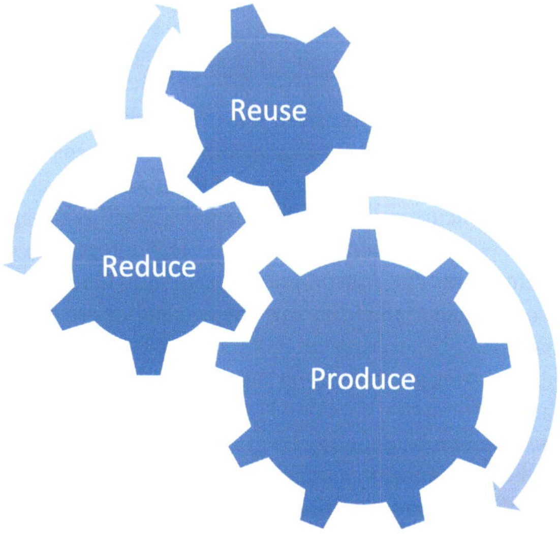

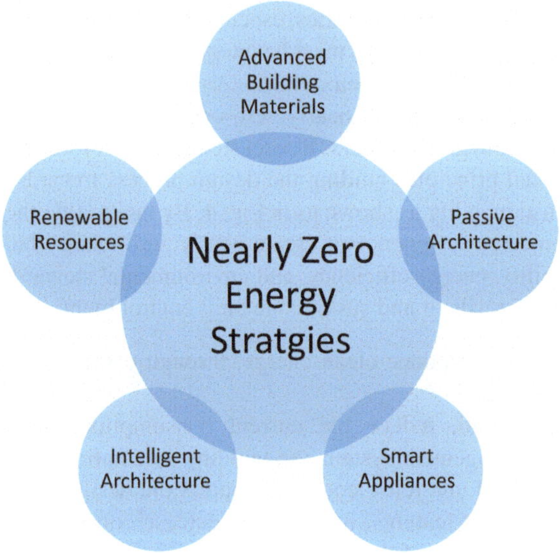

Fig. 5 Main parameters of criteria selection towards nearly zero-energy building design

of decreasing the cooling and heating load through using different techniques like adding atrium, wind catchers, etc. Additionally, intelligent architecture technology can be used to reduce cooling and heating loads through sensors, interactive facades and other smart systems [28].

In the realm of nearly zero-energy building design, a secondary level of parameters further refines the guidelines for architects and designers, encompassing 20 detailed factors crucial for achieving optimal energy performance. Passive architecture, forming the cornerstone of energy-efficient design, is delineated into four secondary levels, including atriums, wind catchers, low-energy features, and considerations of location and climate. These parameters primarily aim to mitigate heating and cooling loads through passive means, harnessing natural elements to optimize thermal comfort within the built environment. Complementing passive strategies, intelligent architecture integrates sophisticated building management systems (BMS), building automation systems (BAS), sensors, and adaptive facades, empowering precise control over indoor environmental conditions and resource management. Concurrently, smart appliances, such as refrigerators, lighting systems, and HVAC systems, leverage advanced technologies to enhance energy efficiency and minimize energy consumption within individual appliances. Moreover, the integration of advanced building materials, spanning recyclable, interactive, intensive, nanomaterials [19], and smart materials, forms a pivotal aspect of nearly zero-energy building design, bolstering structural resilience and thermal comfort while mitigating the impact of external environmental factors. Finally, the utilization of renewable resources, including solar cells, wind turbines, water turbines, and geothermal methods, underscores a commitment to sustainable energy production, with the flexibility to deploy on-site or off-site solutions based on project scale, location, and resource availability. Collectively, these secondary parameters provide a comprehensive framework for designing nearly zero-energy buildings that prioritize energy efficiency, environmental sustainability, and occupant comfort.

There are 34 techniques or strategies that can be integrated into the design of nearly zero-energy buildings at the core level. Considering that each project is unique, it is not imperative to use all of these; rather, it depends upon the specific situation of each project in its special location, as shown as in Fig. 6.

3 Climate-Resilient Strategies and Implications at the City Scale

Climate-resilient strategies at the city scale are imperative in the face of escalating climate change impacts. These strategies encompass a multifaceted approach to mitigating risks and adapting urban environments to changing climatic conditions. At the heart of these strategies lies the recognition of the interconnectedness between various urban systems and the environment. Urban planners and policymakers must prioritize resilience-building measures that address vulnerabilities in infrastructure, transportation, housing, and public health. This entails integrating green infrastructure, such as green roofs, permeable pavements, and urban forests, to mitigate the urban heat island effect and manage stormwater runoff. Additionally, enhancing the resilience of critical infrastructure, such as power grids and water supply systems, is crucial to minimizing disruptions during extreme weather events. Furthermore, promoting social equity and community engagement is paramount in ensuring that climate-resilient strategies are inclusive and benefit all residents, particularly marginalized communities disproportionately affected by climate change. Investing in robust early warning systems, emergency preparedness plans, and disaster response mechanisms is also essential to safeguarding urban populations against climate-related hazards. Moreover, fostering collaboration and knowledge-sharing among cities globally can facilitate the exchange of best practices and innovative solutions for building climate resilience. Ultimately, integrating climate-resilient strategies into urban planning and development processes is essential for creating cities that are adaptive, sustainable, and able to thrive in the face of evolving climate challenges.

Climate-resilient strategies at the city scale must be closely intertwined with efforts to navigate zero-energy buildings towards climate change mitigation and adaptation. These strategies recognize the imperative of reducing greenhouse gas emissions while simultaneously enhancing urban resilience to climate-related risks. Incorporating zero-energy building principles into urban planning and development is integral to achieving this dual objective. By prioritizing energy-efficient design, renewable energy integration, and passive building strategies, cities can significantly reduce their carbon footprint and mitigate the urban heat island effect. Moreover, zero-energy buildings can serve as resilient infrastructure assets, providing reliable energy

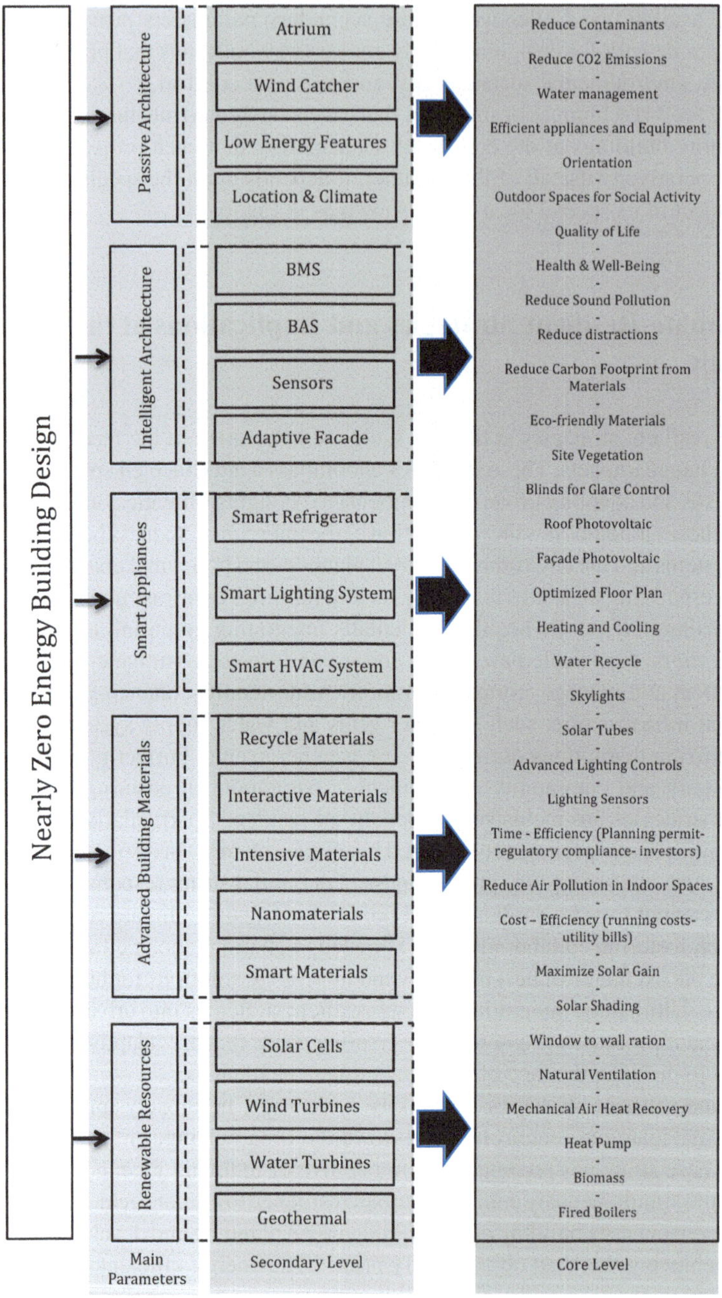

Fig. 6 Nearly zero-energy building design criteria selection (main parameters, secondary level and core level)

supply during extreme weather events and power outages. However, the implications of climate-resilient strategies extend beyond individual buildings to encompass broader urban systems and community resilience. Collaborative efforts are needed to enhance urban green spaces, strengthen infrastructure resilience, and promote equitable access to sustainable housing and energy resources. Furthermore, leveraging technological innovations such as building-integrated photovoltaics (BIPV), energy storage systems, and smart grid technologies can optimize the resilience and efficiency of zero-energy buildings within the urban fabric. Ultimately, integrating climate-resilient strategies with zero-energy building initiatives requires a holistic approach that considers both mitigation and adaptation measures, fosters interdisciplinary collaboration, and prioritizes the well-being of urban residents in the face of climate change impacts.

It is important to break climate actions down further into the areas of transportation, industry, building, energy generation, agriculture, land use and forestry. In this way, it is possible to describe the tasks performed by different industries. Achieving carbon neutrality, for example, requires improving energy efficiency for lower energy demands, decarbonizing the electricity grid, electrifying end uses, using low-carbon fuels and storing carbon. Without specific actions within a given field, this can't be achieved [17].

Many articles have discussed declarations of climate emergency at numerous levels of authority, including governments, municipalities, non-governmental organizations and activists. Worldwide, most governments set the 2030 target without clear visions or policies for the road map for the role of each member of the country in order to accomplish that goal. Therefore, the objective of this research is to identify the factors and principles that should be taken into account by authorities to guide them in implementing this road map, in order for the country to achieve its goals.

When a neighbourhood or city is well planned, it is possible to enhance citizens' everyday lives as well as to encourage them to lead more sustainable and low-impact lifestyles. In addition to transportation, waste, energy and green infrastructures, buildings also need to be 'plugged into' the future plans of the site. By getting this infrastructure right, ambitious climate emergency design targets can be more easily met at the building scale. The microclimate of a particular site is determined by its particular location. Generally, cities are warmer than their surrounding countryside, especially at night, due to their thermal mass. In addition, topography creates opportunities for viewpoints across the landscape, as well as forming barriers. Concepts inspired by the local climate can be very powerful [29].

When referring to climate-resilient implications at the city scale, it's important to mention several strategies that should be taken into consideration for urban resilience, such as sustainable densities, the 15-min city and passive resilience. Sustainable densities, for example, are a topic of debate. There are several factors that determine where a site should be placed, such as the surrounding fabric, the landscape (urban, suburban, rural), the mix of living and working, other services and accessibility to public transportation and alternative modes of transport. A complete picture of all strategies and interventions can be seen as acupunctures of the city, with the help of urban solutions that can influence the mitigation of and adaptation to climate emergencies and which need to be analyzed at a city scale. It has generally been recognized that compact cities with high densities are more sustainable than sprawling cities, owing to the fact that they reduce spread. Those cities that are compact support public transportation, leading to a reduction in car ownership and use for commuting to work. As a result, this reduced fossil-fuel use leads to better air quality, and poor air quality is a major cause of illness and death in all cities [29].

In addition to the fact that cities are experiencing rapid population growth, coupled with high demand for land, forests, water and energy, little attention is being paid to strategies for restoration and prospective sustainability [4, 16]. Providing sustainable ecosystems and supporting population wellbeing in rapidly growing cities has been one of the global agendas of 'urban green', which refers to urban green infrastructure and ecological infrastructure. The urban green infrastructure and/or ecological infrastructure considers elements of land use planning and is used to conserve water resources, create parks, create vegetation cover and conserve natural resources. Green spaces provide environmental benefits by offsetting greenhouse gas emissions, reducing storm water runoff and mitigating global warming. Due to past climate stability, plant species (of grass, shrubs and forests) have high variability, rendering the use of green spaces as an opportunity for sustainability challenging. The importance of green spaces has also been poorly explored, deferring to agriculture, arboriculture and ecological infrastructure for climate change mitigation in cities as part of nature-based solutions. Although the topic and results have received extensive discussion and consensus, implementation is still elusive [12, 16, 23, 25, 27, 35, 36].

The term 'nature-based solutions' refers to a package of practices that are typically natural and increase the ability of people, animals and plants to address challenges [34]. During an era of climate change and extreme phenomena, such as flooding and extreme weather, a nature-based solutions package brings together all previously discussed urban solutions that can be helpful for adaptation to and mitigation of climate change, as shown in Fig. 7.

Ecosystem adaptations to and mitigations of climate change, the reduction of disaster risks before they become problems, water bodies in cities and natural approaches to climate problems are all part of nature-based solutions [22]. Despite the benefits of nature-based solutions in solving environmental problems, the adoption of them is not widespread in practice as part of research and innovation. As a result of funding projects in the Global North (e.g. Europe), the response to nature-based solutions, biodiversity and ecological services has been improved. More jobs have been created and income has increased for such organizations. While the importance of the urban green infrastructure was emphasized in 2009 in this same region, in order to adapt to climate change and restore cities, there was a crucial focus on natural and semi-natural solutions in 2009. Continuing the deployment of nature-based solutions is acknowledged by the European Commission (2019) as contributing towards achieving a net-zero society in the Global North, particularly Europe, by 2050 and ensuring sustainable climate change adaptation and mitigation measures. In order to implement nature-based solutions effectively at all levels, it is important to: involve

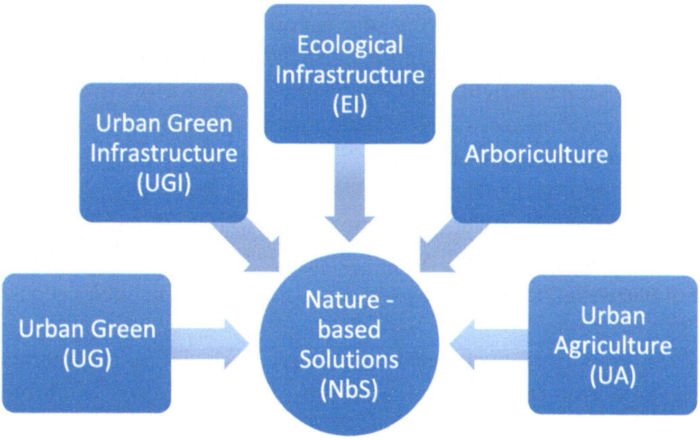

Fig. 7 Nature-based solutions package for climate change mitigation and adaptation

all stakeholders properly; provide guidelines; be uncertain of the value to be delivered; have access to financing sources and/or expertise; and have land and time available [3, 9, 11, 20, 34]. Despite the fact that nature-based solutions have been put into practice in the Global North since 2014, the private sector has been left out of the implementation process by the public sector, which has resulted in many obstacles to the achievement of the goals for nature-based solutions. Public–private enterprises, public/community groups and networking in community gardens, forestry and tourism are the only nature-based solutions investors now present in Europe; purely private firms are not among them [27].

4 Conclusions

In constructing a comprehensive framework for nearly zero-energy strategies aimed at urban climate change adaptation and mitigation, it is essential to assemble the various components of this intricate puzzle. Each element contributes to shaping both micro and macro climates, impacting building scales as well as urban scales. By integrating these diverse puzzle pieces, ranging from passive architectural design principles to renewable energy technologies and advanced building materials, cities can develop holistic strategies that address the multifaceted challenges posed by climate change. Moreover, considering the interconnectedness of these strategies is crucial for achieving synergistic effects and maximizing their effectiveness in reducing greenhouse gas emissions, enhancing energy efficiency, and bolstering urban resilience. Therefore, by assembling the entire puzzle of nearly zero-energy strategies, cities can better navigate the complexities of climate change adaptation and mitigation, ultimately fostering sustainable and resilient urban environments for current and future generations. Taking into account the new concept of neighbourhood, the UN-Habitat urban lab integrates principles across five key city objectives into a holistic approach to neighbourhood design, across sectors (transport, housing, public space, utilities) and across spatial dimensions (neighbourhoods, streets, open public spaces and building units).

There are many approaches that can be used at scale to achieve nearly zero-energy buildings across the world, yet it is also important to note that community engagement, architecture style and identity will also affect these strategies. In the Middle East region in particular, identity plays an important role in society's acceptance and admiration of a building.

This chapter provides an overview for architects of the tools and features of architecture that can be used for adaptation and mitigation as a result of climate change phenomena in buildings and urban scales. To conclude, each player in this mission needs to play their part and coordinate with others to ensure that this integration approach will create climate-adapted neighborhoods that will complement national strategies to combat the extreme phenomenon of climate change.

References

1. Aigwi IE, Duberia A, Nwadike AN (2023) Adaptive reuse of existing buildings as a sustainable tool for climate change mitigation within the built environment. Sustain Energy Technol Assess 56:102945. https://doi.org/10.1016/j.seta.2022.102945
2. Akkose G, Akgül Ç, Gürsel Dino İ (2021) Educational building retrofit under climate change and urban heat island effect. J Build Eng 40(102294)
3. Albert C, Brillinger M, Guerrero P, Gottwald S, Henze J, Schmidt S, Ott E, Schröter B (2020) Planning nature-based solutions: principles, steps, and insights. Ambio 50(8):1446–1461. https://doi.org/10.1007/s13280-020-01365-1
4. Amoah-Antwi C, Kwiatkowska-Malina J, Thornton SF, Fenton O, Malina G, Szara E (2020) Restoration of soil quality using biochar and brown coal waste: a review. Sci Total Environ 722:137852. https://doi.org/10.1016/j.scitotenv.2020.137852
5. Andrić I, Koc M, Al-Ghamdi SG (2019) A review of climate change implications for built environment: impacts, mitigation measures and associated challenges in developed and developing countries. J Clean Prod 211:83–102. https://doi.org/10.1016/j.jclepro.2018.11.128
6. Attia S (2018) Regenerative and positive impact architecture, 1st ed. Springer, Cham. https://doi.org/10.1007/978-3-319-66718-8
7. Bektaş Y, Sakarya A (2023) The relationship between the built environment and climate change: the case of Turkish Provinces. Sustainability 15(2):1659. https://doi.org/10.3390/su15021659
8. Castaño-Rosa R, Pelsmakers S, Järventausta H, Poutanen J, Tähtinen L, Rashidfarokhi A, Toivonen S (2022) Resilience in the built environment: key characteristics for solutions to multiple crises. Sustain Cities Soc 87:104259. https://doi.org/10.1016/j.scs.2022.104259
9. Cohen-Shacham E, Walters G, Janzen C, Maginnis S (eds) (2016) Nature-based Solutions to address global societal challenges. Gland, Switzerland: IUCN. xiii + 97 pp. In: International Union for Conservation of Nature. https://doi.org/10.2305/IUCN.CH.2016.13.en. International Union for Conservation of Nature and Natural Resources. https://portals.iucn.org/library/sites/library/files/documents/2016-036.pdf. Accessed 17 June 2024
10. Dino IG, Akgül CM (2019) Impact of climate change on the existing residential building stock in Turkey: an analysis on energy use, greenhouse gas emissions and occupant comfort. Renew Energy 141:828–846. https://doi.org/10.1016/j.renene.2019.03.150

11. European Commission (2015) Towards an EU research and innovation policy agenda for nature-based solutions and re-naturing cities: final report of the Horizon 2020 expert group on 'Nature-based solutions and re-naturing cities': (full version), Publications Office. In European Commission. European Commission, Directorate-General for Research and Innovation. https://ec.europa.eu/newsroom/horizon2020/items/24388/en. Accessed 17 June 2024
12. Feltynowski M, Kronenberg J (2020) Urban green spaces—an underestimated resource in third-tier towns in Poland. Land 9(11):453. https://doi.org/10.3390/land9110453
13. Garcia-Fernandez B, Omar O (2023) Sustainable performance in public buildings supported by daylighting technology. Sol Energy 264:112068. https://doi.org/10.1016/j.solener.2023.112068
14. Groezinger J, Boermans T, John A, Seehusen J, Wehringer F, Scherberich M (2014) Overview of member states information on NZEBs working version of the progress report—final report. In: Ecofys 2014 by order of: European Commission. http://www.nezeh.eu/assets/media/PDF/2014_nzeb_progress_report49.pdf. Accessed 17 June 2024
15. Hamdy M, Carlucci S, Hoes PJ, Hensen JL (2017) The impact of climate change on the overheating risk in dwellings—a Dutch case study. Build Environ 122:307–323. https://doi.org/10.1016/j.buildenv.2017.06.031
16. Hansen R, Rall E, Chapman E, Rolf W, Pauleit S (2017) Urban green infrastructure planning: a guide for practitioners. Green Surge. http://greensurge.eu/working-packages/wp5/. Acessed 17 June 2024
17. He BJ, Sharifi A, Feng C, Yang J, Prasad D, Jupesta J, Pignatta G (2022) Climate emergency, actions and environmental sustainability. In: Advances in science, technology and innovation, pp 1–6. https://doi.org/10.1007/978-3-031-12015-2_1
18. IPCC (2014) Climate change 2014: impacts, adaptation, and vulnerability. Part A: global and sectoral aspects. In: Contribution of working group II to the fifth assessment report of the intergovernmental panel on climate change [Field CB, Barros VR, Dokken DJ, Mach KJ, Mastrandrea MD, Bilir TE, Chatterjee M, Ebi KL, Estrada YO, Genova RC, Girma B, Kissel ES, Levy AN, MacCracken S, Mastrandrea PR, White LL (eds)] Cambridge University Press, Cambridge, United Kingdom and New York, NY, USA, 1132 pp (nd)
19. Ibrahim MA, Hanafi MA, Omar OM (2010) Nanoarchitecture and global warming. Alexandria University
20. Kabisch N, Korn H, Stadler J, Bonn A (2017). Nature-based solutions to climate change adaptation in urban areas. Springer. http://books.google.ie/books?id=QtpCDwAAQBAJ&printsec=frontcover&dq=Kabisch,+N.,+Korn,+H.,+Stadler,+J.,+Bonn,+A.+(2017).+Nature based+solutions+to+climate+change+adaptation+in+urban+areas:+linkages+between+science,+policy+and+practice.+(Vol.+337):+Springer+Cham&hl=&cd=2&source=gbs_api
21. Kaitouni SI, Pfafferott J, Jamil A, Ahachad M, Brigui J (2023) A holistic digital workflow methodology to shifting towards net zero energy urban residential buildings in a semi-arid climate. Sol Energy 263:111959. https://doi.org/10.1016/j.solener.2023.111959
22. Kooijman ED, McQuaid S, Rhodes ML, Collier MJ, Pilla F (2021) Innovating with nature: from nature-based solutions to nature-based enterprises. Sustainability 13(3):1263. https://doi.org/10.3390/su13031263
23. Kreft H, Jetz W (2007) Global patterns and determinants of vascular plant diversity. Proc Natl Acad Sci USA 104(14):5925–5930. https://doi.org/10.1073/pnas.0608361104Lucy
24. Lawler JJ, Spencer B, Olden JD, Kim SH, Lowe C, Bolton S, Beamon BM, Thompson L, Voss JG (2013) Mitigation and adaptation strategies to reduce climate vulnerabilities and maintain ecosystem services. Clim Vulnerab 315–335. https://doi.org/10.1016/B978-0-12-384703-4.00436-6
25. Lee B, Binns T, Dixon AB (2010) The dynamics of urban agriculture in Hanoi, Vietnam. Open Edition Journal. http://journals.openedition.org/factsreports/464. Accessed 17 June 2024
26. NASA (2023) Climate change: vital signs of the planet. https://climate.nasa.gov/%C2%A0/. Accessed 17 June 2024
27. Nassary EK, Msomba BH, Masele WE, Ndaki PM, Kahangwa CA (2022) Exploring urban green packages as part of nature-based solutions for climate change adaptation measures in

rapidly growing cities of the Global South. J Environ Manag 310:114786. https://doi.org/10.1016/j.jenvman.2022.114786
28. Omar O (2018) Intelligent building, definitions, factors and evaluation criteria of selection. Alex Eng J 57(4):2903–2910. https://doi.org/10.1016/j.aej.2018.07.004
29. Pelsmakers S, Group TF, Hoggard A, Kozminska U, Donovan E (2022) Designing for the climate emergency. Riba Publishing. http://books.google.ie/books?id=_5LRzgEACAAJ&dq=Kozminska,+S.+P.+E.+D.+A.+H.+U.+(2022).+Designing+for+the+climate+emergency+RIBA+Publishing&hl=&cd=1&source=gbs_api
30. Polly B, Kutscher C, Macumber D, Schott M, Pless S, Livingoo B, Van Geet O (2016) From zero energy buildings to zero energy districts. In: Proceedings of the 2016 American council for an energy efficient economy summer study on energy efficiency in buildings, Pacific Grove, CA, USA, pp 21–26. https://www.aceee.org/files/proceedings/2016/data/papers/10_566.pdf
31. Reinhart CF, Davila CC (2016) Urban building energy modeling—a review of a nascent field. Build Environ 97:196–202. https://doi.org/10.1016/j.buildenv.2015.12.001
32. Rienow A, Kantakumar LN, Ghazaryan G, Dröge-Rothaar A, Sticksel S, Trampnau B, Thonfeld F (2022) Modelling the spatial impact of regional planning and climate change prevention strategies on land consumption in the Rhine-Ruhr metropolitan area 2017–2030. Landsc Urban Plan 217:104284. https://doi.org/10.1016/j.landurbplan.2021.104284
33. Saifudeen A, Mani M (2024) Adaptation of buildings to climate change: an overview. Front Built Environ 10. https://doi.org/10.3389/fbuil.2024.1327747
34. Seddon N, Mace GM, Naeem S, Tobias JA, Pigot AL, Cavanagh R, Mouillot D, Vause J, Walpole M (2016) Biodiversity in the anthropocene: prospects and policy. Proc Royal Soc Biol Sci 283(1844):20162094. https://doi.org/10.1098/rspb.2016.2094
35. Sun S, Jiang Y, Zheng S (2020) Research on ecological infrastructure from 1990 to 2018: a bibliometric analysis. Sustainability 12(6):2304. https://doi.org/10.3390/su12062304
36. Taylor L, Hochuli DF (2017) Defining greenspace: multiple uses across multiple disciplines. Landsc Urban Plan 158:25–38. https://doi.org/10.1016/j.landurbplan.2016.09.024
37. De Wilde P, Coley D (2012) The implications of a changing climate for buildings. Build Environ 55:1–7. https://doi.org/10.1016/j.buildenv.2012.03.014
38. Zięba Z, Dąbrowska J, Marschalko M, Pinto J, Mrówczyńska M, Leśniak A, Petrovski A, Kazak JK (2020) Built environment challenges due to climate change. IOP Conf Ser Earth Environ Sci 609(1):012061. https://doi.org/10.1088/1755-1315/609/1/012061

Enhancing Climate Resilience Through Urban Landscape Design: Strategies for Green Infrastructure and Adaptation

Reza Farhadi, Hadi Soltanifard, and Behdad Alizadeh

Abstract Climate change presents urgent global challenges that demand immediate action to mitigate its impacts. Urban landscape design has emerged as a promising approach to addressing climate change and its consequences. This chapter focuses on the relationship between urban landscape planning and green infrastructure design to promote climate resilience and adaptation. Urban Green Infrastructure (UGI) planning is a strategic approach to developing multifunctional networks of blue and green areas that can deliver various environmental, social, and economic advantages while improving cities' climate resilience. Community participation in urban planning, fostering social cohesion, and creating inclusive spaces are emphasized as key pillars that can withstand climate change impacts. Practical initiatives, such as community gardens and neighborhood networks, are explored as effective means of enhancing community resilience. It is necessary to argue how green infrastructure planning can be applied to integrate planning strategies into community-adaptive planning. This is to cope with urban climate change and promote local resilience. To address climate change effectively, it is essential to understand the variety of types of green infrastructure, their implications, and ecological design. Adopting the 'traditional' grey infrastructure or technologically-based solutions is preferable. The planning process includes a number of structured adaptive solutions derived from ecological planning and climate change adaptation in order to facilitate multiple planning responses at the same time. This will enhance planning flexibility in practice. The challenges and prospects of urban landscape design in the context of climate change adaptation. In order to fully realize the potential of urban landscape design in addressing climate change, it emphasizes the importance of interdisciplinary collaborations,

R. Farhadi (✉)
Department of Human Geography and Planning, Faculty of Geography, University of Tehran, Tehran, Iran
e-mail: reza.farhadi@ut.ac.ir

H. Soltanifard
Faculty of Geography and Environmental Sciences, Hakim Sabzevari University, Sabzevar, Iran

B. Alizadeh
Department of Horticulture and Landscape Architecture, Ferdowsi University of Mashhad, Mashhad, Iran

B. He et al. (eds.), *Urban Climate and Urban Design*, Urban Sustainability,
https://doi.org/10.1007/978-981-96-1521-6_7

policy support, and financial investment. An emphasis is placed on nature-based solutions and community involvement when designing resilient and sustainable cities. It is a valuable contribution for scholars, urban designers, professionals, and policymakers interested in climate change adaptation and urban planners. As a final point, we discuss adaptation strategies in urban landscape design.

Keywords Climate change · Adaptation strategies · Urban landscape · Sustainable cities · Resilience

1 Introduction

Climate change has become a major challenge for human settlements, the physical environment, and natural ecosystems around the world. The crisis is marked by increasing global temperatures, rising sea levels, more frequent extreme weather events such as floods, storms, droughts, and the spread of diseases and pandemics [14]. Over the past few decades, poorly planned urbanization has rapidly increased and expanded. This has brought about a major challenge for cities in terms of both climate mitigation and adaptation. In this regard, urban areas are significantly known as the main contributor to climate change through concentrating high population density, built infrastructure, and socio-economic interactions. For instance, the fast-paced growth and development contribute significantly to the emission of greenhouse gases that have an adverse impact on the environment.

According to a United Nations report, urban areas are responsible for 71–76% of global CO_2 emissions from energy use, primarily transport, industry, and buildings [4]. Moreover, to meet the energy needs of the world's population, urban areas have become one of the largest energy consumers globally. However, this high demand for energy, particularly in developing countries, can lead to an increase in global temperatures even if efforts are made to stabilize climate change at 2 °C [25]. Nevertheless, cities are among the most vulnerable areas due to climate change at the local and global levels. Therefore, as a global phenomenon, climate change is the primary threat to the quality of life in the current century. To address this challenge, we need immediate and effective mitigation strategies while also implementing urgent adaptive measures to combat the increasing impact of climate change.

The fight against climate change can be broadly classified into two categories, namely mitigation and adaptation. Mitigation involves gradually decreasing the emission of greenhouse gases that cause global warming. Adaptation also aims to minimize the vulnerability of environmental, social, and economic systems and to enhance their resilience to climate change. While there are a variety of adaptation and mitigation measures available to tackle climate change, no single solution can effectively address the issue on its own [20]. In the recent years, there has been an increasing number of multidisciplinary approaches proposed to address the challenge of landscape and urban planning and design. These fields are defined differently based on geographical properties, goals, objectives, and perspective [24]. One of the most integrated approaches is "Green Infrastructure" (GI) which involves incorporating urban green elements into the planning, design, and implementation of cities to mitigate the effects of climate change. Generally, the infrastructure of green spaces, water, and built environment systems, such as forests, wetlands, parks, green roofs, and walls, build up GI [9, 10, 19]. This approach has been extensively used in diverse areas such as urban and regional planning and landscape architecture. However, due to its wide-ranging applicability, several planning initiatives have been developed, making it difficult to define and execute GI planning. Despite this challenge, GI planning has the potential to improve urban resilience by offering a comprehensive understanding of the interactions between ecological and societal systems [24]. Therefore, GI implementation can help strengthen urban ecosystems and provide various benefits to humans through adaptation against climate change.

This chapter explores the connection between urban landscape planning and green infrastructure design to promote climate resilience and adaptation. It highlights the importance of community participation, fostering social cohesion, and creating inclusive spaces as key pillars to withstand climate change impacts. The chapter also examines practical initiatives such as community gardens and neighborhood networks as effective means of enhancing community resilience. The application of green infrastructure planning can integrate planning strategies into community-adaptive planning. This is necessary to cope with urban climate change and promote local resilience. It is important to understand the different types of green infrastructure, their implications, and ecological design. Traditional grey infrastructure or technologically-based solutions are less preferable. The planning process should include structured adaptive solutions derived from ecological planning and climate change adaptation to facilitate multiple planning responses at the same time. This will enhance planning flexibility in practice.

In addition, the chapter discusses the challenges and prospects of urban landscape design in the context of climate change adaptation. It emphasizes the importance of interdisciplinary collaborations, policy support, and financial investment to fully realize the potential of urban landscape design in addressing climate change. Nature-based solutions and community involvement are emphasized when designing resilient and sustainable cities. Finally, the chapter discusses adaptation strategies in urban landscape design, making it a valuable contribution for scholars, urban designers, professionals, and policymakers interested in climate change adaptation and urban planning.

2 Relationship Between Landscape Ecology Approach and Green Infrastructure

In 2002, Sandstrom introduced the term GI to broaden the scope of green spaces beyond recreation. GI encompasses multiple purposes such as biodiversity preservation, city structure, cultural identity, environmental quality, and biological solutions to technical and engineering problems [21, 26]. Later, Tzoulas and his colleagues [35] aimed to provide a more accurate depiction of the concept, including its constituent elements and defining their desired scale within an interconnected system [35]. In the last decade, scholars have made several attempts to provide a comprehensive definition of the GI concept, encompassing various aspects of relevant scale, spatial patterns, functions, and planning. Among these, the European Union has provided a holistic definition of GI that aims to conserve, restore, and enhance natural resources in order to prevent biodiversity loss and enable ecosystems to provide essential services to people [12]. Strategically, this concept refers to a planned system of natural and semi-natural areas, along with other environmental features, that are designed and managed to provide a wide range of ecosystem services in urban, suburban, and rural areas [21]. To make cities more resilient in the face of climate challenges, the EU strategy for implementing green infrastructure focuses on the following objectives:

- Stormwater Management: Green infrastructure helps to absorb and store rainwater, which decreases the risk of flooding and reduces the pressure on urban drainage systems.
- Heat Island Reduction: Urban areas are usually warmer than rural areas due to the heat-absorbing properties of buildings and pavement. Trees and green spaces, which are part of green infrastructure, can help mitigate the urban heat island effect by providing shade and cooling the air.
- Air Quality Improvement: Trees and vegetation play a vital role in decreasing air pollution by absorbing harmful gases and filtering particulate matter.
- Biodiversity Preservation: Green infrastructure provides habitat for wildlife, promoting biodiversity in urban areas.
- Community Well-being: Access to green spaces has been proven to improve mental health, encourage physical activity, and foster social interaction among communities [12].

However, the concept of GI has been widely adopted in the various scale of urban planning to promote ecological connectivity between landscape elements. This is achieved through the incorporation of "green spaces," "green systems," or "green structures" that enhance the structure of urban areas and provide access to natural resources in predominantly built-up regions [18, 30]. Therefore, GI involves the use of vegetation, soils, and natural processes to manage water and create healthier urban environments. This can range in scale from designing sites with features like rain gardens and green roofs to regional planning that involves conservation of large tracts of open land. When combined with gray infrastructure, an interconnected network of green infrastructure can help enhance community resilience by increasing water supplies, reducing flooding, combating the urban heat island effect, and improving water quality [11].

From a Landscape Ecology perspective, the planning of urban Green Infrastructure (GI) is directly influenced by the scale and spatial distribution of landscape elements. At the regional scale, the GI strategy is mainly used as a conservation planning approach to enhance the connectivity of green hubs by utilizing the ecological network of natural resources in urban areas. Meanwhile, at the urban scale, it mostly emphasizes multi-functionality and distribution of ecosystem services such as water management and urban heat reduction by integrating the GI strategy into urban planning practice and giving ecological resources the same priority as other infrastructure, such as roads and communications. Therefore, recent GI planning approaches can be categorized as conservation planning, restorative planning, and landscape planning, depending on the objectives of the planning process [5, 13, 17, 24, 31].

Increasing Urban Climate Resilience by Incorporating Nature-Based Strategies

As the impacts of climate change rapidly escalate, the need to revolutionize sustainable planning in urban zones becomes not just important, but alarmingly urgent. The potential for climate-resilient urban development is not just inspiring, but a critical necessity, especially when we consider that a staggering 70% of CO_2 emissions are generated in cities [23]. In this context, incorporating nature-based solutions (NBS) emerges as a beacon of hope and a key strategy for enhancing urban climate resilience (Fig. 1).

Challenges of implementing NBS in cities:

- Limited available space in dense urban environments.
- Upfront costs for implementation and maintenance.
- Lack of awareness or understanding of the benefits among decision-makers.
- Resistance to changing established infrastructure and development practices.

Benefits of Incorporating NBS

The benefits of incorporating NBS are not just limited to climate resilience. They encompass a wide range of advantages, including improved air quality, reduced urban heat island effect, enhanced biodiversity, and recreational opportunities. These benefits paint a vivid picture of a more sustainable and livable urban environment. One of the most reassuring aspects of incorporating NBS is its long-term cost savings compared to conventional grey infrastructure. This is achieved through reduced energy needs and efficient stormwater management, providing a solid financial foundation for sustainable urban planning. The improved human health and well-being from increased access to nature further reinforces the confidence in the effectiveness of NBS. Increased climate change adaptation and mitigation through carbon sequestration, stormwater management, and temperature regulation.

Characteristics of Effective NBS Strategies

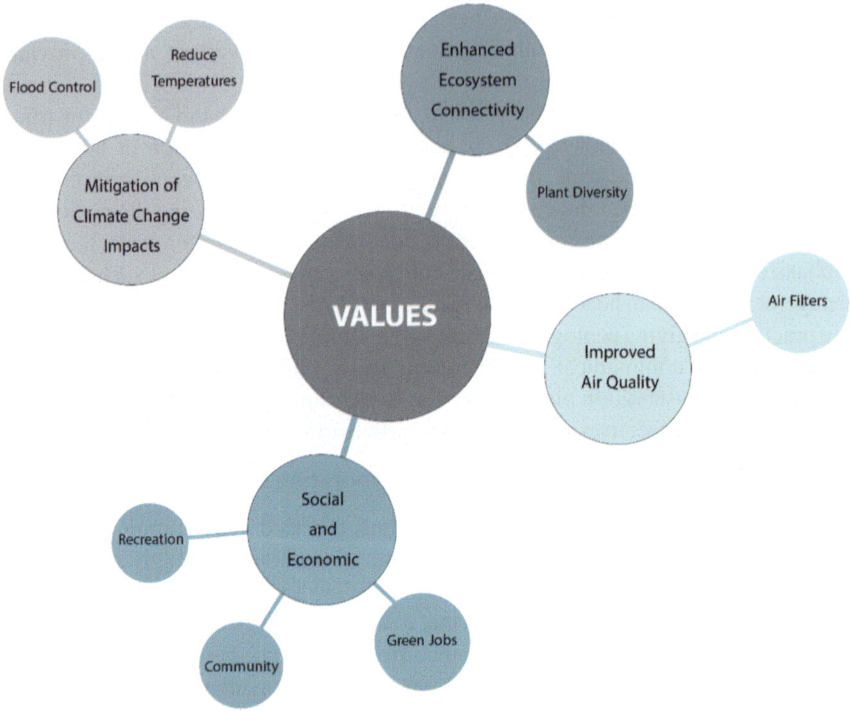

Fig. 1 The values of nature-based strategies (NBs)

- Layered, interconnected natural systems like urban forests, green roofs, and bioswales that work together.
- Integration with existing grey infrastructure like roads and buildings.
- Tailored to local climate, ecology, and socioeconomic context.
- Inclusive design processes that engage diverse stakeholders.
- Monitoring and adaptive management to ensure long-term functionality.

The model of a micro-ecosystem subject to factorial manipulation, which can produce a rapid response, is one of the critical approaches to analyzing the long-term effects of climate change on the ecosystem [2]. NBS promote multifunctional resilience in urban environments as an alternative to conventional grey infrastructure [27]. Urban forests, for instance, can reduce ambient temperatures by 2–4 °C according to modelling. Toronto mandated green roofs to capture 85% of stormwater from extreme precipitation events [29]. As a result of such NBS, ecosystem connectivity is enhanced. For instance, biodiversity assessments show that cities with over 20% vegetated cover have 30% more diverse plant species [2, 3]. However, with climate change expected to intensify by up to 4 °C by 2100, the strategic integration and interconnection between natural systems becomes essential and urgent for cities to build resilience.

3 Ecological Strategy and Urban Planning Adaptability

It is essential for climate resilience that urban planning centres on ecological foresight. Despite projections that heatwaves and flash floods will become more intense, the specifics will depend on the trajectory of emissions [6]. As a result, adaptation requires integrated strategies that provide flexibility in the face of uncertainty. It has been demonstrated that climate-sensitive designs with wetlands buffers mitigate the effects of heat islands and regulate flood risk simultaneously [34]. Cities can better navigate climate uncertainty by embedding such multifunctional NBS within adaptive frameworks.

Natural climate adaptation techniques provide practical, cost-effective solutions. Citizens who are concerned about NBS integration can advocate for its integration by:

- Promoting the implementation of climate-friendly green infrastructure by lobbying city officials.
- Taking part in community greening initiatives.
- Consultation with experts regarding the most effective way to capture water in the backyard and balcony.

As part of the urban fabric, green infrastructure contributes to developing multi-functional climate resilience [16]. Modelling studies have demonstrated that strategically planted urban forests have the potential to reduce surface temperatures by 2–4 °C [36]. Several cities, including Toronto, are now mandating green roofs capable of retaining up to 85% of stormwater during extreme precipitation events. Recent biodiversity assessments reveal that towns with a vegetated cover of more than 20% have more diverse plant species [29]. The benefits of their ecosystem services and cost savings over traditional grey infrastructure are evident in cities that have invested in NBS recovery within five to seven years. A multifunctional green infrastructure that extends beyond visual enhancement must be explored to build urban resilience. It describes various natural and hybrid components harmoniously woven into urban landscapes as well as expansive urban canopies and bio-diverse vegetative systems, such as living roofs and porous surfaces [22]. As a result of dissecting ecological design principles, this section emphasizes the importance of ecosystem interconnectivity for urban climate change mitigation and adaptation [37].

A sustainable urban plan must stem from ecological awareness and foresight, as it is the lifeblood of urban resilience. This chapter offers a refined perspective on adaptive strategic frameworks crucial for urban areas to withstand climate volatility. It argues that such strategies enable cities to adapt flexibly to the changing climate narrative by enhancing resilience with the attribute of adaptability. Incorporating NBS within urban planning paradigms is critically appraised for its potential to yield sustainable, flexible, and visually appealing urban settings [8]. This segment investigates the intricate challenges climate change poses on urban design and the opportunities that arise from addressing such issues (Fig. 2). It delves into the prospects enabled by cross-disciplinary cooperation, foundational policy development, and

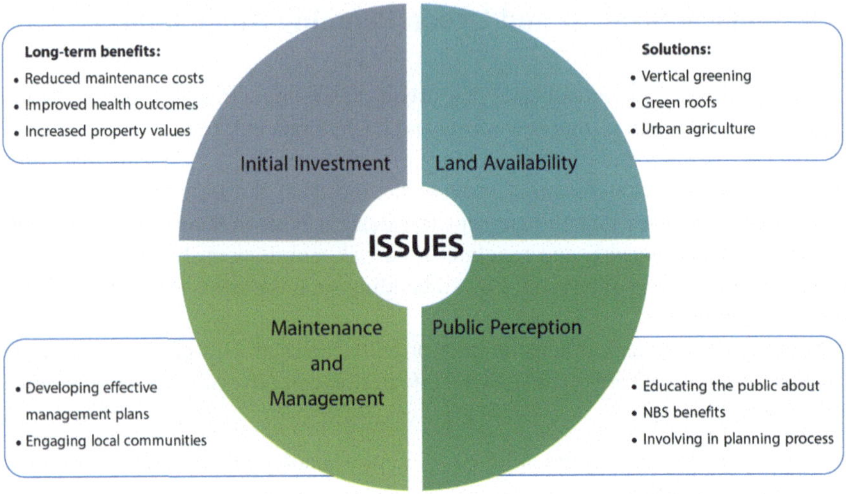

Fig. 2 Issues

visionary financial tactics. The complex nature of climate change calls for an integrated approach, where professionals across various fields are united in the ambition to develop innovative and enduring solutions. Collaboration and policy action are critical in fostering nature-based approaches for climate adaptation in urban environments. Communities, designers, academics, and legislators must work together to achieve successful adaptation. Developing and executing impactful strategies requires collaborative dynamics that facilitate a cohesive understanding of adversities and prospects. An adequate policy infrastructure must support incentives and regulations that support NBS integration within urban planning initiatives.

4 Community Engagement and the Vision for Sustainable Urbanity

Climate change, resource constraints, and socioeconomic inequalities have made urban sustainability an imperative in the twenty-first century [7]. Achieving urban sustainability requires transforming cities on an environmental, economic, and social level. Diverse urban stakeholders and communities must be involved in comprehensive and inclusive processes. Engaging the community involves involving, empowering, and participating in local decisions affecting their communities and cities it is a crucial strategy to create sustainable and resilient urban systems. To develop policies, programs, and projects that meet the needs of communities, municipalities can utilize residents' knowledge, priorities, and resources. Through this grassroots, bottom-up approach, innovative and context-appropriate solutions that are not apparent to external experts can often be developed [1]. It is increasingly recognized that community engagement and public participation are crucial to achieving sustainable urban development. Local stakeholders must be mobilized as partners in the planning and governance of urban areas due to the 1992 Rio Earth Summit. Subsequent global conferences and agreements have reaffirmed this message. In addition to incorporating inclusive, participatory processes that harness the ideas and assets of all segments of urban society, leading sustainability frameworks such as the Eco2 Cities Model emphasize inclusive, participatory approaches [32]. Cities must engage their communities effectively to diagnose sustainability challenges from a community perspective, identify assets and capacity gaps, mobilize grassroots action, and align solutions with local values and priorities. It encourages behavior change and ongoing participation in sustainability initiatives by increasing public ownership. A metropolitan vision that incorporates local needs into sustainable urban transformation can be achieved through inclusive engagement [32]. Various strategies are available for community engagement, including participatory budgeting, grassroots toolkits, deliberative polls, crowdsourcing platforms, social media campaigns, and co-planning workshops (Fig. 3). The use of these techniques helps translate the broad vision for urban sustainability into context-specific policies and programs that meet the needs of diverse constituents by sharing knowledge and co-creating solutions. Developing environmentally resilient, socially inclusive, and economically vibrant cities requires collaborative visioning and problem-solving processes [28].

Community engagement is essential to the success of sustainable urbanization. Cities must partner with empowered local stakeholders to achieve the 21st-century vision of livable, sustainable urban centres. Developing and implementing NBS procedures requires the active participation of the community. Community engagement plays a crucial role in fostering resilient, sustainable urban environments. In it, we examine community-driven ownership and stewardship's role in sustaining adaptation efforts. A holistic and culturally resonant approach is achieved by integrating local insights and practices into NBS. For urban zones to be transformed

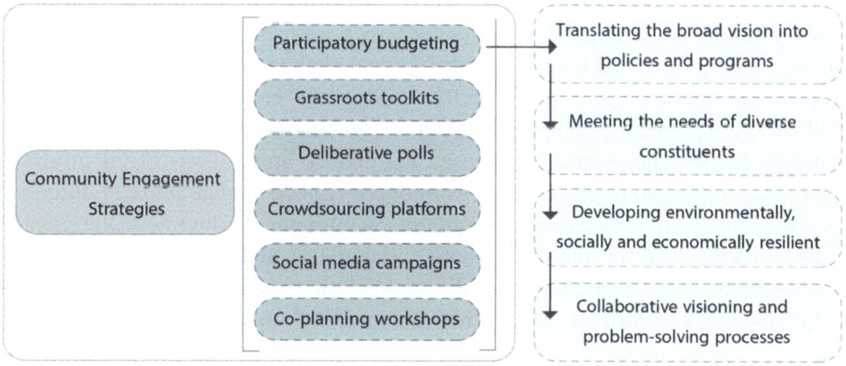

Fig. 3 Community engagement strategies flowchart

through NBS, financial commitment is necessary. NBS execution and perpetuation rely on monetary backing, which deals with the obstacles of accumulating funds for extensive projects and explores progressive financial conduits like public–private mergers and eco-centric bonds. NBS initiatives are effectively and sustainably implemented through targeted investments, acknowledging the long-lasting fiscal benefits of resilient cities. This section describes how pioneering technologies are used to monitor and evaluate NBS outcomes, given the surge in smart city innovations. Data processing, remote satellite analysis, and sensor networks are all used to consider green infrastructure in urban ecosystems [33]. As part of NBS's efforts to ensure robustness and long-term success, technological applications are evaluated concerning their ability to improve urban adaptability and refine the paradigms of strategic decision-making. To conclude, the work emphasizes the importance of integrating NBS into urban landscapes to respond effectively to the complex challenges of climate change. The challenges and opportunities highlight the need for interdisciplinary collaboration, comprehensive policy support, and discerning financial commitments [15]. With community involvement, technological integration, and ecologically based adaptation strategies, urban panoramas can be transformed into resilient, sustainable, and visually appealing habitats. It represents the convergence of a future-oriented shift, aiming for urban growth that is symbiotic with nature, setting the foundation for climate-resilient and vibrant cities [15].

5 Conclusion

In conclusion, the chapter has shed light on the critical importance of enhancing climate resilience through urban landscape design. Climate change presents unprecedented challenges to our environment, society, and economy, making it imperative to adopt proactive measures to mitigate its impacts and adapt to its changing conditions. The chapter emphasized the dual approach of mitigation and adaptation as essential strategies for addressing the multifaceted challenges of climate change. A central theme of the chapter was the role of Green Infrastructure (GI) in promoting climate resilience through the integration of green elements into urban planning and design. GI offers a versatile and effective strategy for enhancing urban resilience by incorporating green spaces, water systems, and built environment structures into the urban fabric. The chapter showcased the diverse applications of GI in urban landscapes, highlighting its potential to mitigate climate change impacts, improve environmental quality, and enhance community well-being. Furthermore, the chapter underscored the need for interdisciplinary collaboration and innovative solutions in urban planning and design to address the complexities of climate change. By prioritizing sustainable practices, green infrastructure, and climate resilience in urban landscapes, policymakers, planners, designers, and stakeholders can create more resilient and sustainable cities that are better equipped to withstand the challenges of a changing climate. As we look towards the future, it is crucial for all stakeholders to work together to implement sustainable practices and resilient strategies that prioritize environmental protection and community well-being. By embracing green infrastructure and climate-resilient design principles, we can build healthier, more sustainable cities that are prepared to face the challenges of climate change and ensure a brighter future for generations to come. Through collective action and strategic interventions, we can create a more resilient urban landscape that fosters environmental sustainability, social equity, and economic prosperity.

References

1. Adamson D, Bromiley R (2013) Community empowerment: learning from practice in community regeneration. Int J Publ Sector Manag 26(3):190–202. https://doi.org/10.1108/ijpsm-08-2011-0105
2. Alizadeh B, Hitchmough J (2019) A review of urban landscape adaptation to the challenge of climate change. Int J Clim Change Strateg Manag 11(2):178–194
3. Alizadeh B, Hitchmough JD (2020) Impact of simulated climate change on slug grazing in designed plant communities. Appl Ecol Environ Res 18(3)
4. Almulhim AI, Cobbinah PB (2023) Urbanization-environment conundrum: an invitation to sustainable development in Saudi Arabian cities. Int J Sustain Dev World Ecol 30(4):359–373
5. Anguelovski I, Irazábal-Zurita C, Connolly JJT (2019) Grabbed urban landscapes: socio-spatial tensions in green infrastructure planning in Medellín. Int J Urban Reg Res 43(1):133–156
6. Boateng EA, Asibey MO, Cobbinah PB, Adutwum IO, Blija DK (2023) Enabling nature-based solutions: innovating urban climate resilience. J Environ Manag 332:117433
7. Coaffee J, Therrien MC, Chelleri L, Henstra D, Aldrich DP, Mitchell CL, Participants et al (2018) Urban resilience implementation: a policy challenge and research agenda for the 21st century. J Conting Crisis Manag 26(3):403–410
8. De Ridder DW (2019) Considering nature-based solutions for place-branding: a spatial planning perspective. Doctoral dissertation, North-West University, South Africa
9. Demuzere M, Orru K, Heidrich O, Olazabal E, Geneletti D, Orru H, Faehnle M et al (2014) Mitigating and adapting to climate change: multi-functional and multi-scale assessment of green urban infrastructure. J Environ Manag 146:107–115
10. Derkzen ML, Van Teeffelen AJA, Verburg PH (2017) Green infrastructure for urban climate adaptation: how do residents' views on climate impacts and green infrastructure shape adaptation preferences? Landsc Urban Plan 157:106–130
11. EPA US (2016) Green infrastructure and climate change: collaborating to improve community resiliency
12. European Commission (2011) Communication from the commission to the European Parliament, the Council, the European economic and social committee and the committee of the regions youth opportunities initiative. European Commission Brussels, Belgium
13. Forman RTT (1995) Land mosaics: the ecology of landscapes and regions. Cambridge University Press
14. Gao J, Sun Y, Liu Q, Zhou M, Lu Y, Li L (2015) Impact of extreme high temperature on mortality and regional level definition of heat wave: a multi-city study in China. Sci Total Environ 505:535–544
15. Hakwendenda PC (2022) Climate change resilience and adaptation strategies in ecologically fragile urban mining communities: a review of existing research and practice. Int J Environ Clim Change 38–73. https://doi.org/10.9734/ijecc/2022/v12i1030771
16. Hansen R, Olafsson AS, Van Der Jagt AP, Rall E, Pauleit S (2019) Planning multifunctional green infrastructure for compact cities: what is the state of practice? Ecol Ind 96:99–110
17. Lanzas M, Hermoso V, de-Miguel S, Bota G, Brotons L (2019) Designing a network of green infrastructure to enhance the conservation value of protected areas and maintain ecosystem services. Sci Total Environ 651:541–550

18. Liu H, Shen Y (2014) The impact of green space changes on air pollution and microclimates: a case study of the Taipei metropolitan area. Sustainability 6:8827–8855. https://doi.org/10.3390/su6128827
19. Matthews T, Lo AY, Byrne JA (2015) Reconceptualizing green infrastructure for climate change adaptation: barriers to adoption and drivers for uptake by spatial planners. Landsc Urban Plan 138:155–163
20. Pachauri RK, Allen MR, Barros VR, Broome J, Cramer W, Christ R et al (2014) Climate change 2014: synthesis report. Contribution of Working Groups I, II and III to the fifth assessment report of the intergovernmental panel on climate change. IPCC
21. Pamukcu-Albers P, Ugolini F, La Rosa D, Gruadinaru SR, Azevedo JC, Wu J (2021) Building green infrastructure to enhance urban resilience to climate change and pandemics. Landsc Ecol 36(3):665–673
22. Pitman SD, Daniels CB, Ely ME (2015) Green infrastructure as life support: urban nature and climate change. Trans R Soc S Aust 139(1):97–112
23. Ragazzo L (2021) Urban development and the world bank: historical perspective and present-day programs
24. Ramyar R, Ackerman A, Johnston DM (2021) Adapting cities for climate change through urban green infrastructure planning. Cities 117:103316
25. Rao ND, Min J, Mastrucci A (2019) Energy requirements for decent living in India. Brazil and South Africa. Nat Energy 4(12):1025–1032
26. Sandström UG, Angelstam P, Khakee A (2006) Urban comprehensive planning–identifying barriers for the maintenance of functional habitat networks. Landsc Urban Plan 75(1–2):43–57
27. Santiago Fink H (2016) Human-nature for climate action: nature-based solutions for urban sustainability. Sustainability 8(3):254
28. Shalowitz MU, Isacco A, Barquin N, Clark-Kauffman E, Delger P, Nelson D, Wagenaar KA et al (2009) Community-based participatory research: a review of the literature with strategies for community engagement. J Dev Behav Pediatr 30(4):350–361
29. Slater KR, Ventura J, Robinson JB, Fernandez C, Dutfield S, King L (2022) Assessing climate action progress of the City of Toronto. Build Cities 3(1)
30. Sturiale L, Scuderi A (2019) The role of green infrastructures in urban planning for climate change adaptation. Climate 7(10):119
31. Suppakittpaisarn P, Larsen L, Sullivan WC (2019) Preferences for green infrastructure and green stormwater infrastructure in urban landscapes: differences between designers and laypeople. Urban For Urban Green 43:126378
32. Suzuki H, Dastur A, Moffatt S, Yabuki N, Maruyama H (2010) Eco2 cities: ecological cities as economic cities. World Bank Publications.
33. Tayefi Nasrabadi M (2022) How do nature-based solutions contribute to urban landscape sustainability? Environ Dev Sustain 24(1):576–591
34. The IPCC Sixth Assessment Report on Climate Change Impacts (2022) Popul Dev Rev 48(2):629–633. https://doi.org/10.1111/padr.12497
35. Tzoulas K, Korpela K, Venn S, Yli-Pelkonen V, Kaźmierczak A, Niemela J, James P (2007) Promoting ecosystem and human health in urban areas using green infrastructure: a literature review. Landsc Urban Plan 81(3):167–178
36. Wang Z, Li Y, Song J, Wang K, Xie J, Chan PW, Di Sabatino S et al (2022) Modelling and optimizing tree planning for urban climate in a subtropical high-density city. Urban Clim 43:101141
37. Yocom K (2014) Principles of ecological landscape design. Ecol Restor 32(2):215–216. https://doi.org/10.3368/er.32.2.215

Prioritizing Energy Efficiency and Low-Carbon Approaches: Climate Adaptation Strategies in Urban West Asia

Akram Ahmed Noman Alabsi

Abstract The rising demand for energy and the accompanying economic pressures, combined with the negative effects of greenhouse gas emissions and the resulting climate change, necessitate immediate action to mitigate future consequences. As a result of rapid urbanization fueled by population growth and economic development, many West Asian nations have neglected environmental considerations and sustainable strategies. As a consequence, complex environmental challenges have emerged, and the region faces significant risks from climate change, which threatens ecosystems, economies, and infrastructure dependent on natural resources. This research aims to develop standards and strategies for climate adaptation in West Asian nations that prioritize energy efficiency and a low-carbon approach. Examining the influence of region-specific climatic factors and exploring traditional local technologies used for climate effect control, as well as evaluating their compatibility with modern approaches, are essential components. The research methodology employed here focuses on learning from the past to improve energy utilization practices in modern architecture. In addition, this study provides a comprehensive review of prior research that examined the development and improvement of techniques and strategies to improve the thermal performance of buildings, thereby advancing energy efficiency. The research outcomes address the existing gaps in climate adaptation standards and guidelines for thermal comfort in West Asia, demonstrating the potential for significant energy savings in future housing compared to existing homes. The projected energy savings for the five climate zones of West Asia are approximately 80% during winter and 60% during summer. In addition, the study presents green design strategies and energy-efficient methods tailored specifically for West Asian arid regions with high temperatures. However, additional research is required to simulate the energy performance of each technology in order to fully comprehend their energy efficiency potential.

A. A. N. Alabsi (✉)
College of Architecture and Urban Planning, Fujian University of Technology, Fuzhou 350118, China
e-mail: AlabsiAkram@fjut.edu.cn; sallam_akram@hotmail.com

B. He et al. (eds.), *Urban Climate and Urban Design*, Urban Sustainability,
https://doi.org/10.1007/978-981-96-1521-6_8

Keywords Climate adaptation · Energy efficiency · Low-carbon approach · Traditional technologies · West Asia

1 Introduction

West Asia is positioned within the expanse bridging Central Asia, Africa, and Southeast Europe. Commonly recognised as the Middle East, the region excludes Egypt geographically, though culturally it is deemed a Middle Eastern country. Characterised by vast arid expanses, West Asia boasts access to major water bodies such as the Black Sea, the Persian Gulf, the Arabian Sea, the Red Sea, the Caspian Sea, the Mediterranean Sea, and the Aegean Sea. Despite marked topographic diversity, the climate in the region is broadly delineated into two primary categories: the predominant hot-dry climate and the Mediterranean climate [2]. Climatic conditions in Western Asia exhibit considerable variation based on season and geography, succinctly summarised as hot and dry, with moderate winters featuring occasional rainfall. Exceptions occur in mountainous regions where deserts give way to steppe landscapes, notably in northern Iraq, northern Iran, and eastern Turkey, experiencing harsh winters. The Arabian Peninsula registers some of the most extreme and arid conditions globally. Climate studies in the 21st century present compelling evidence of global warming, evident in peak temperatures observed in Iran's inland areas during summer and heightened spring snowfall in southeastern Turkey correlating with rising temperatures.

Climate, alongside social and cultural considerations, exerts a profound influence on architectural and urban configurations. It stands as a pivotal factor shaping form, height, population density, and spatial organisation, reflecting the trajectory of civilisations [18]. The last century witnessed a transformative shift in global urban and architectural trends, however, Western Asia did not witness an equivalent diversity in interpreting the impact of cosmic changes on architecture, as observed in the West. Disparities arise from adopting urban and legislative systems incongruent with traditional urban and architectural patterns, neglecting traditional methodologies, and the advent of modern building materials and techniques associated with Western models. Traditional architecture in Western Asia is cha racterised by simplicity, spontaneity, and direct harmony with specific needs and environmental conditions. Urban growth unfolded organically, evolving with family expansion within the frameworks of religious principles and local cultural norms, including land conservation and the impact of laws, customs, and other societal systems governing relationships [7]. The environment in West Asia, shaped by these laws, principles, and local customs, is regarded as an organised rather than a pre-planned system, developing gradually in accordance with local and societal values. While organised environments, despite their noble intentions, have proven ineffective in managing responsibility for organisation and administration after usurping these powers from direct beneficiaries, leading to a dispersion of responsibility and environmental degradation.

This chapter will scrutinise and analyse the impact of climate adaptation on energy performance and green design, examining adaptation experiences in traditional architecture. The analysis will encompass several buildings and cities as case studies, alongside an exploration of the adaptation impact on each type of climate in the West Asia region. The objective is to identify the most effective and influential climatic elements in energy-efficient design decisions, leading to the formulation of sustainable adaptation principles at both urban and building levels. This approach aims to promote the most energy-efficient and environmentally friendly designs based on climate adaptation.

2 Climatic Conditions of West Asia

West Asia, exhibits a complex climatic mosaic shaped by its diverse topography and surrounding bodies of water. Over millennia, human activities such as deforestation, grazing, and watercourse manipulation have transformed the landscape [11]. This transformation, coupled with rapid population growth and political tensions, heightens vulnerability to climate variability and change. The region's climate classification, under the widely accepted Köppen system, primarily identifies as Hot Desert (BWh), characterised by scorching daytime temperatures and scarce rainfall, covering approximately 80% of the area [17]. However, notable variations exist due to local geography. For instance, along the 40° N latitude, a swift transition occurs from desert (BWh) through steppe (BSh) to cooler highland climates (H) within a mere 400 kms [2]. Coastal and mountainous features significantly influence microclimates, orographic effects enrich precipitation in the Taurus and Zagros Mountains, sustaining river flow vital for Mesopotamia. The Mediterranean influence extends into the region's northern reaches, bringing wet winters and dry summers (Csa). Conversely, the south experiences the Indian summer monsoon's impact, influencing climate dynamics [7]. Noteworthy landscapes include the Sarawat Mountains in the west, the expansive deserts of Najd, and the Rub Al–Khali to the southeast. These geographical features profoundly shape West Asia's climate regime, contributing to its distinct arid and semi-arid conditions, where water scarcity remains a pressing challenge amidst a growing population and escalating water needs (see Fig. 1). Applying the Kuben classification of climatic zones, West Asia is discernibly divided into three major climatic regions: (1) Desert climate zones: (Desert Hot-Arid climate—Coastal hot-arid climate), (2) Mediterranean climate zone, (3) Semi-Arid climate zone.

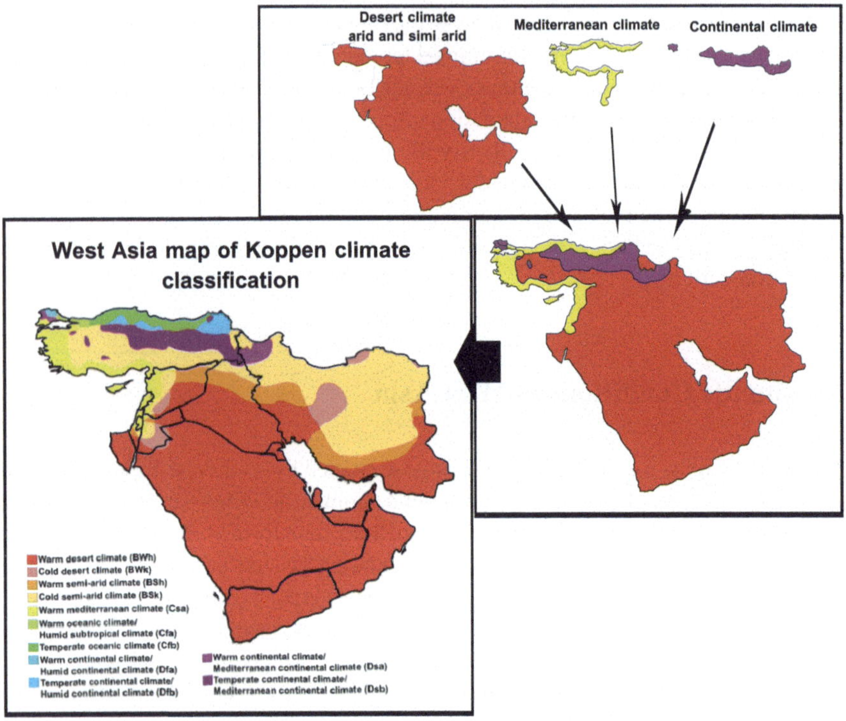

Fig. 1 Climate classifications of West Asia. (*Source* Kuben classification Edited by the author)

3 Methodology

The methodology employed in this study for evaluating climate adaptation strategies and their impact on energy efficiency in residential buildings across West Asia integrates both inductive and deductive methods. Drawing from prior research and global findings, this study undertakes a specific regional investigation to address challenges related to adaptation in traditional buildings. The research methodology emphasises historical insights to inform contemporary energy utilization practices in modern architecture. To assess climate adaptation strategies, the study conducts a comprehensive review of existing research, focusing on techniques to enhance the thermal performance of buildings and advance overall energy efficiency. This involves analysing region-specific climatic factors and evaluating traditional technologies alongside modern approaches for their effectiveness in achieving energy efficiency and a low-carbon footprint. The methodology will involve case studies of various buildings and cities across West Asia to analyse the impact of climate adaptation on energy performance and green design. It aims to identify influential climatic elements in energy-efficient design decisions, proposing sustainable adaptation principles at urban and building levels. Figure 2 illustrates the methodological

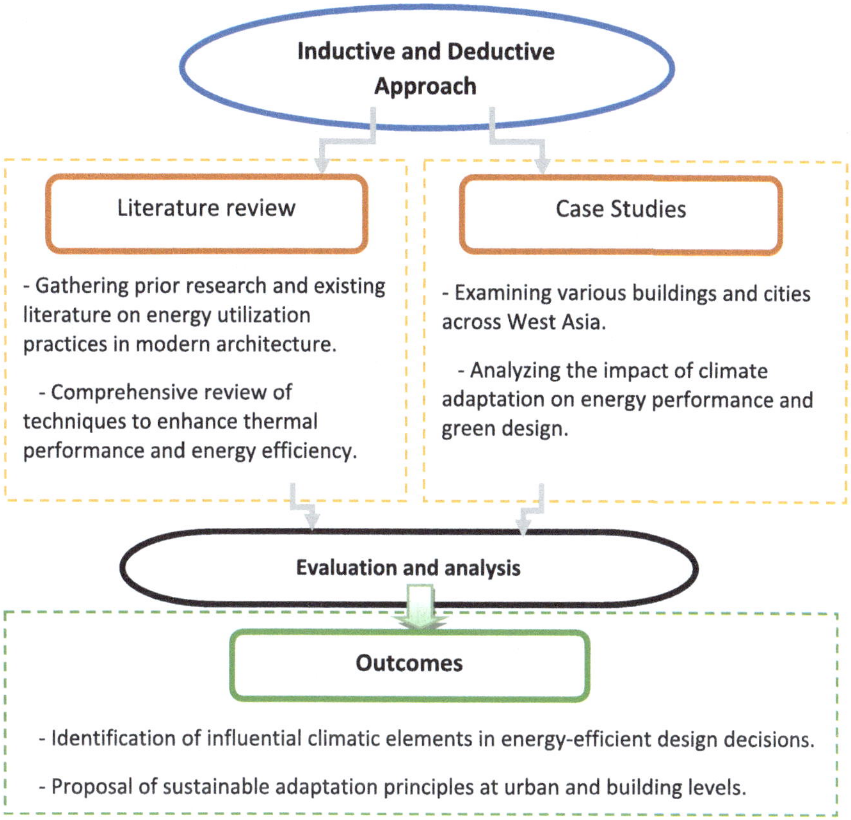

Fig. 2 Methodological framework of the study

framework, showcasing the integration of historical insights, regional studies, and global research findings to develop effective climate adaptation strategies prioritising energy efficiency and low-carbon approaches in West Asia.

3.1 Case Study Selection

The cities selected for this study fall within the geographical scope commonly referred to as West Asia. The selection criteria encompass various parts of the region, including the north, south, west, and east. Another criterion pertains to the antiquity of the cities, encompassing both ancient and historical aspects, with a mix of traditional and modern buildings. The third criterion involves the significance of the city, considering factors such as population size and degree of urbanization. Based on these criteria, the chosen cities include Jeddah (Saudi Arbia), Shibam Hadramout

(Yemen), Yazd (Iran), Sana'a (Yemen), Baghdad (Iraq), Damascus (Syria), and Dubai (UAE).

4 Examination of West Asia Case Studies

4.1 Environmental Performance and Climate Adaptation According to Different Climates in West Asia

Since there is insufficient information about the historic building methods of the selected boomtowns, this research and analysis are broadened to cover the climate zones in which they lie. The goal of this analysis is not to create a list of all existing building types and principles of climate-adapted architecture for an overview or compendium, abut to compile the data in order to select potential principles that can be used for the building envelope [1]. The examples show traditional building methods, some of which more than a thousand years old. Some are rarely used today, others have been preserved into modern times. The core question of this section is the potential use of such traditional principles in modern building construction; and their application to energy efficiency and green design in particular [5]. The enduring simplicity and harmony of traditional architecture with environmental conditions highlight valuable insights that can inform contemporary approaches to sustainable building strategies.

4.1.1 Hot Dry Climate Zones

(1) Planning context

In the planning of hot arid regions, a compact layout is imperative, minimizing the lengths of roads and corridors (see Fig. 3). Paved footpaths assume a pivotal role, serving not only as pedestrian thoroughfares but also affording natural protection against direct sunlight, radiant heat, solar glare, dust, and wind. The segmented lines of pedestrian corridors, intentionally configured to restrict wind movement within the space, contribute to effective shading [12]. The architecture of coastal settlements in West Asia, particularly within desert locales, is shaped by the humid climate, socio-cultural life, and the availability of local construction methods and materials [16].

Characterised by high-density planning with closely situated or attached buildings, these areas showcase shaded narrow alleys, known as Sikka. These alleys, designed to harness sea breezes in coastal zones or mitigate dusty winds inland, often align from north to south, enabling prevailing north winds to traverse through. Houses with an 'introvert plan' are prevalent, featuring rooms around a central courtyard open to the sky [7]. This arrangement addresses social conditions, ensuring

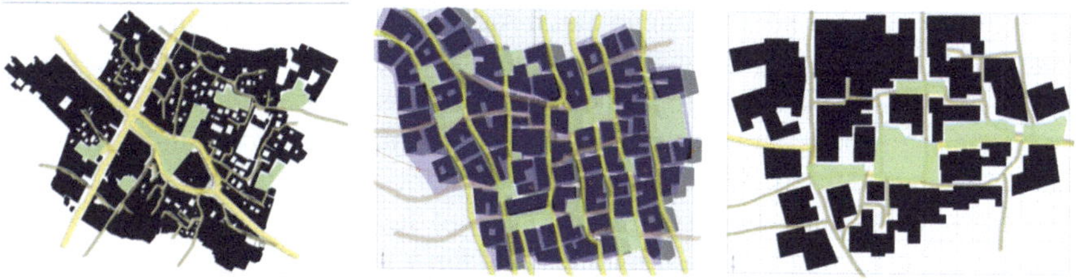

Fig. 3 Compacted planning in the traditional cities of West Asia. (*Source* by Author)

privacy for extended family members and spatial adaptability. The courtyard acts as a wind generator, promoting air circulation as hot air ascends and is replaced by cooler air from surrounding rooms. Furthermore, wind towers are utilised to passively cool occupants, introducing variation to the district's skyline and enhancing its visual allure.

(2) Buildings layout

To optimally address the challenges of hot and dry climates, building design must focus on shaping structures that naturally provide shade and limit sun exposure. Incorporating embedded shapes proves effective in reducing surface exposure to intense external conditions during the day, preventing overheating [14]. At night, the same design facilitates heat dissipation through radiation and convection. The layout should strategically encourage cool breezes to traverse building openings, enhancing interior cooling. In hot and humid climates, the diffusion-form design becomes particularly effective. The creation of an introverted model or patio design offers multifaceted advantages, especially in hot and dry climates. Responding to environmental impacts, spatial areas within buildings need to harness desirable factors like cool breezes while shielding against undesirable elements such as hot winds and dust [15]. The Al-Hosh (meaning courtyard in Arabic) stands as an exemplary sustainable design, harmonizing climatic, social, and cultural requirements. Desert engineering in West Asia evolves from three key factors: the dry/wet hot climate, socio-cultural dynamics, and locally available building methods and materials. Characterized by high-density plans, buildings in these areas are closely connected, intersected by shaded alleyways like Sikka. These alleys strategically channel sea breezes in coastal zones or mitigate dusty winds inland. Houses with a "start plan," featuring rooms around a central courtyard, showcase adaptability to the extended family's changing requirements. This design not only ensures privacy but also acts as a wind generator, fostering constant air circulation and creating a pleasant living environment. The integration of wind towers adds visual diversity to the skyline while facilitating passive cooling for occupants.

4.1.2 Hot Dry Climate Case Studies

(1) House features for climatically adaptive traditional design in Saudi Arabia

Traditional residential architecture in Saudi Arabia, specifically Jeddah city, demonstrates many climatic adaptations [13]. Influenced by cross-cultural exchanges with Middle-Eastern, Asian, and European traders, as well as architectural influences from Egypt, the indigenous architecture evolved to optimize thermal comfort. Notable features include tall, airy structures up to five stories high, strategically designed for wealthy merchants. The settlement pattern of Jeddah's old town showcases a climate-responsive layout (see Table 1). Buildings are spaced apart to allow air circulation, while narrow, shaded streets mitigate solar exposure. Streets oriented north–south maximize shading, with bends and projecting Mashrabiyas (screened windows) further reducing solar gain. Intersection spaces, heated by intense sunlight, drive convection currents that ventilate the narrower streets. House designs incorporate passive features like Mashrabiyas for shading and airflow control [3]. The architecture prioritizes natural ventilation and shade, adapting ingeniously to Saudi Arabia's hot-humid climate. Such adaptations highlight the synergy between spatial organization, construction techniques, and material selection in traditional Hejazi architecture [8].

(2) House features for climatically adaptive traditional design in Iraq

Choose the appropriate dimensions and squares of the human scale and climatic factors and the needs of people. Allocate the plazas or courtyards per group of buildings. Allocate courtyards inside most of the houses. Gardens planning in a rectangular shape about a longitudinal axis. Use waterfalls in the gardens and fountains in the inner courtyard of houses [12]. The cultivation of a roof, as in some cities, such as the city of Sana'a. Use colorful mosaics and the use of stained glass. The use of wood and decorated wood which manmade. The use of engineering units known as arabesque decorative Arabic (see Table 2).

(3) Traditional house features for climatically adaptive design in Dubai

Table 1 Features of Jeddah houses Saudi Arabia

The Jeddah house features for climatic adaptive traditional design	Mashrabiyas details
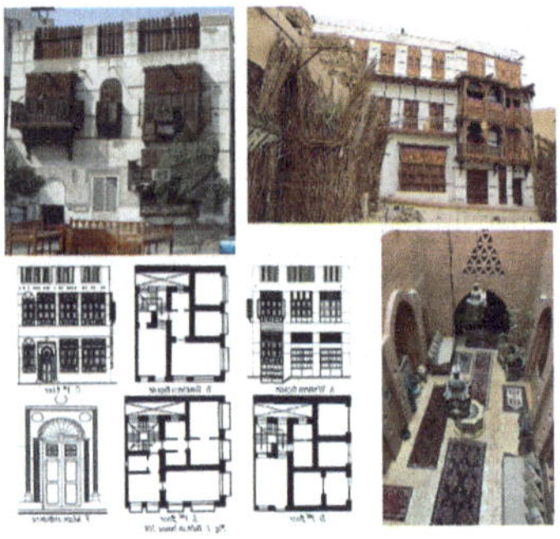	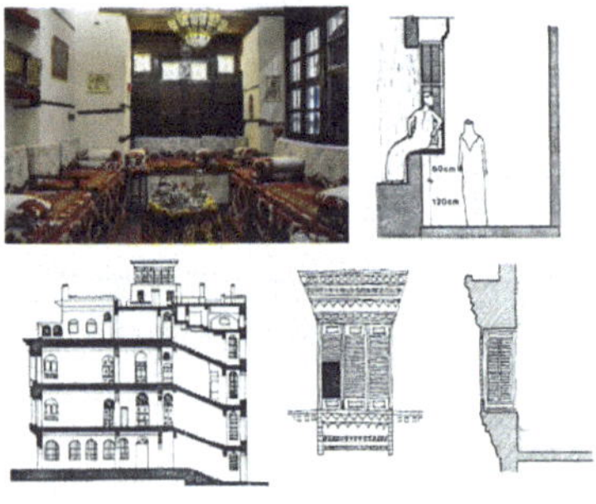 (*Source* Bagader [6] edited by author)

Table 2 Features of traditional Iraqi houses

The Baghdad house features for climatic adaptive traditional design	Traditional houses in Basra–Iraq
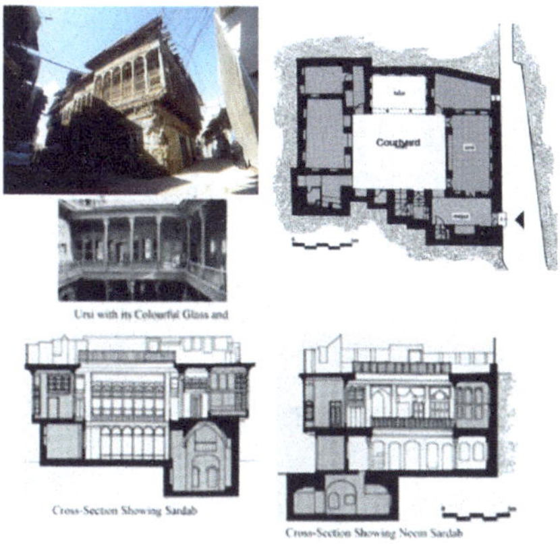	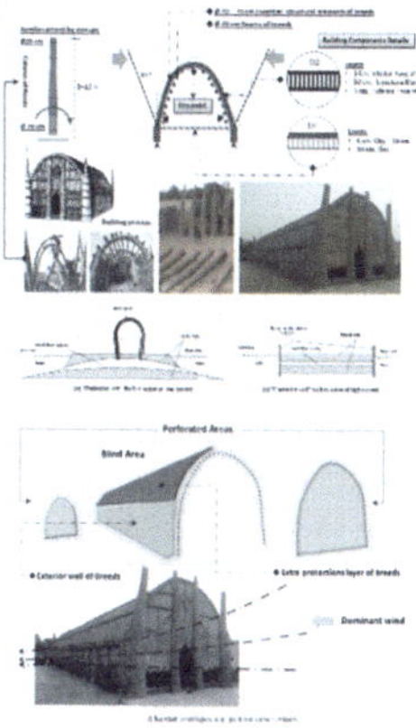
(*Source* Ayyash [4] edited by author)	(*Source* Jawad and Al-Yousif [12] edited by author)

In Dubai (see Table 3), traditional house design reflects meticulous climate adaptations for the region's hot summers. Builders crafted solid structures with minimal exterior openings, aligning with social values and traditions. The orientation, North–South, maximizes exposure to wind and cross ventilation while minimizing solar radiation. Evaporative cooling from creeks aligns with courtyards, contributing to energy efficiency [9]. Construction materials, including mud, coral stones, and palm trunks, offer insulation, and roofs with bamboo and palm leaves ensure thermal efficiency. Bastakia houses demonstrate a temperature difference, showcasing the effectiveness of thick walls. Wind towers and air pullers enhance natural ventilation, while various daylight openings, including traditional windows and wind towers, promote cooling strategies. Overall, Dubai's traditional houses ingeniously integrate design elements to adapt to the challenging climate, emphasizing sustainability and comfort.

(4) Traditional houses feature climatic adaptations in Iran

Traditional Iranian houses exhibit ingenious climatic adaptations. Characterized by dense structures with joint walls, narrow curved alleys prevent unwanted winds and offer defense. The use of accessible materials like clay and mudbrick, with high heat capacity, ensures effective insulation. The houses feature thick walls, high ceilings, and inward-facing openings, fostering an introspective design (see Table 4). Central courtyards, deeper than the building, incorporate fountains and trees to lower temperatures [7]. Wind towers, integral to Persian architecture, enable natural ventilation. The number and design of openings in windcatchers depend on prevailing winds, with the addition of fountains creating a refreshing ambiance. Strategic orientation

Table 3 Features of traditional Dubai houses

Traditional residence in the old Bastakiya area—Dubai	A residential house of Saeed bin Maktoum Al Maktoum–Dubai
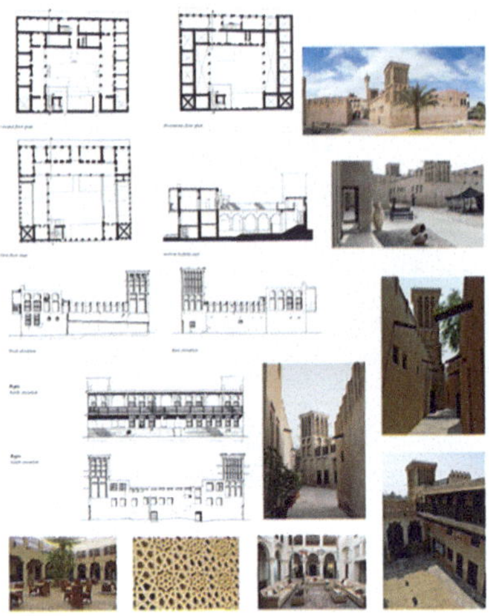 (*Source* Damluji [9] edited by author)	

shields houses from sandstorms, showcasing the synergy of form, materials, and orientation in adapting to the Iranian climate [18].

(5) Climate adaptation features in traditional tower houses in Yemen

Traditional tower houses in Yemen showcase climate adaptation through thoughtful design elements. In cities like Sana'a and Shibam Hadramout (see Table 5), the layout prioritizes spaces for human comfort and privacy while aligning with ecological balance. Traditional construction methods, particularly earth construction, offer sustainable advantages, ensuring a promising future. Vertical extension aids in temperature regulation, with shading on vertical surfaces enhancing efficiency. Optimal orientation, especially in southern facades, maximizes exposure to sunlight throughout the year. The use of local materials, such as stone and mud bricks, with high thermal capacity contributes to effective thermal transmission. Controlled air movement, strategic placement of windows, and consideration of solar access further highlight the meticulous climate adaptation features in these traditional Yemeni tower houses [1].

4.1.3 The Most Effective Climate Adaptation Technologies in Hot Dry Areas of West Asia

In West Asia's hot arid regions, traditional architecture employs ingenious climate adaptation technologies to combat severe heat. Strategies include clustering buildings

Table 4 Features of traditional Iranian houses

Traditional Abbasian house in Yazd–Iran	Traditional house in Kashan–Iran
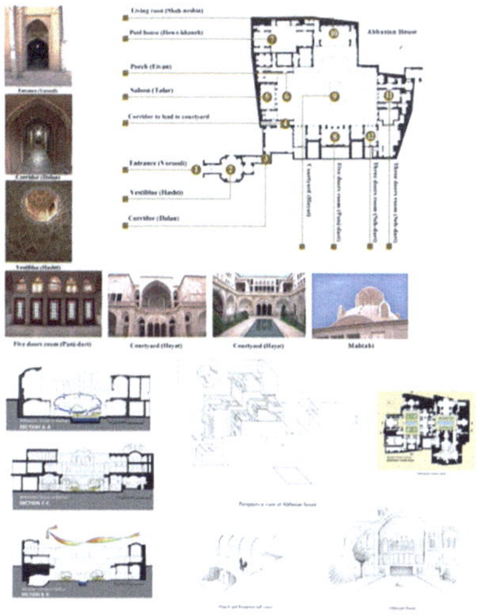 (*Source* Eskandari [10] edited by author)	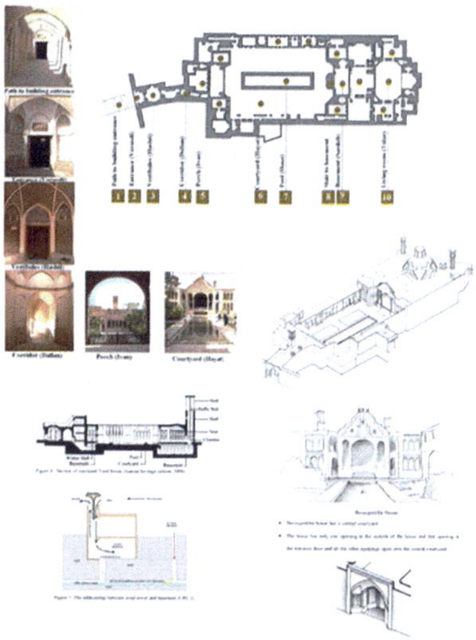

Table 5 Features of traditional Yemen houses

Traditional tower houses in Sana'a–Yemen	Traditional tower houses in Shibam–Yemen
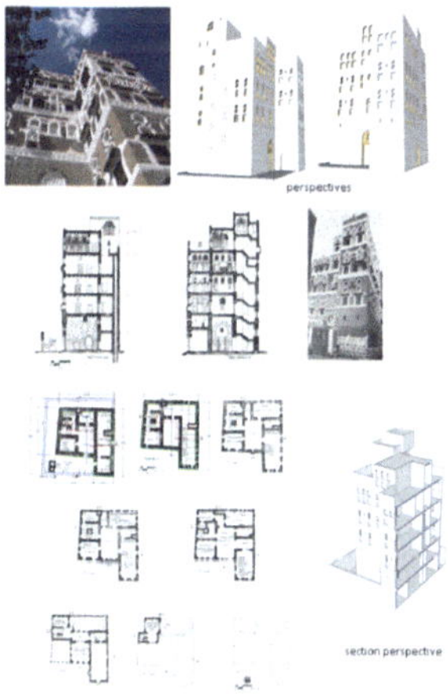 (*Source* by the authors)	

to minimize exposure, limiting openings to reduce heat gain and dust penetration, and constructing thick adobe walls for thermal mass. Creating cool microclimates with courtyards, trees, and water features further enhances comfort. Sensible traditions involve adapting lifestyles to climate cycles, utilizing cool basements during hot afternoons and rooftop sleeping in cool nights. Passive cooling systems, such as wind towers, catchers, curved roofs, cisterns, and ice makers, showcase the region's historical expertise. Climate adaptation techniques for cooling encompass preventive measures, modifying heat acquisition, and heat dissipation strategies [10]. Landscaping, building shape, and solar control fall under preventive techniques, while modifying heat acquisition involves storing heat in building structures. Heat dissipation methods leverage environmental heat sinks like radiative cooling using the sky, evaporative cooling with air and water, and thermal cooling with air and soil. These proven techniques significantly contribute to reducing the cooling load of buildings, offering sustainable and effective solutions for hot, dry climates in West Asia [12].

(1) Cooling strategies for building

Implementing cooling techniques in traditional architecture has proven highly effective, significantly reducing building cooling loads. These strategies encompass three main categories: Preventive Techniques: To shield against solar and heat gains, measures include strategic landscaping, utilization of outdoor spaces, thoughtful building forms, layouts, and external finishes, solar control through shading, thermal insulation, and management of internal heat gains. Modulation of Heat Gain Techniques: Focusing on the building's ability to store heat, this strategy aims to diminish peak cooling loads by delaying heat discharge. This modulation of internal temperature occurs over an extended period. Heat Dissipation Techniques: Relying on natural means to expel excess heat, this approach demands two key conditions: a suitable environmental heat sink and efficient thermal coupling with sufficient temperature differentials. Techniques involve radiative cooling using the sky, evaporative cooling through air and water, and convective cooling utilizing air and ground soil.

These proven techniques showcase the effectiveness of traditional architectural cooling methods, offering sustainable solutions for reducing the overall cooling burden on buildings.

(2) Thermal performance

In addressing thermal performance, effective control of solar gain is pivotal for building heat management. The design and arrangement of solid and glazed walls/roof play a crucial role in mitigating solar gain, particularly in hot climates. In hot dry regions, minimizing large openings, especially on west and east orientations due to low solar altitude, is essential. Conversely, hot humid climates benefit from sizable openings for cool breezes, necessitating shading through extended porches. Shading treatments vary by orientation: south-facing openings require horizontal shading, while west and east orientations need vertical shading. Employing egg-crate shading devices is recommended in very hot climates to counter various solar radiation components. Additionally, manipulating the solar-optical properties of glazed surfaces proves instrumental in reducing heat gains. These strategies underscore the significance of thoughtful design interventions for effective thermal regulation.

(3) Climate adaptation strategies

In West Asia, climate adaptation strategies in traditional urban and architectural designs focus on compact layouts to combat intense solar radiation. Key tactics include increasing built area density, narrow alleys, elevating buildings while narrowing alleys for optimal shading, and strategically designing public squares and covered alleys. Architectural features like courtyards, air-puller walls, and traditional windows provide extensive shading, while coastal settlements align alleys for passive cooling through convective and ventilative methods. The urban and architectural approaches work synergistically to create sustainable, energy-efficient designs that prioritize both urban planning and building features in the region.

5 Evaluation of Climate Adaptation Strategies in Terms of Energy Efficiency and a Low Carbon Approach

Efficient and climate-adaptive cooling systems have the potential to enhance energy efficiency and reduce operational costs, albeit with potential additional construction costs. Minimizing heat gains through adaptive architectural design, such as site selection, building form, and shading techniques, is crucial. Traditional climate adaptation technologies in West Asia's hot, arid regions demonstrate ingenious methods to combat severe heat. These include clustering buildings to minimize exposure, limiting openings to reduce heat gain and dust penetration, and utilizing thick adobe walls for thermal mass. Creating cool microclimates with courtyards, trees, and water features further enhances comfort. Effective cooling strategies for buildings encompass preventive techniques, modulation of heat gain, and heat dissipation methods, utilizing natural means such as radiative cooling and evaporative cooling. Additionally, optimizing thermal performance through thoughtful design interventions, including control of solar gain and shading treatments, is pivotal for effective heat management. Climate adaptation strategies in traditional urban and architectural designs emphasize compact layouts, narrow alleys, and strategic public squares to combat intense solar radiation. These approaches prioritize sustainable, energy-efficient designs that harmonize urban planning with building features. This can be achieved by integrating passive systems and a small mechanical cooling system, if needed. The research emphasizes a holistic approach, blending natural climate design with modern technologies, creating dynamic building envelopes suited for dry-hot areas. These dynamic envelopes, combining traditional principles with smart technology, offer cost-effective construction while promoting resilience and adaptability. The relationship between climate factors and adaptive building systems is dynamic and reciprocal, influencing each other. Achieving thermal equilibrium in hot, dry areas requires a higher thermal balance and a climate-adapted design methodology, ultimately reducing the reliance on mechanical heating and cooling. The study underscores the importance of a comprehensive vision in urban development, integrating originality, stability, and openness to new technologies. It proposes that this approach not only ensures continuity and communication for the future but also leads to thermally efficient building casings, promoting high energy efficiency in the face of climate variations.

6 Concluding Remarks

Rapid urbanization and the heightened need for habitats, driven by an expanding urban population and an augmented lifestyle, demand increasing energy consumption annually. The predominant reliance on non-renewable power generation globally significantly contributes to greenhouse gas emissions, notably carbon dioxide. Mitigating energy demand through building efficiency proves economically more viable than generating equivalent energy through coal or nuclear means. Strategies to markedly reduce greenhouse gas emissions in the construction sector fall into three categories: (1) planning and design, (2) building envelope, materials, and equipment selection, and (3) additive technologies. The concept of establishing "low-energy, low-carbon architecture" centres on increased dependence on passive and low-energy systems. Technical realisation involves meticulous design encompassing built-in configurations, solar controls, daylighting strategies, envelope materials, and on-site renewable systems. Building energy efficiency directly correlates with the effective thermal and climatic performance of all components. Enhancing design efficiency aligned with climate adaptation emerges as a unique avenue for achieving requisite energy efficiency, concurrently yielding multiple benefits. The research outlined in this study makes a substantial contribution towards addressing critical issues linked with urbanization, energy demand, and climate change in West Asia. By examining the interplay between climate adaptation strategies and energy efficiency, this work provides innovative perspectives on sustainable architectural and urban design practices. It stresses the importance of integrating traditional knowledge with modern technologies to create climate-resilient building solutions, specifically tailored for the arid environments of West Asia. A key aspect of this research is the identification and evaluation of region-specific climatic factors and traditional technologies, forming the basis for developing energy-efficient building standards. Moreover, the study highlights the economic feasibility of low-energy, low-carbon architecture, underlining the potential for significant reductions in greenhouse gas emissions through strategic planning, design, and material selection. This involves positive interaction with climate data and the environment, minimising environmental impact, and proactively addressing existing environmental issues. Climate adaptation extends beyond alignment with current conditions, requiring continuous dynamic modification to accommodate changing climatic and weather conditions, ensuring consistent thermal comfort and fostering a symbiotic relationship with the living environment.

Acknowledgements The author expresses heartfelt gratitude to Fujian University of Technology for financial support (startup research project GY-Z22072) and the Urban and Rural Ultra-Low Energy Consumption and Low-Carbon Building Technology Research Institute for unwavering support in this research.

References

1. Alabsi AAN, Song D, Zheng L (2016) Traditional solutions in climate adaptation and low energy buildings of hot-arid regions in west Asia. Urban Dev Stud 23(1):1–7
2. Alabsi AAN, Wu Y, Koko AF, Alshareem KM, Hamed R (2021) Towards climate adaptation in cities: indicators of the sustainable climate-adaptive urban fabric of traditional cities in West Asia. Appl Sci (Switzerland) 11(21). https://doi.org/10.3390/app112110428
3. Alzahrani A (2022) Understanding the role of architectural identity in forming contemporary architecture in Saudi Arabia. Alex Eng J 61(12):11715–11736. https://doi.org/10.1016/j.aej.2022.05.041
4. Ayyash AB (2015) Sustainability in vernacular architecture of Iraq and its applicability in modern residential buildings in the Iraq region
5. Baboli FBM, Ibrahim N, Sharif DM (2015) Design characteristics and adaptive role of the traditional courtyard houses in the moderate climate of Iran. Procedia Soc Behav Sci 201:213–223. https://doi.org/10.1016/j.sbspro.2015.08.170
6. Bagader M (2017) Climate adaptability in the Hejazi traditional architecture. Int J Herit Archit Stud Repairs Maint 1(4):683–693. https://doi.org/10.2495/ha-v1-n4-683-693
7. Baghaiepoor M, Jovanovic G, Stanimirovic M (2019) Climate adapted houses in Iran: hot, cold and humid climate. Facta Universitatis - Ser Archit Civil Eng 17(4):429–443. https://doi.org/10.2298/fuace180721025b
8. Bardeesi B, Muhrji D, Gutub G, Al-Ahdal H, El-Shorbagy AM (2022) The lost identity of Jeddah. Civil Eng Archit 10(5):2180–2186. https://doi.org/10.13189/cea.2022.100536
9. Damluji SS (2006) The architecture of the United Arab Emirates. Garnet Publishing Ltd.
10. Eskandari P (2011) Analysis of traditional Iranian houses of Kashan, Iran in terms of space organization and access design. Eastern Mediterranean University
11. Evans JP, Smith RB, Oglesby RJ (2004) Middle east climate simulation and dominant precipitation processes. Int J Climatol 24(13):1671–1694. https://doi.org/10.1002/joc.1084
12. Jawad I, Al-Yousif K (2008) Reduction cooling load using intelligent envelope system. Iraqi J Archit 14. https://www.researchgate.net/publication/313309130
13. Kamal MA, Warsi TR, Nasir O (2023) Islamic principles as a design framework for the urban system: for the environmental concern and sustainable development. J Islam Archit 7(4):699–712. https://doi.org/10.18860/jia.v7i4.21187
14. Manzano-Agugliaro F, Montoya FG, Sabio-Ortega A, García-Cruz A (2015) Review of bioclimatic architecture strategies for achieving thermal comfort. In: Renewable and sustainable energy reviews, vol 49. Elsevier Ltd., pp 736–755. https://doi.org/10.1016/j.rser.2015.04.095
15. Mohamed ASY, Elmeligy DA, Azmy NY (2020) Eco-adaptive architecture through the bioclimatic design in historical Arab regions. EQA Int J Environ Qual 39:32–51. https://doi.org/10.6092/issn.2281-4485/10484
16. Mohamed Gamal Abdelmonem A, Arch M, Loehlein G, Gamal Abdelmonem M (2007) Sustainability in traditional houses in the UAE potentials and improvement of buildings abilities sustainability in traditional houses in the UAE potentials and improvement of buildings abilities. In: The second international conference of Dubai conservation

17. Patlakas P, Stathopoulos C, Flocas H, Kalogeri C, Kallos G (2019) Regional climatic features of the Arabian Peninsula. Atmosphere 10(4). https://doi.org/10.3390/ATMOS10040220
18. Taheri J, Moghadam TT, Taheri S, Safari MK, Eslami F (2022) Assessment of passive design strategies in traditional houses of Sabzevar, Iran. J Cult Herit Manag Sustain Dev 12(4):570–592. https://doi.org/10.1108/JCHMSD-10-2020-0145

Advanced and Emerging Design Models and Tools

Cooling Pakistan: Exploring Historical Wisdom and Modern Strategies for Passive Cooling in Pakistan

Shanzae Naeem, Aisha Aziz, and Baojie He

Abstract The urban population is increasing rapidly in Pakistan especially in large cities as people are migrating from Rural to Urban areas for better life opportunities. This phenomenon has far-reaching effects on the region's climate. Lahore being the second largest city in the country, has undergone a massive increase in its urban population; and the urban population of the world will increase almost 75% by 2030 due to urbanization. This increase in population is resulting in increased resource consumption within the urban context, which in turn affects the urban climate, giving rise to phenomena such as the UHI (urban heat island) effect. The climate of Lahore is semi-arid therefore, the problems mentioned above increase the consumption of energy for cooling purposes in buildings. In this chapter, we will discuss the operational energy increase with the expansion of urbanization and its negative effects on the urban climate. The increase in air conditioning due to hot weather in Lahore has caused a massive surge of almost 60% in electricity consumption. These energies release GHG Emissions which play a key role in global warming and climate change. Moreover, Carbon dioxide (CO_2) is the most significant among GHGs, contributing more than 70% of atmospheric concentrations. So, it is vital to control the production of GHG to take a step towards net zero carbon emissions as Pakistan is one of the signatories of the 2015 Paris Agreement. This problem however can be solved by adopting Passive evaporative cooling techniques. Introduced to this part of the region

S. Naeem · A. Aziz
University of Engineering and Technology, Lahore, Pakistan

B. He (✉)
Centre for Climate–Resilient and Low–Carbon Cities, School of Architecture and Urban Planning, Key Laboratory of New Technology for Construction of Cities in Mountain Area, Ministry of Education, Chongqing University, Chongqing 400045, China
e-mail: baojie.he@cqu.edu.cn

School of Architecture, Design and Planning, The University of Queensland, Brisbane 4072, Australia

Institute for Smart City of Chongqing University in Liyang, Chongqing University, Liyang 213300, Jiangsu, China

CMA Key Open Laboratory of Transforming Climate Resources to Economy, Chongqing 401147, China

B. He et al. (eds.), *Urban Climate and Urban Design*, Urban Sustainability,
https://doi.org/10.1007/978-981-96-1521-6_9

by the Mughals, this array of different strategies helped Mughal-era designers create comfortable indoor living conditions. It will also be beneficial for the sustainability of the urban climate of Lahore and if opted in contemporary architectural designs, can help reduce electricity consumption resulting in controlling carbon emissions.

Keywords Urban climate · Carbon emissions · Energy consumption · Operational energies · Passive cooling techniques · Passive evaporation technique

1 Introduction

Sustainability has become a course of discussion around the globe recently. Within the context of climate, it becomes a vital approach for architectural design to counter the effects of global warming. Data indicates that Pakistan is the fastest growing country in terms of urbanization in South Asia. This phenomenon has led to a number of problems, mainly climate change and global warming. Figures indicate that the use of energy consumption is increasing drastically to provide thermal comfort in indoor spaces. As Pakistan is one of the 194 parties that joined hands for the Paris Agreement of 2015, climate-responsive architecture could be the key to a sustainable future.

Pakistan is a developing country; its urban population is also rapidly increasing. In the previous decades, 10% population was living in urban areas, whereas now it is estimated that by 2030 it will be 75% [1]. The climate of Pakistan is mostly semi-arid, so, this huge population has a drastic effect on the urban weather pattern of the city and the residents have to depend heavily on mechanical cooling systems/air conditioning in the built structures to achieve thermal comfort. In hot climate areas, about 40–50% of the energy resources are utilized for cooling purposes. Whereas some scholars claim it to be up to 60%. In this chapter, we will delve into the semi-arid climate zone of Pakistan, with a particular focus on Lahore, the country's second-largest city situated within the province of Punjab. Buildings in Lahore are generally built with brick walls having low to moderate insulation, putting all the responsibility of cooling any space on active cooling methods, thus increasing its carbon footprint. The idea of carbon footprint is closely related to the concept of embodied and operational energies of a building. To minimize the carbon footprint, it is required to conserve the embodied energies and curtail the operational energies. In the next few pages, we will look into the data and numbers of energy consumption in Lahore and its effects on building design and the built environment. But firstly, it is important to understand the concepts of embodied and operational energy within the context of a building as it promotes resource efficiency, reducing environmental impact, and fostering long-term sustainability. By minimizing energy consumption throughout a building's lifecycle, architects can create resilient structures that mitigate climate change impacts and contribute to a more sustainable urban environment.

1.1 Embodied Energies

The embodied energy of a building is the sum of all the energy required to extract, process, deliver, and install the materials when a building is constructed. This fact also paves the way for the conservation of old structures through adaptive reuse to preserve the embodied energy. Mostly, these embodied energies are divided into the whole process of building construction [8].

1.2 Operational Energies

Operational energy of the building can be defined as the energy expenditure for the functionality of the building throughout its life span which provides comfortable living conditions. It is the energy utilized for the purpose of lighting, heating, cooling, HVAC, and other electricity appliances such as dishwashers, refrigerators, etc. [13]. Utilization of cooling equipment is included in operational energies which in return are producing greenhouse gases (GHGs) [9].

2 Climate Conditions

Pakistan is geographically located between 24–37°N latitudes and 62–75°E longitudes in the western zone of South Asia. Pakistan is the only country in the whole world to have a distinctive range of altitudes from sea level to the second-highest mountain peak. This unique characteristic gives it an enormous variation in climate across its different areas. The climate of Pakistan is generally arid, characterized by hot summers and cold winters. A large part of Pakistan is arid and semi-arid, whereas some areas are hyper-arid in the lower southern half of the country. Lahore is in the semi-arid zone.

2.1 Koppen Climate Classification of Lahore

According to Koppen Climate Classification for Lahore, its climate lies under "BSh". BSh is semi-arid tropical steppe climates where average annual temperature is more than 180 °C. These climates are usually found in the neighborhood of hot desert and need humidity.

2.2 *Bioclimatic Chart of Lahore*

This chart (Fig. 1) for determining the satisfactory climatic conditions for living was discovered by the Olgyay's Brother in 1950s. This chart shows the relationship of different climatic elements to each other. The human thermal comfort zone where no external resources are needed of heating or cooling is shown in the middle of the chart. The procedure is to collect temperature data of all months of the year and then plotting it according to the average humidity. The thermal comfort zone diagram of Lahore shows the climate conditions in all 12 months in Lahore. The months of January (54°F, 46%), February (59°F, 41%) and December (56°F, 49%) needs heating. The month of March (69°F, 29%) and November (66°F, 41%) are comfortable months which do not need any heating or cooling. The months of May (88°F, 20%) and June (93°F, 22%) needs humidity. The months of April (79°F, 23%) and October (77°F, 33%) needs humidity and wind whereas the months of July (88°F, 50%), August (88°F, 57%) and September (85°F, 43%) needs winds only. This shows that total number of hot months that require active cooling are seven and out of it four needs humidity as well.

3 Electricity Consumption

According to researchers, Pakistan is at a vulnerable condition as the consumption of electricity is increasing considering rising temperature trends. Only 16% of the rural areas are connected to the electricity grid where as 85% of the urban areas consume electricity directly from grid. Pakistan is at lack of 10–15% of electricity demand.

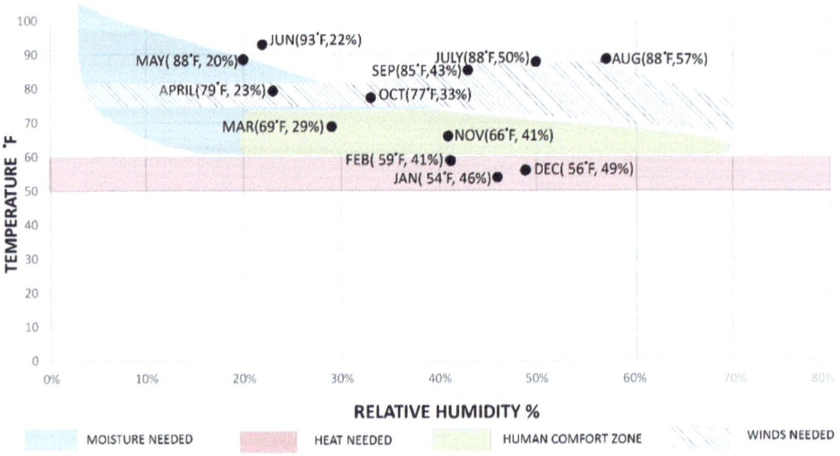

Fig. 1 The Y-axis represents the temperature, and the X-axis represents relative humidity

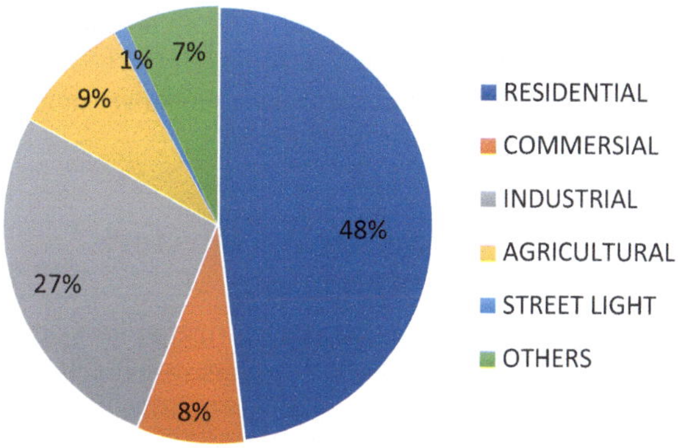

Fig. 2 Electricity consumption in different sectors

According to the analysis of 2018 (Fig. 2), residential zone consumes almost half of the total percentage 48% of the electricity usage. Pakistan's electricity demand will be 192.64 TWh, 346.31 TWh, 600.08 TWh, and 2299.59 TWh of electricity by 2025, 2030, 2035, and 2050 respectively [3].

3.1 Increased Trend of Airconditioners in Lahore

As shown in Fig. 2, the residential zone consumes the maximum electricity in Lahore. It varies periodically, according to the climate. As explained earlier (Fig. 1), the semi-arid nature of Lahore's climate requires the use of air conditioners for cooling purposes for almost seven months annually. A study published in 2017 in the Pakistan Journal of Meteorology, showed the number of household electrical appliances. According to the statistics obtained, out of 300 households, only 70 were not using air conditioners for cooling their living spaces while others were using up to 5 air conditioners. These results indicate that more than 75% of the households are following the increased trend of using air conditioners in Lahore city [9].

3.2 Greenhouse Gas Emissions from Air Conditioners

Air conditioners are a huge source of greenhouse gas emissions both directly and indirectly in the urban environment. The refrigerants in the air conditioners remove heat from the environment, but their leakage is a direct emission of GHG. On the other hand, the production of electricity to run these air conditioners is a cause of indirect emissions of GHGs. The most famous refrigerants used in air conditioners are Chlorofluorocarbons (CFCs) known as R12 and hydrochlorofluorocarbons (HCFCs) known as R22. According to the Montreal protocol both the refrigerants used in the air conditioners are considered harmful to the ozone layer of the urban environment and gave the policy of depleting out both R12 and R22 by 2040. As a substitute for these refrigerants, hydrofluorocarbon (HFCs) became popular in cooling systems but many of these refrigerants also have global warming potential (GWP) [6].

4 Climate Responsive Architecture

Before the advent of mechanical refrigeration systems and these toxic refrigerants, buildings were built as climate-responsive in function. The concept of climate responsive means that the building's architectural design, planning, and facades are designed by incorporating passive design strategies for different climatic conditions. These design strategies helped maintain the temperature of the buildings. The most efficient techniques were the "Passive Cooling/heating techniques" according to climatic conditions. It will decrease the consumption of electricity and production of GHG's.

4.1 Passive Cooling Techniques

The concept of passive cooling essentially means, achieving comfortable temperature in the internal spaces of a building, using zero operational energy. This implies the use of numerous passive cooling techniques. These techniques are used around the globe according to the climatic conditions of the region such as natural ventilation, Insulation and shading elements, building orientation, Natural Landscaping, Adding thermal mass to the structure, Passive Evaporative Cooling.

As Lahore has a semi-arid climate, water bodies such as fountains in the exterior, huge courtyards, various shading devices, vegetation, wind towers, jharokhas, air vents, domes, etc. have been a choice for the designers and architects. But the most effective of all is the use of water bodies as a passive cooling strategy. Earlier, water bodies were used as a symbol of beauty in this part of the region. The Mughals with their ingenuity in architectural and urban design, brought this technique to its zenith. Mughals incorporated water bodies to increase the thermal comfort of the interior spaces [2].

5 Passive Evaporative Cooling

The basic scientific principle at work is the reduction of air temperature through the evaporation of water within an airstream. The water vapors raise the level of humidity in the air, which in turn makes the air cooler and decreases the temperature of that area. There are numerous ways to achieve thermal comfort through this particular method. The oldest and simplest type is the use of direct evaporative cooling technique [2].

5.1 Direct Evaporative Cooling Technique

This method is based on the simple yet effective technique of channeling direct air from the exterior of the building through a water body which in turn cools down the air by evaporation, hence making the interior of the building comfortable. Using direct evaporative cooling was an important feature of the Mughal architecture and landscape. One of the major functions of Mughal gardens was to enhance the micro-climate. The design elements such as water fountains and ponds etc. were provided in the middle of the courtyards. Air would pass over the cool water ponds causing evaporation which resulted in heat absorption by the water, making the air humid, which reduced the temperature in the internal spaces. Water has a very high thermal storage capacity. For every gram of water that evaporates, it is estimated that around 2500 Joule heat energy is consumed [2].

5.1.1 Implementation of Direct Evaporation in Mughal Architecture

Whether the Mughals were aware of the above-stated scientific fact is unclear. But what is evident from the architectural heritage they left, is an ingenious use of this natural phenomenon. Water fountains at the Centre of the courtyard were used by the Mughals frequently as a design element with water channels attached, taking water into internal spaces. Nahar-i-Bahisht (canal of paradise) in the Red Fort Delhi, is an example of a water channel that passes through internal spaces of the building to modify the internal environment [2]. These channels inside the fort opened into water fountains. To maximize the surface area of the water, for evaporation of the internal air, aeration of the drinking water, and ornamentation. Similarly, direct evaporative cooling can be seen in the Shish Mahal Pavilion at the Lahore Fort. The Pavilion consists of a central pool and many shallow circular pools at different locations. The courtyard is subdivision into four quadrants from where four narrow water channels are distributed. The basement of the Shish Mahal also consists of underground wind tunnels and openings, through which the process of direct evaporation takes place.

5.1.2 Direct Evaporation in Modern Projects: Pearl Academy Jaipur India

This project is situated in India, Jaipur, and has been made a part of this study because the climate of Lahore and Jaipur both lie under the BSH category of Koppen climate classification (Fig. 3). In the Jaipur project water body is located in the basement, which is erected through pillars. The air from the exterior enters the building by passing through these pillars and over the water body, thus decreasing its temperature. The cool air is then circulated throughout the building before it exits [10].

There are some other methods to introduce direct evaporative cooling techniques in buildings such as mashrabiya's, wind towers, passive downdraught evaporation,

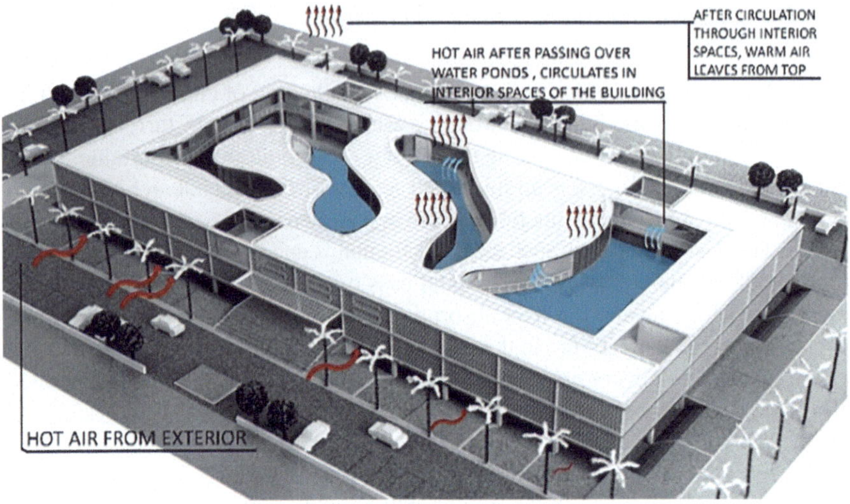

Fig. 3 Direct evaporation through Pearl Academy Jaipur

evapcool systems, etc. In the following discussion, we will be discussing some existing projects to understand how these techniques work in real-time.

5.1.3 Mashrabiya

The word mashrabiya is of Arabic origin and have been used extensively in traditional Islamic architecture (Fig. 4). These were wooden screens/windows that provided shade from the sun and allowed cool wind to enter the building, incorporating porous jars of water. The concept was similar to that of Mughal water ponds in the courtyard. The air from the exterior after passing through the mashrabiya's passes over the porous jars. Heat is then transferred into the water and vapors are added into the air making the air cool. This technique was used in Pakistan, Egypt, and Iran. The size, shape, and material of mashrabiya differ in different regions. In one of the modern projects, this technique is implemented in Pearl Academy of Fashion India.

5.1.4 Wind Towers

One of the eminent techniques used by the designers historically was called wind towers or wind catchers. Wind catchers, traditionally known as 'malqafs', catches the air due to difference in air pressure. The air flows from a high-pressure area (outside the building) to a low-pressure area (inside the building). There are many different types of wind towers. The basic type is generalized into four categories, based on the number of sides with openings, which means a tower could be either one, two, Four, six, or eight-sided. These towers could also be circular in plan. The location

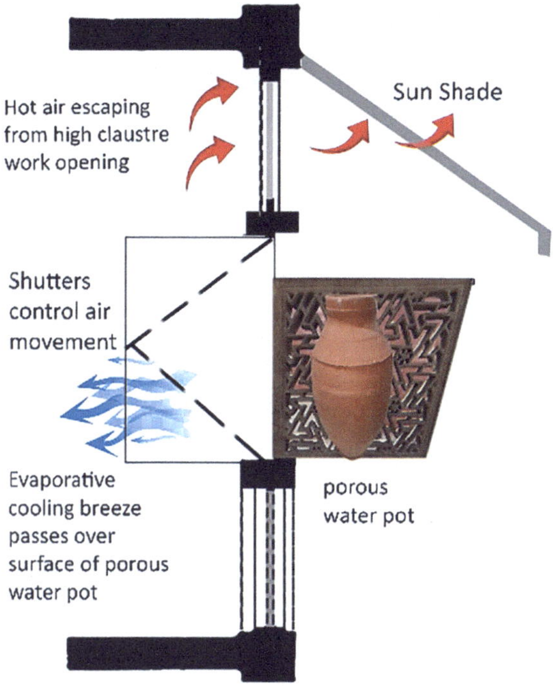

Fig. 4 Mashrabiya used in Islamic architecture

and number of these openings in the tower depend on wind direction. Four, six, and eight-sided wind towers are bigger in size to catch the maximum amount of air from outside. The cylindrical wind towers are the latest type of wind towers. They are the most effective type of towers compared to the other three categories of wind towers. Bahadori was the pioneer of this type. He referred to those wind towers as, the "wind towers with wetted columns, and the wind towers with wetted surfaces" [5].

5.1.5 Wind Towers with Wetted Columns

In this type of wind tower, it consists of unglazed ceramic conduits. Air from the exterior enters the wind tower, and is sprayed by an inbuilt water spraying pipe. The air temperature lowers down and then cool air through the conduits enter the interior of the building. Excess water is than stored at the bottom of the tower in the form of water pool. This pool consists of a small pump that regulates the same water again and again. This type of passive evaporation cooling technique is very effective for hot semi-arid climates like Lahore. Its installation is also not very expensive.

5.1.6 Wind Towers with Wetted Surfaces

This type of wind towers is specially designed for areas where the speed of air flow is slow such as in Lahore. The surface of the wind tower consists of a series of straws or cellulose pads. These cellulose pads are designed at the aperture of the wind tower and are kept wet by spraying water on them. Air passes through these wet pads and as a result its temperature decreases due to increase in humidity in the air. Cool air than passes through two chambers, and as cool air is heavier it enters into the interior of the room. It circulates around the room and after changes the temperature of the room hot air exits from the internal space from top apertures of the building [5].

5.1.7 Passive Downdraught Evaporation

In the late 1980s, Passive Downdraught Evaporative Cooling was discovered. The Tower comprises of single or multiple towers equipped with a water spraying nozzles, placed on the top. This innovation consists of replacing the wetted pads with rows of atomizers (nozzles, which produce an artificial fog by injecting water at high-pressure) (Fig. 5). During continuous spray of water, droplets descend through the tower thus creating conditions close to saturation along its length. The cool air is heavier and enters into the adjacent spaces. The concept is based on the relatively large amount of energy required to convert water from its liquid to gaseous. This leads to the movement of air from a zone of high pressure, where the air is hot and less dense (top of the tower) to a zone of lower pressure, where air is colder and denser (bottom of the tower). The concept of passive downdraught evaporative cooling is similar to that of wind towers with wetted columns.

5.1.8 Torrent Research Centre

This technique was first used at a pharmaceutical research laboratory in Ahmedabad, at a large scale in 1998. The climate of Ahmadabad, according to Koppen climate lies under BSh, similar to that of Lahore. An experiment carried out in 1998 showed that the temperature of ground floor achieved by PDEC was 27 °C and that of first floor was 29 °C when the exterior temperature was maximum 38 °C. It was due to this technique that, it was reported that the building used 64% less electrical energy in the first year. 72% of the central building has achieved human comfort conditions using Passive Downdraft Evaporative Cooling [7].

5.1.9 Evapcool System for Direct Evaporation

This passive evaporative cooling system using porous ceramic components that can be integrated into the building envelope. It is mostly designed inside the wall of the building or the roof. The cavity of the wall or roof consists of a ceramic evaporator.

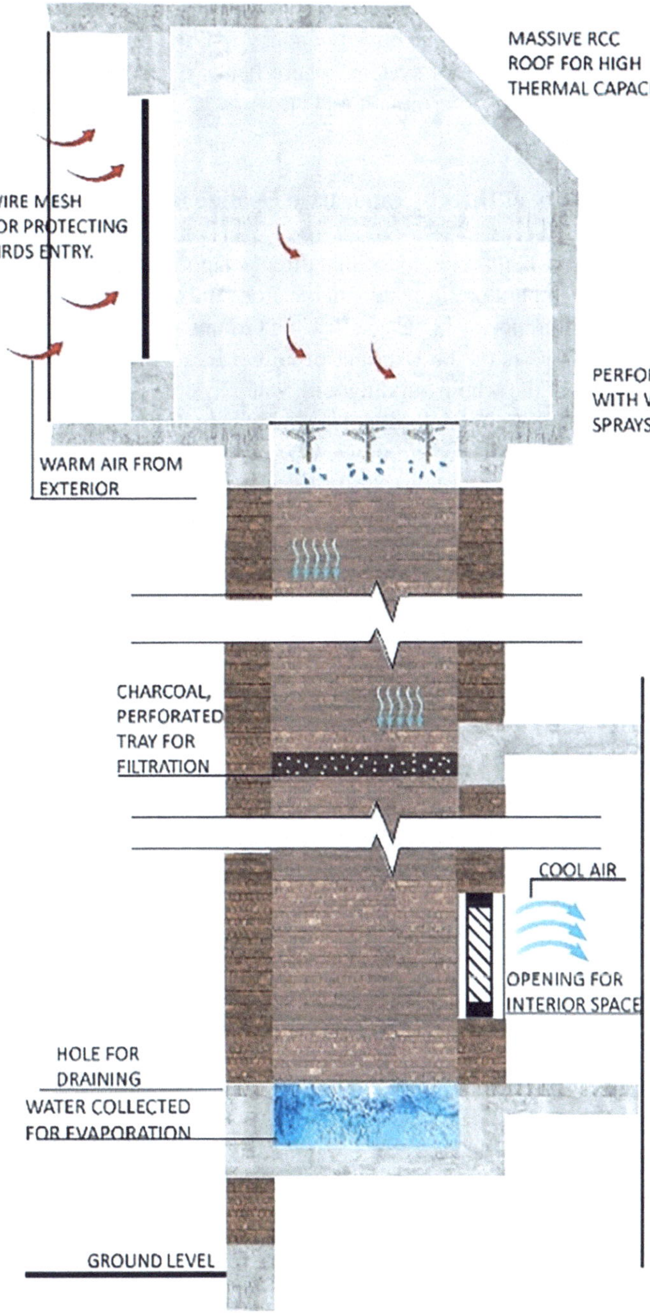

Fig. 5 Passive downdraught evaporation

The air in contact with the wet surface of the ceramic evaporators becomes heavier and cooler. It then gets delivered by negative buoyancy into the indoor spaces. Another option for integration is the roof system, where banks of ceramic evaporators are integrated into a wind-catcher termination [11].

5.1.10 Applacability of Direct Evaporative Cooling in Lahore

According to data above, it is concluded that direct evaporative cooling can be beneficial for reducing temperature in residential sector. The study of Torrent Research (5.1.8) shows the difference of 11 °C and 9 °C on Ground and first floor respectively by using PDEC. PDEC is the best variant of Direct Evaporative Cooling as it can reduce temperature of the whole building with water system installed at one corner. The effectiveness of PDEC is influenced by wind direction and wind speed. The impact of wind direction can be minimized by designing four faced wind towers facing each side on top of the house in Lahore to catch winds coming from any direction. The effectiveness of PDEC is reduced by strong winds as it creates disturbance at the top of the inlet [4]. Quantitative research conducted by the author of this chapter questioned architects that why this technique is not used as design element in their design. Most architects questioned the costing of the construction, but it is researched that a PDEC building could be built for 6% less than an airconditioned building. Less water is consumed as the same water regulates again and again [7].

5.1.11 Risk of Direct Evaporative Cooling

Risk related to this technique such as hardening of water, microbiological contamination of water can be addressed by taking certain design measurements such as UV filters, regular maintenance etc. As it is not possible to completely depend on the passive means of cooling in all hot months in Lahore, this technique can be coupled with chilled water-cooling coils at top of the air shaft [7].

5.2 *Indirect Evaporative Cooling*

This technique utilizes the process of conduction of heat, without changing the humidity by the transfer of heat energy. Conduction is the process by which heat energy is transmitted through collisions between neighboring atoms or molecules. Indirect evaporative cooling includes many techniques such as evaporative roof cooling and thermal storage roof.

5.2.1 Evaporative Roof Cooling

Among different passive evaporative techniques, this strategy focuses on the heat absorption by the roof of the building. According to research, the roof alone can be responsible for up to 50% of the heat load. There are many ways to insulate the roof from getting heat up. The use of water as a thermal mass is the most common.

5.2.2 Principle of Evaporative Roof Cooling

In evaporative roof cooling, the space is cooled with the help of a water pond on the roof. The process of radiation occurs during daytime to transfer heat energy of water to the interior spaces through reinforced slabs and thus acts like the radiant cooling system. The concept of radiation can be defined as a process where heat waves are emitted that may be absorbed, reflected, or transmitted through a colder body, the water pond on the roof in this case. Heat transfers from high concentration area to low concentration area so at night the concept of "Sky cooling" effect is achieved. The cooling of the interior space is carried out through the process of evaporation of the water on the roof pond as the water acts as a heat sink/thermal buffer. The water absorbs the heat from the interior spaces and transfers the heat to the environment by the process of evaporation. Evaporative roof cooling proved to reduce 2 degrees in hot and humid areas. Roof ponds work using both the concept of evaporative cooling at night and radiation cooling during day time. Previous literature works conclude that 19 different variants of roof ponds are present, which use the concept of evaporation.

(A) **Open Roof Pond with and Without Spray**

Open roof pond without has a flat concreate slap roof with recommended depth of 30 cm for a water pond (Fig. 6). During day time heat is absorbed by the water which helps in maintaining the interior temperature cool as water acts as a buffer. The phenomenon of evaporation takes place when water reaches its maximum capacity of absorbing heat and starts evaporating, causing the "cooling effect". At night the ambient temperature is cooler outside, so the heat energy moves from higher concentration to lower. The water is cool, absorbs heat energy from the interior through radiation and then emits it into the atmosphere. The only difference in open roof pond with spray is that water is sprayed on the roof pond throughout the day. It increases the evaporative cooling and heat absorption process. The recommended radius of water droplet is 0.05–0.1 cm, with flow rate 1.0–1.5 Vol/h.

(B) **Roof Ponds with Movable Insulations**

Roof ponds with movable insulations can be with continuous spray throughout the day (Fig. 7). It can be only at night time or without spray. These ponds have movable insulation panels made up of polystyrene, polyurethane, Styrofoam or corrugated metal sheet. These movable sheets can be fold-able or sliding with highly reflective

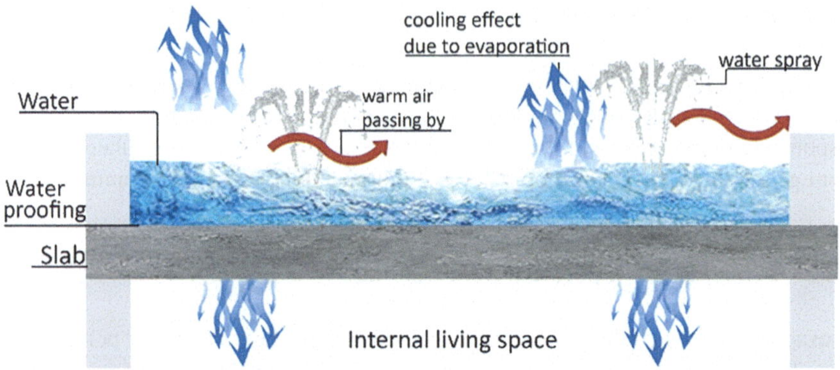

Fig. 6 Open roof pond with water spray

opaque material to reflect solar radiations. This insulation helps prevent the water from getting hot during daytime and at night when removed from over the pond its temperature is further lowered by ambient environment temperature and the process of evaporation which helps the movement of indoor heat energy to radiate through the reinforced slab into the water.

(C) **Walkable Roof Ponds**

These roof ponds have two variants, "roof pond with insulation embedded within the pond", and "walkable roof pond with night water circulation" (Fig. 8). Roof pond with insulation embedded within the pond have slightly slanted reinforced slab to drain rainwater with a waterproof layer on top of it. Third layer is composed of pebbles 2 cm thick, after which 5–10 cm thick insulation panels made up of polystyrene are placed with gaps to allowing "thermo symphonic circulation". Above it is another layer of pebbles of around 12 cm. In this layer of pebbles, a float is incorporated which makes sure that water doesn't flow above 3 cm above the insulated panels. In winters, water is removed through hole from the lower most layer. In "walkable roof

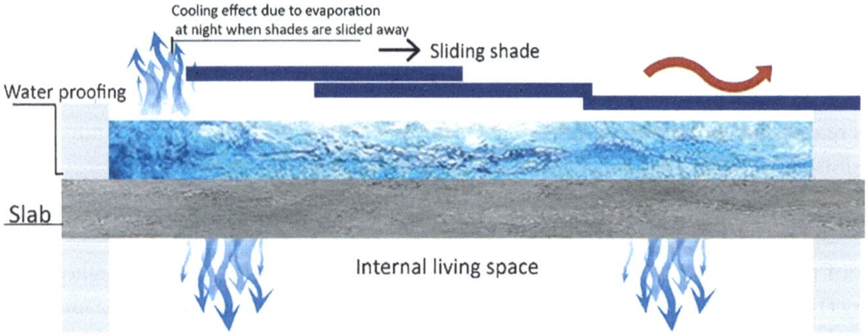

Fig. 7 Roof pond with moveable insulation

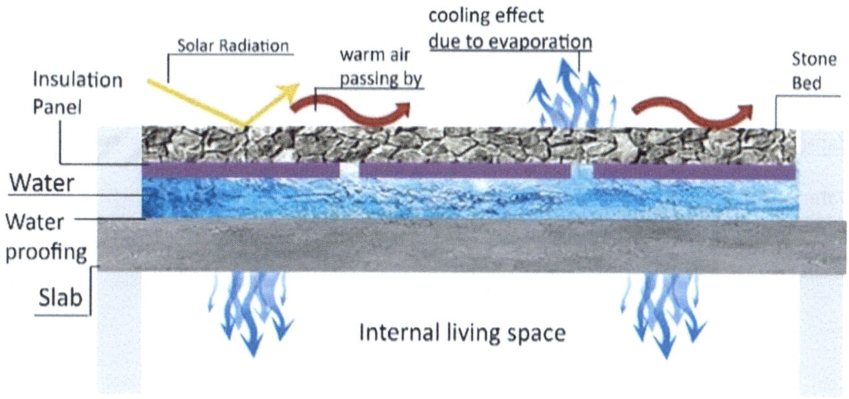

Fig. 8 Walkable roof pond

pond with night water circulation" there is only one layer of pebbles imbedded in water pond, above which there is a layer of insulation covered by thin tiles to walk on. At night time water is circulated on the tiles for the process of evaporation.

(D) **Roof Ponds with Gunny Bags**

Roof ponds with gunny bags have two variants, "Roof Pond with wet gunny bags (RPWGB) and shaded roof pond with wet gunny bags (shaded RPWGB)" (Fig. 9). Roof ponds with wet gunny bags have depth of 15 cm. Insulation layer or a mesh or grid with polystyrene strips or other floatable materials are attached underneath the gunny bags. The process of cooling takes place through the concept of evaporation, convection and nocturnal radiation. In shaded roof pond with wet gunny bags, a shading panel is placed above the gunny bags with a gap so that air can pass through.

(E) **Shaded Roof Ponds**

These type of roof ponds has louver plates on the pond which may be fixed or operable (Fig. 10). These louvers in operable type remain closed during daytime to avoid solar radiations whereas remain open during night time so that cooling of water

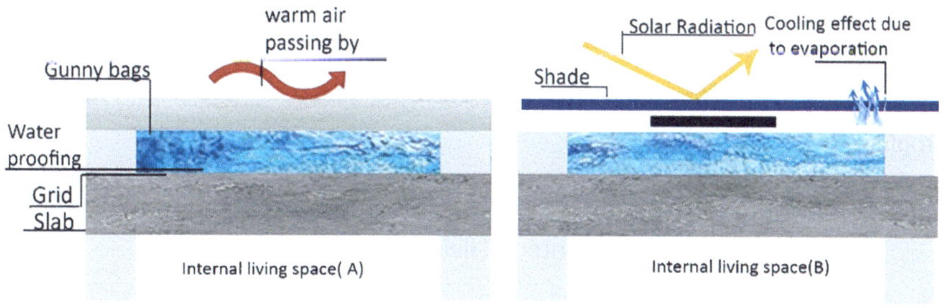

Fig. 9 Roof pond with wet gunny bags (**a**), shaded RPWGB (**b**)

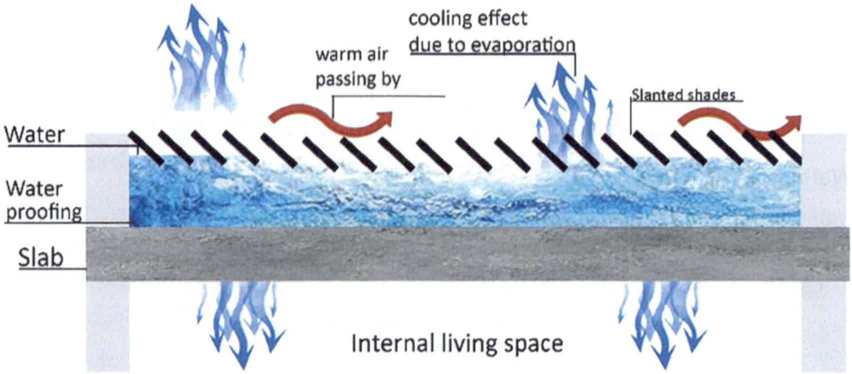

Fig. 10 Shaded roof pond

takes place through the process of evaporation and radiation. In the fixed louver type, the louvers are set in a position to avoid direct solar radiations. In addition to louvers, water sprays are also added to enhance the effect of evaporation.

(F) **Ventilated Roof Ponds**

These type of roof ponds are covered from top with a reflective metal insulation (Fig. 11). The water pond on top of concrete slab is ventilated by apertures with a gap of 0.3 m for the purpose uninterrupted airflow. Depth of the pool should be around 9 cm. Fans can be added on one side (inlet) or both inlet and outlet to enhance the effect of evaporation.

(G) **Closed Roof Ponds**

This type of roof ponds always remains closed from top. These are known as "evapo-reflective roof pond" (Fig. 12). This roof pond consists of pebbles inside the water which is separated by unventilated air gap from the roof. The roof is made up of aluminum plate painted with "titanium-based pigment" to reflect solar radiations.

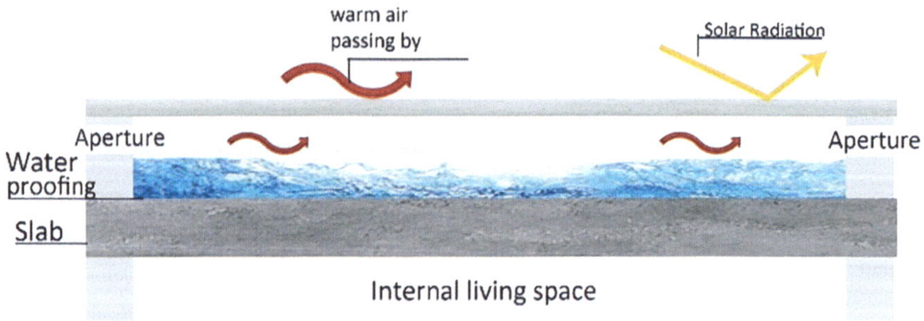

Fig. 11 Ventilated roof pond

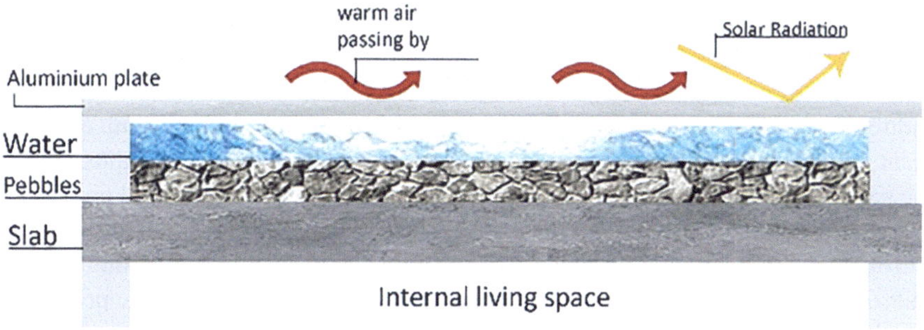

Fig. 12 Closed roof pond

At night, temperature of the water is higher vapors does not leave the system but condenses and fall into the water pond again [12].

5.2.3 Applacability of Indirect Evaporative Cooling in Lahore

In a stimulation study by Tang and Etzion found that implementation of RPWGB reduces 7 °C from a space and shaded RPWGB performs even better. According to different articles, RPWCB performs better than ventilated and open roof pond without spray. It is shown that RPWGB, performs similar, slightly better or worse than pond with moveable insulation as discussed in three different articles. RPWGB performs better than roof pond with gunny bags. Whereas shaded RPWGB performs better than all other variants of roof ponds. It concludes that out of many indirect evaporative variants, RPWGB and shaded RPWGB would be the most suitable for the BSh climate of Lahore.

5.2.4 Risk of Indirect Evaporative Cooling

There are some concerns regarding using roof ponds on roof tops. If proper maintenance is not carried out there is a high risk of leakages. These ponds increase the roof weight and effect its accessibility.

6 Conclusion

Despite its many opportunities, urbanization has become an immense challenge globally. The increase in the embodied and operational energies which in turn increases the carbon foot print, thus producing toxic gasses. So, it is imperative to cut down these energies especially the excessive use of air conditioners, in order to make a positive effort towards achieving the net-zero carbon emission. Lahore's climate lies under the

"BSh" category and due its semi-arid nature, therefore, Passive Evaporative Cooling is considered the most suitable method. It not only helps in decreasing the consumption of electricity but, also the production of GHG's. There are numerous Direct and Indirect Evaporative methods to achieve passive evaporative cooling in built environment. Some of these have been developed over centuries, and can be adapted to suit the modern needs. The above points have briefly shed light on the natural process of evaporation and movement of air due to difference in pressure and temperature of the surroundings of the built environment. As Lahore needs introduction of humidity in four out of seven hot summer months it is essential that water bodies are proposed, which add moisture through evaporation. Converting the hot, dry air stream into cold and humid air. This technique of adding moisture to air, is called Direct Passive Evaporative Cooling. There are many types of Direct Evaporative Cooling techniques such as adding ponds in courtyards, Mashrabiya, wind towers but the most useful technique and efficient one is Passive Downdraught Evaporative cooling. These techniques combine natural processes along with some mechanical means to induct water in the system. It decreased the need to use air conditioners, which is a major source of toxic greenhouse gas emissions. In a nutshell, this discussion on passive evaporative cooling methods in Lahore exemplifies a crucial intersection between sustainable architecture and climate change mitigation. By elucidating the effectiveness of appropriate cooling techniques within the region's semi-arid climate, this chapter offers practical solutions to reduce energy consumption and curb greenhouse gas emissions in the built environment to foster resilience to rising temperatures but also contributes to broader efforts towards achieving net-zero carbon emissions. By prioritizing sustainability and climate adaptation in urban development, architects and policymakers can play a pivotal role in creating healthier, more resilient cities that mitigate the impacts of climate change.

References

1. Abbasi NQ, Jabeen A, Khalid M (2022) Urban generosity: strategies for rejuvenation of historic site-Mughal Tombs of Shahdara. Pak Lang Humanit Rev 6:282–295
2. Ali A (2012) Passive cooling and vernacularism in Mughal buildings in North India: a source of inspiration for sustainable development. Int Transact J Eng Manage Appl Sci Technol 4(1):15–27
3. Ali M, Iqbal MJ, Sharif M (2013) Relationship between extreme temperature and electricity demand in Pakistan. Int J Energy Environ Eng 4:1–7
4. Alshenaifi M, Sharples S (2019) A parametric analysis of the influence of wind speed and direction on the thermal comfort performance of a passive downdraught evaporative cooling (PDEC) system–field measurements from a Saudi Arabian library. In: Presented at the IOP conference series: earth and environmental science, IOP Publishing, p 012042
5. Dehghani-sanij AR, Soltani M, Raahemifar K (2015) A new design of wind tower for passive ventilation in buildings to reduce energy consumption in windy regions. Renew Sustain Energy Rev 42:182–195
6. Dong Y, Coleman M, Miller SA (2021) Greenhouse gas emissions from air conditioning and refrigeration service expansion in developing countries. Annu Rev Environ Resour 46:59–83

7. Ford B (2001) Passive downdraught evaporative cooling: principles and practice. ARQ Archit Res Q 5:271–280
8. Jackson M (2005) Embodied energy and historic preservation: a needed reassessment. APT Bull J Preserv Technol 36:47–52
9. Khan W, Siddiqui S (2017) Estimation of greenhouse gas emissions by household energy consumption: a case study of Lahore, Pakistan. Pak J Meteorol 14(27):65–83
10. Sarswat G, Kamal MA (2015) Passive cooling through natural ventilation techniques in green buildings: inspirations from the past. Civ Eng Environ Technol 2(2):169–173
11. Schiano-Phan R (2004) The development of passive downdraught evaporative cooling systems using porous ceramic evaporators and their application in residential buildings (Doctoral dissertation, Architectural Association)
12. Sharifi A, Yamagata Y (2015) Roof ponds as passive heating and cooling systems: a systematic review. Appl Energy 160:336–357
13. Surekha B, Jagadish K, Hegde M (2018) Operational energy in buildings. Int J Sustain Build Technol 1:8–18

The Urban Design Climate Workshop: Integrating Mitigation and Adaptation Strategies to Confront Urban Climate Change

Jeffrey Raven, Mattia Federico Leone, and Giovanni Nocerino

Abstract 21st century urban design must configure densely occupied urban settlements that offset undesirable local outcomes through climate-sensitive urban design. To bridge between climate science and climate action, policymakers/stakeholders need tools and methods to identify, configure and evaluate evidence-based Urban Climate Factors at urban and local scales. The goal is to configure densely occupied urban settlements that offset undesirable local outcomes through climate-sensitive strategies. The challenge is configuring these districts to reduce the impact of increased urban heat island (UHI) and flooding due to the changing climate while enhancing an equitable, high-quality, low-carbon lifestyle. Enabling transformative climate action in cities requires expanding on the traditional influence and capabilities of urban planning and urban design. In response, we have developed a climate-driven planning process called the *Urban Design Climate Workshop* whose urban climate factors overlap with stakeholder-driven performance indicators.

Keywords Urban design · Climate · Adaptation · Mitigation · Urban design climate workshop · Architecture · Morphology

1 Introduction

Drawing on evidence-based Urban Climate Factors, the *Urban Design Climate Workshop* process was outlined in the Urban Planning and Urban Design Chapter of the *Second Assessment Report on Climate Change and Cities* [16] and refined in the *Third Assessment Report on Climate Change and Cities* [18]. Building on years of UDCW experience, the *Architecture, Urban Design and Planning for Climate Action in the Third Assessment Report on Climate Change and Cities* (ARC3.3) evaluates

J. Raven (✉)
School of Architecture and Design, New York Institute of Technology, New York City, USA
e-mail: jraven@nyit.edu

M. F. Leone · G. Nocerino
Department of Architecture, Università di Napoli Federico II, Napoli, Italy

B. He et al. (eds.), *Urban Climate and Urban Design*, Urban Sustainability,
https://doi.org/10.1007/978-981-96-1521-6_10

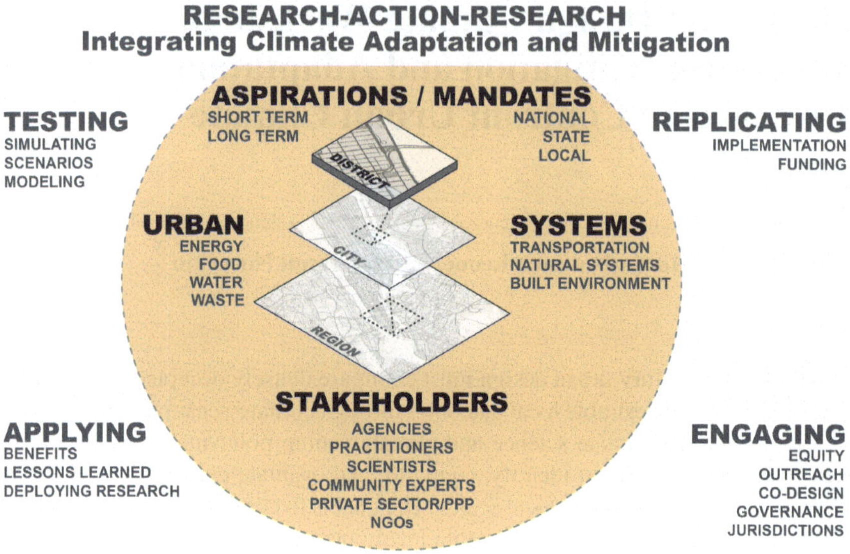

Fig. 1 Framework for climate-resilient urban design (*Source* Raven et al. [18])

research, field testing, and validation of urban systems, built environment models, and frameworks integrating climate mitigation and adaptation (Fig. 1). It includes environmental justice, synergies between research and practice, innovative tools, capacity building, and a clear roadmap for climate action. It also considers cross-sectoral impact, planning processes, global case studies, and persistent research gaps, suggesting opportunities for improvement in the UDCW process. With UDCW sites in diverse global cities, the planning process strives to balance ambitious city visions with ensuring marginalized communities' access to services. UDCWs have been conducted in cities such as New York, Paris, Naples, Rio de Janeiro and Durban, South Africa, since 2015.

2 UDCW Process and Phasing

The UDCW methodology is structured around sequential and iterative phases [13–16, 18] that guide the project's development through a multi-disciplinary and multi-scale approach. These phases are implemented with the assistance of UCCRN multidisciplinary experts, community experts, and other stakeholders. The process integrates knowledge-sharing and co-design actions with urban decision-makers and local communities, alongside the use of computational design tools to simulate and control key indicators influencing building and open space performance under climatic stress conditions.

The UDCW process aims to explore potential "Climate Synergies" aligned with stakeholder priorities, urban system specificities, and planning/design opportunities related to the study area and the targeted program. The four phases of the UDCW process are illustrated in the following sub-sections, not strictly following a chronological order but being interchangeable based on specific needs [18].

3 Urban Climate Factors

Urban Climate Factors [16], a crucial tool in the UDCW process, have been updated in the *Third Assessment Report on Climate Change and Cities* [18] based on a series of UDCW activities (2018–2023). They serve as a foundational tool throughout UDCW phases for planning and urban design. Mitigation and adaptation solutions form part of a knowledge-sharing process explicitly linking stakeholder priorities to climate benefits. This framework facilitates the mapping of co-benefits for climate action within the local context.

The four Urban Climate Factors can be summarized as follows (Fig. 2).

Efficiency of Urban Systems: More efficient urban systems reduce anthropogenic (human-caused) emissions from urban systems including energy, food, transportation and waste. Reducing the impact of polluting vehicles, manufacturing, construction, and waste heat from buildings are critical priorities.

Form and Layout: Urban districts should be designed to provide cooling and ventilation that reduces the need for energy use and allows citizens to cope with higher temperatures, while enabling cities to better manage rainfall extremes. Design that

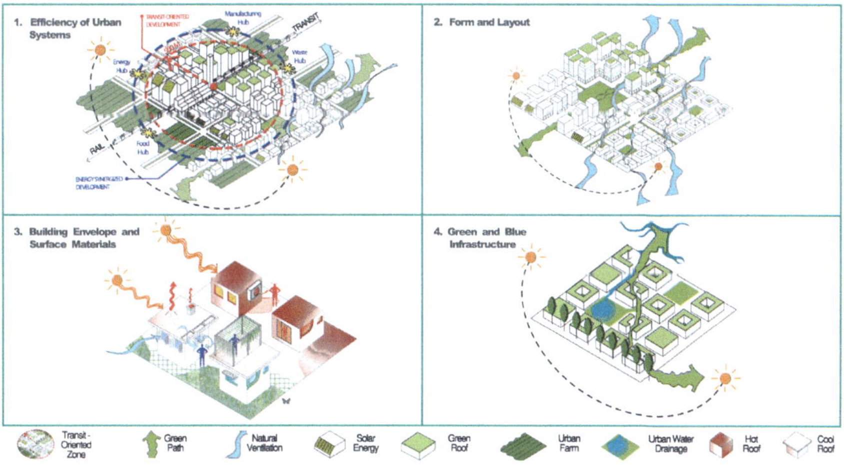

Fig. 2 Urban climate factors (*Source* Raven et al. [18]. Adapted from Raven et al. [16])

integrates climate considerations, natural systems and compact urban form can result in attractive and healthy microclimates.

Building Envelope and Surface Materials: Selecting low heat capacity construction materials and reflective coatings can improve building performance by managing heat exchange at the surface. Air conditioning-reliant buildings are often isolated from their neighbourhood microclimate. One approach would be to define a wider range of acceptable indoor temperatures, by allowing buildings to be better connected to healthier, outdoor microclimates.

Green and Blue Infrastructure: Increasing vegetative cover can simultaneously lower outdoor temperatures, building cooling demand, rain and floodwater runoff, and pollution, while sequestering carbon. Investments in pedestrian and cycling corridors, particularly when integrated with parks, water bodies and other natural systems, can reduce carbon emissions, enhance carbon sequestration, capture stormwater and, perhaps most effectively, cool cities through evapotranspiration and shading [18].

4 UDCW Process

4.1 Stakeholder Priorities

UDCW collaborative and participatory activities aim to assess the needs and expectations of stakeholders and local communities in a co-production process that fosters ownership and inclusivity. The engagement methods employed include collaborative mapping and co-design sessions. UDCW phases include:

Workshop Preparation and Introductory Session. Organizers introduce the topic of climate mitigation and adaptation in broad terms, illustrated by a large board with four Urban Climate Factors. Stakeholders present relevant urban plans and projects and provide documentation.

Priorities mapping session. Stakeholder breakout session, where tables led by facilitators discuss and generate lists of priorities; ***not*** specifically related to climate.

Study Area Borders and Site Visit session. The study focus area is established. A site visit, led by local hosts, provides direct information and confronts topics from the introductory session.

Climate Session. UDCW organizers describe factors driving local climate, from heat to flooding to GHG emissions linking those to the specificity of Urban Climate Factors in the study area.

Evaluating Win–Win Synergies. Stakeholders affix their Priorities to the large Urban Climate Factors board and discuss the relationship between each Priority and Urban Climate Factors.

Collaborative Mapping. The synergies between Community Priorities and Urban Climate Factors are mapped in consideration of key social, economic, and environmental drivers.

4.2 Urban Systems

Policy mandates and stakeholder priorities (see Fig. 1) drive the interaction between urban systems and climate [17]. This shapes the mitigation and adaptation design process. The urban systems encompass energy, food, water, waste, transportation, natural systems, and the built environment. The climate profile of a study area includes factors such as heat, flooding, and greenhouse gas emissions. Development scenario modelling illustrates urban climate, microclimate, and energy profiles. The outputs present potential mitigation and adaptation outcomes based on local climate projections [13–16, 18].

4.2.1 Indicators, Metrics and Simulation Tools

Off-the-shelf mapping tools are utilized to generate and illustrate urban systems and climate outcomes, including heat hazards, flood-prone zones, and greenhouse gas (GHG) emissions. Three-dimensional computational design tools simulate climate and energy drivers at various scales, from district to sub-district to building. Analytical parametric design tools are integrated into the UDCW simulation toolkit, providing user-friendly options for planners and designers to analyze climate change projections guided by Urban Climate Factors (Fig. 3). Geographic Information Systems (GIS) tools perform two-dimensional analyses, while Algorithm-Aided Design (AAD) tools perform three-dimensional modeling.

Drawing from the UDCW taxonomy for input data requirements from district to sub-district scales, development scenarios are evaluated through the lens of integrated climate adaptation and mitigation outcomes. These quantitative climate indicators are augmented with quantitative and qualitative assessment for social, economic, and environmental co-benefits. A three-dimensional modeling configurator tool supports a co-design process between practitioners and local authorities, providing metrics from design alternatives at the concept phase of project development [13].

4.2.2 Geographic Information Systems (GIS)

Geographic Information Systems (GIS) seamlessly integrate varied information (including those derived from remote sensing sources) into a single analytical geospatial model [9]. Innovative GIS-based technologies have the potential to assist urban planners address and even anticipate the complexity of contemporary urban growth [9] as they allow urban planning processes to be data driven more effectively

Urban Climate Factors	Indicators	Metrics
Efficiency of Urban Systems 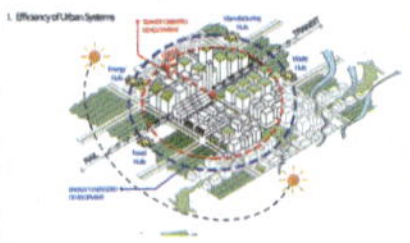	Stationary energy Waste Food Transport Manufacturing Waste heat Circular economy indicators GHG emissions per capita	Average building energy consumption [kWh/m2y] Total energy consumption [kWh/m2y] Renewable energy production potential [kWh/y] GHG Emissions [tCO2e/y] Hospitalization costs [€] Mortality rate increase [%] Mean Radiant Temperature (TMRT) [°C] Universal Thermal Climate Index (UTCI) [°C] Apparent Temperature [°C] Vehicle Kilometers Traveled (MKT) Municipal solid waste generation per capita
Form and Layout 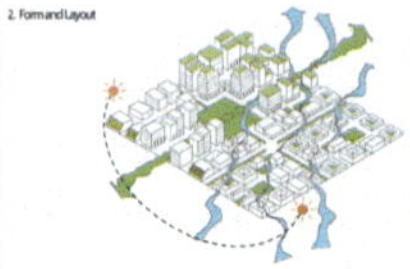	Massing diagrams Wind-Sun impacts Sky view factor Land surface temperature Thermal comfort analysis Heat wave hazards Heat wave impacts Energy profile	Mean Radiant Temperature (TMRT) [°C] Universal Thermal Climate Index (UTCI) [°C] Apparent Temperature [°C] Land Surface Temperature [°C] Energy Use Intensity [kWh/m2y] Renewable energy production potential [kWh/y]
Building Envelope and Surface Materials 	Radiation analysis Building envelope rating Thermal comfort analysis Heat wave hazard Heat wave Impact Energy profile	District energy consumption [kWh/m2y] Building energy consumption [kWh/m2y] Renewable energy production potential [kWh/y] Carbon Footprint of Building Materials [tCO2e/ y] Indoor Predicted Mean Vote [PMV] Mean Radiant Temperature (TMRT) [°C] Universal Thermal Climate Index (UTCI) [°C] Apparent Temperature [°C] Run-off [-]
Green and Blue Infrastructure 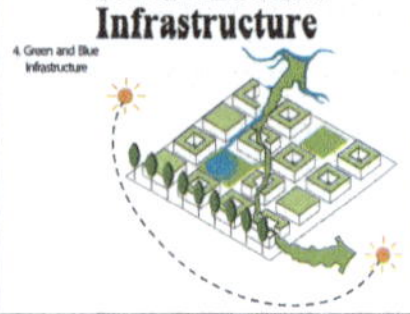	Tree canopy cover Biodiversity Heat wave hazard Heat wave Impact Flood Hazard Flood Impact	Tree canopy cover % City Biodiversity Index (CBI) Carbon storage potential from vegetation [tCO2e/y] Run-off [-] Flood probability index [-] Building damage (structure/content) [€/$] Road infrastructure damage (cleaning/repairing) [€/$]

Fig. 3 Urban climate factors, indicators and metrics (*Source* Raven et al. [18])

than conventional techniques. The UDCW GIS tools include a set of algorithms programmed using PostgreSQL and Python, and designed to support evidence-based planning in relation to urban scale issues (Fig. 4). To represent and analyze urban areas, the algorithms are based on a 2.5D model [1], which combines the two-dimensional features of land cover with selected three-dimensional information, from elevation and shading to sky-view factor. The tool can analyze current and projected urban climate conditions and design scenarios. At the current state of development, the models allow the assessment of the health and economic impact of heat waves and floods (pluvial and coastal), the effect of temperature variations on energy consumption and production, and the CO_2 absorption potential from vegetation cover. The tools are based on validated models, such as ENVI-MET [20] for land surface temperature associated with different cover materials, and SOLWEIG [5, 6] for the calculation of Mean Radiant Temperature. Impact assessments incorporate models such as Gasparrini et al. [3] for estimated mortality rate increase due

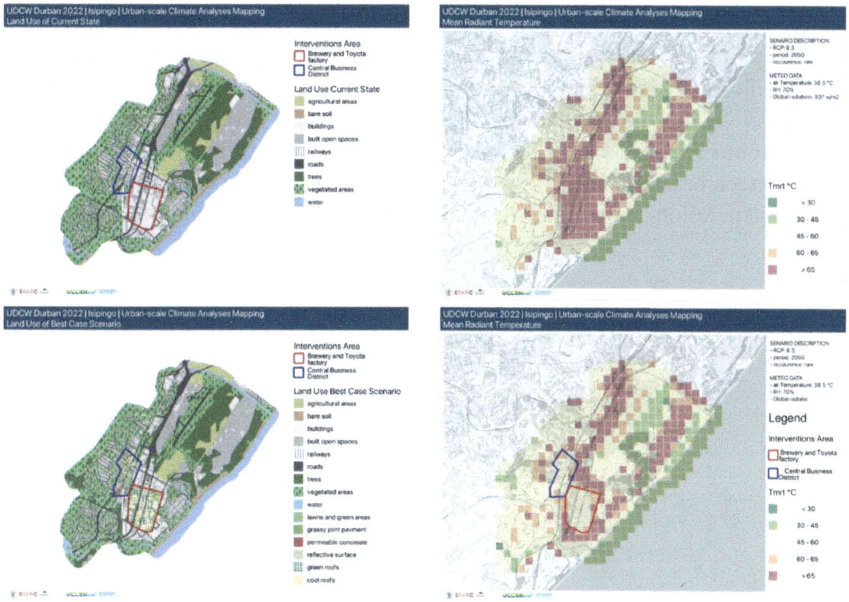

Fig. 4 GIS tool applied in Durban. Tmrt analysis of current state versus 2050 design scenario (*Source* Raven et al. [18])

to heat waves, and Huizinga et al. [4] for estimated economic damage from floods on buildings and transport infrastructure.

4.2.3 3D Modelling

Algorithm Aided Design (AAD) tools are increasingly used in architecture and urban design practice for their ability of integrating simulation components within conventional 3D modelling software (see Figs. 5 and 6). These tools run complex calculation models, manage data, and assess impacts from planning and design decisions in real time during project development. While these tools are flexible, the accuracy of the results is directly proportional to the amount of information that the designer chooses to integrate in the workflow, according to an "easy to start, hard to master" process logic. In particular, the modular nature of algorithmic design tools allows the tools to be enriched with ever-different elements and data [20], which, in the case of simulations configure results that can be more or less accurate according to requirements. This computational methodology is therefore efficient in different phases of the design process, from the conceptual to the final design. The UDCW 3D modelling tools offer comprehensive, quantitative, and spatial analysis of the climate-resilient potential of planning and design scenarios at the neighbourhood scale. It assesses adaptation benefits related to heat wave and flood events and measures the energy and carbon intensity of different design solutions. The process requires detailed information about building envelope and HVAC solutions (both for existing buildings and targeted designs for new or retrofitted buildings), as well as technical specification for outdoor paving materials and vegetation types. The modelling builds upon validated tools such as Ladybug [7, 8] and i-Tree [12]. It incorporates novel custom Python components to assess additional environmental performances based on green building rating systems such as LEED [10] and use EleFront [21] for metadata generation and management, and Urbano [2] for GIS interoperability.

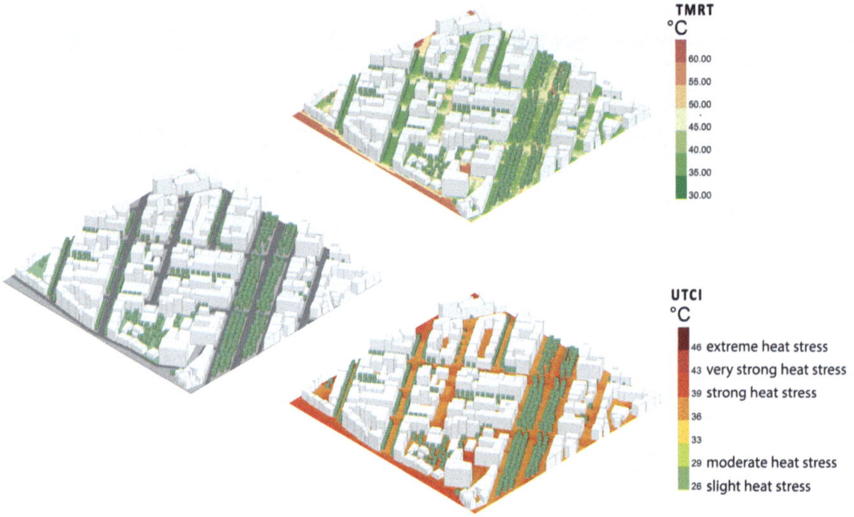

Fig. 5 3D modelling AAD tools in Lisbon. TMRT and UTCI analysis of current state scenario (*Source* Nocerino and Leone [11])

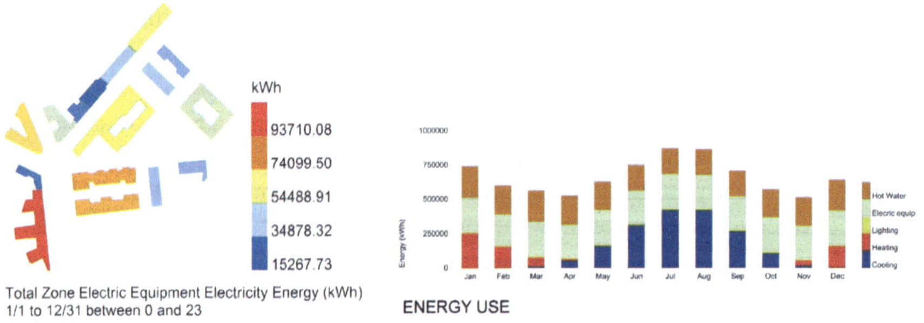

Fig. 6 Applied 3D modelling AAD tools to calculate energy consumption at neighbourhood scale (*Source* Nocerino and Leone [11])

4.2.4 Neighbourhood Configurator

Through the management of simple 3D objects and several parameters, the Configurator tool aims to create a simple and intuitive modelling environment. It generates urban scenarios and verifies their performance in relation to qualitative and quantitative indicators. This tool is intended to simplify the comprehension of climate modelling and consequently climate resilient planning and design. It assesses key performance indicators for climate-resilience from the conceptual stage of design, including building massing.

The Configurator tool consists of a series of components programmed in Grasshopper (the Visual Programming Language integrated in Rhinoceros). It starts from a certain set of parameters, to generate simplified and "blocky" 3D elements, easily handled in the Rhino environment by non-expert users. Once all the elements have been generated, it is possible for the user to combine them in the Rhinoceros working environment to generate a simplified and schematic model of an open space, monitoring all the essential information in real time. At the current state of development, the Configurator can generate five typologies of elements: terrains, buildings, trees, roads and canopies. The configurator uses the input attributes associated with each object inside the 3D model to generate the outputs of the UCDW 3D modelling tools (see Fig. 3). The tool can be used in the beginning to support the design process.

The 3D models generated using the configurator can be used as input for the other 3D modelling tools in order to obtain 3D visualisation of current city conditions and proposed design concepts, with assessment of key indicators related to climate-benefits (mitigation and adaptation) and co-benefits (social, economic and environmental). The algorithm consists of five components: Ground Generator, Building Generator, Tree Generator, Road Generator and Canopy Generator (Fig. 7). For each of these components the algorithm requires a set of inputs to generate the elements to compose the urban space. These components are still under development, with plans to test the tool with further modules.

The current version of the UDCW approach and toolkit has undergone development, testing, and improvement within several recent research projects in Europe and the US, including H2020 "CLARITY," H2020 "ESPREssO," Erasmus+ "UCCRN_edu," Horizon Europe "KNOWING," Horizon Europe "UP2030," and the National Science Foundation—Belmont Forum RCN "City-as-Lab."

4.3 Co-design

Urban transformation pathways are shaped by demographic projections, socio-economic considerations, and driven by stakeholder aspirations or policy mandates [19]. The Co-Design process aims to illustrate how urban design, planning, and development address the nexus between climate hazards, urban systems, and low-carbon development [13]. UDCW participants draw from research and cross-sectoral

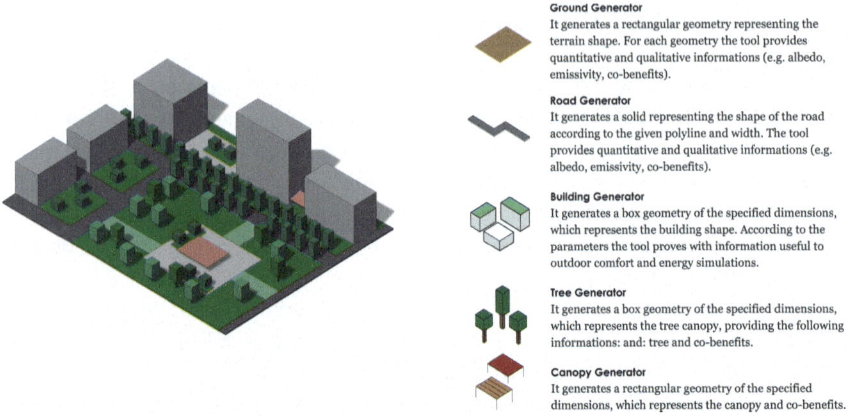

Fig. 7 Output generated by the neighborhood configurator (*Source* Nocerino and Leone [11])

stakeholders to engage cities whose vulnerable neighborhoods face comprehensive rezoning or development pressures (Fig. 8).

1. **Urban Systems' Users**. The population scenarios consider future "Business as Usual" scenarios for the UDCW project area. This hypothetical scenario is based on proposed development, projects in the development pipeline, rezoning plans and "market driven" full build-out assumptions. The Urban Systems Inventory and Footprint considers Current and Future scenarios, based upon user groups and population projections. The interconnected Synergies-Adjacencies columns illustrate spatial synergies and flows of people, goods, services.
2. **Scales**. The same synergies and adjacencies analysis confronts aspirations and policy mandates across multiple scales: Building, district, city, region, national and even international. These synergies are identified between urban systems connected across these spatial scales.
3. **Programming Design Typologies, Simulation and Modeling**. Emerging site planning and building typologies and metrics responding to climate aspirations. These include mixed use, residential, commercial, institutional, manufacturing, infrastructure facilities, energy hubs, food hubs, waste hubs, water treatment, intermodal hubs.
4. **Urban Climate Factors**. Study Area evaluation according to urban form and urban function principles for configuring climate resilient neighborhoods. These include Efficiency of Urban Systems, Form and Layout, Building Envelope/ Surface Materials and Green–Blue Infrastructure.
5. **Site Opportunities and Constraints**. Analyzing and configuring interconnected microclimates and urban systems within the study area to achieve reduced energy loads, cleaner air and enhanced civic life. Considerations range from climate analysis to community culture, future activity nodes and goods/services distribution hubs.

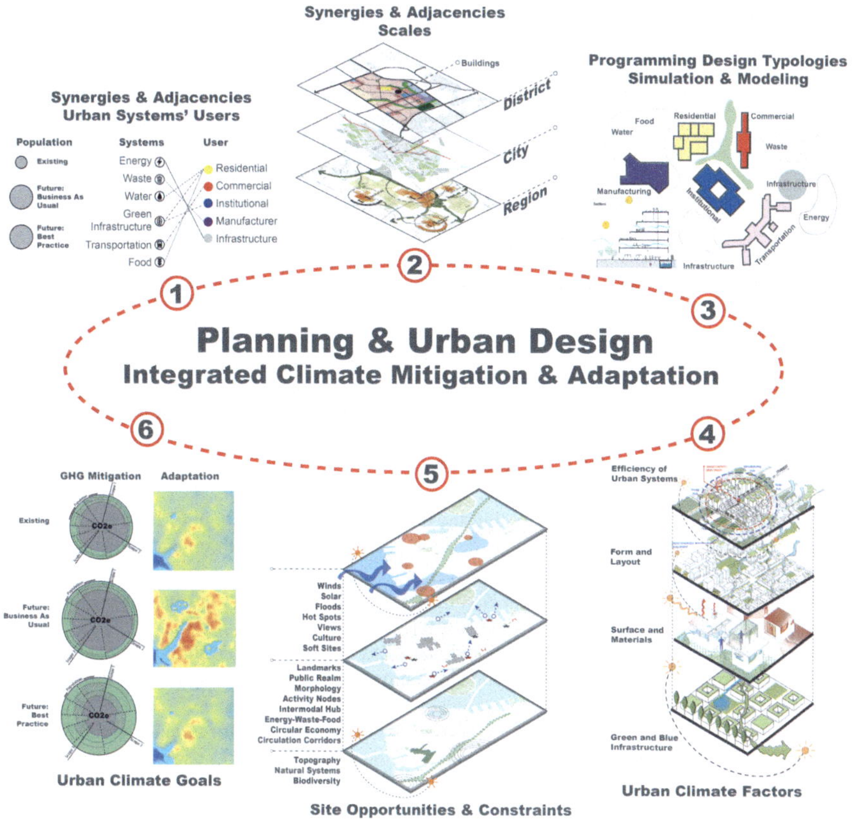

Fig. 8 Synergies and Adjacencies to explore within the UDCW co-design process (*Source* Raven et al. [18])

6. **Urban Climate Goals**. Current and future scenarios for integrated climate mitigation and adaptation. Considerations include GHG emission, heat and flooding outcomes. In addition to existing conditions, scenarios also included are future baseline and best practice projections.

5 Conclusion

Strengthening the classification of case study prototypes from which to draw lessons learned and identify research gaps is a priority. These feedback loops can form the basis for iterative improvements moving forward.

References

1. Costamagna E (2014) Geographic information science (GIS) 3D. In: Encyclopedia of quality of life and well-being research. Springer, Netherlands, pp 2512–2521
2. Dogan T, Yang Y, Samaranayake S, Saraf N (2020) Urbano: a tool to promote active mobility modeling and amenity analysis in urban design. Technol Archit Des 4(1):92–105
3. Gasparrini A, Armstrong B, Kenward MG (2010) Distributed lag non-linear models. Stat Med 29(21):2224–2234
4. Huizinga J, De Moel H, Szewczyk W (2017) Global flood depth-damage functions: methodology and the database with guidelines. Publications Office of the European Union, Luxembourg. https://doi.org/10.2760/16510. JRC105688. ISBN: 978-92-79-67781-6
5. Lindberg F, Grimmond CSB (2011) The influence of vegetation and building morphology on shadow patterns and mean radiant temperatures in urban areas: model development and evaluation. Theoret Appl Climatol 105(3–4):311–323
6. Lindberg F, Onomura S, Grimmond CSB (2016) Influence of ground surface characteristics on the mean radiant temperature in urban areas. Int J Biometeorol 60(9):1439–1452
7. Mackey C, Roudsari MS (2018) The tool(s) versus the toolkit. Humanizing digital reality. Springer, Singapore, pp 93–101
8. Mackey C (2019) The tool(s) vs toolkit: the hierarchical kit. In: Naboni E, Havinga LC (eds) Regenerative design in digital practice: a handbook for the built environment. Eurac, Bolzano, IT, p 11
9. Maliene V, Grigonis V, Palevičius V, Griffiths S (2011) Geographic information system: old principles with new capabilities. Urban Des Int 16:1–6
10. Nocerino G, Leone MF (2023) Computational LEED: computational thinking strategies and visual programming languages to support environmental design and LEED credits achievement. Energy Build 278:112626
11. Nocerino G, Leone MF (2024) WorkerBEE: a 3D modelling tool for climate resilient urban development. In: International symposium: new metropolitan perspectives. Springer, Cham, pp 16–26
12. Nowak DJ (2021) Understanding i-Tree: 2021 summary of programs and methods. US Department of Agriculture, Forest Service, Northern Research Station
13. Raven J, Braneon C, Rosenzweig C (2021) Embedding climate change in urban planning and urban design in New York City. In: Ren C, McGregor G (eds) Urban climate science for planning healthy cities. Biometeorology, vol 5. Springer, Cham. https://doi.org/10.1007/978-3-030-87598-5_3
14. Raven J (2021) Urban design climate workshops. In: Dash J (ed) Climate and policy, encyclopedia of climate change, vol 3. World Scientific Publishing Co. Inc. ISBN: 9789811209321
15. Raven J, Braneon C, Esposito M (2020) Urban design climate workshop: Gowanus, Brooklyn. Urban Land Institute, Washington, DC. https://knowledge.uli.org/urban-design-climate-workshop-gowanus
16. Raven J, Stone B, Mills G, Towers J, Katzschner L, Leone M, Gaborit P, Georgescu M, Hariri M (2018) Urban planning and urban design. In: Rosenzweig C, Solecki W, Romero-Lankao P, Mehrotra S, Dhakal S, Ali Ibrahim S (eds) Climate change and cities: second assessment report of the urban climate change research network. Cambridge University Press
17. Raven J (2011) Cooling the public realm: climate-resilient urban design resilient cities, vol 1. Springer, pp 451–463. In: Otto-Zimmermann K (ed) Cities and adaptation to climate change - proceedings of the global forum 2010, series: local sustainability, vol 1. XLIV, 576 pp
18. Raven J, Leone MF et al (2025). Architecture, Urban Design and Planning for Climate Action. In: Rosenzweig C, Solecki W et al (eds) Climate Change and Cities: Third assessment report of the urban climate change research network. Cambridge University Press
19. Reckien D, Lwasa S, Satterthwaite D, McEvoy D, Creutzig F, Montgomery M, Schensul D, Balk D, Khan I (2018) Equity, environmental justice, and urban climate change. In: Rosenzweig C, Solecki W, Romero-Lankao P, Mehrotra S, Dhakal S, AliIbrahim S (eds) Climate change

and cities: second assessment report of the urban climate change research network. Cambridge University Press, New York, pp 173–224
20. Simon H, Bruse M (2020) Modelling of microclimates. Eco-design of buildings and infrastructure: developments in the period 2016–2020, 4, 318
21. Van Der Heijden R, Levelle E, Reise M (2015) Parametric building information generation for design and construction. In: Computational ecologies: design in the anthropocene-35th annual conference of the association for computer aided design in architecture, pp 417–429

Understanding Urban Heat Islands in Tunisian City Centers: An Integrated Approach

Messaoudi Abir

Abstract A comprehensive study of urban climate is necessary to better understand the presence of heat islands in each part of the city centers of Sousse, Tunis, and Sfax. Temperature has a direct impact on energy consumption and the well-being of residents. However, despite the rapid and continuous urbanization of cities, few studies have specifically focused on the effects of urban heat islands (UHI). In this study, we aim to analyze the urban heat island effect in the city centers of Sousse, Tunis, and Sfax by examining the relationships between land surface temperature (LST) and spatial characteristics, composition, and configuration of impervious and natural spaces. We investigate how intensive urbanization, evapotranspiration (IDWI), urban morphology, and the radiativity of building materials influence the formation of urban heat islands in these specific contexts. To conduct this in-depth analysis, we will utilize geospatial data and various approaches, including statistics, urban landscape metrics, and spatial analysis techniques. These methods will help us understand how the characteristics of the urban landscape and urban development choices contribute to the formation of urban heat islands. Several indices will be considered, including the Normalized Difference Vegetation Index (NDVI), the Normalized Difference Built-up Index (NDBI), and land surface temperature (LST). Correlation analyses will enable us to quantify the strength and direction of the relationships between different indices and urban climate parameters. For example, we will assess whether there is a positive correlation between the density of impervious spaces (measured by the NDBI) and land surface temperature. Similarly, we will examine the correlation between the NDVI and land surface temperature to evaluate the mitigating role of green spaces in reducing urban heat islands. By employing these statistical methods, we will obtain precise and quantitative results regarding the relationships between indices and urban climate parameters. These findings will help us understand the mechanisms of urban heat islands in the city centers of Sousse, Tunis, and Sfax. Key questions to be addressed include the extent and intensity of urban heat islands in these city centers, the correlation between impervious space density and land surface temperature, and the role of green spaces in mitigating

M. Abir (✉)
Research Laboratory Governance, Ambiance and Development of Cities (GADEV Lab), ENAU, University of Carthage, Tunis, Tunisia
e-mail: abir.messaoudi@isteub.u-carthage.tn

B. He et al. (eds.), *Urban Climate and Urban Design*, Urban Sustainability,
https://doi.org/10.1007/978-981-96-1521-6_11

urban heat islands. This information will be essential for guiding urban planning and development decisions, aiming to create more resilient and sustainable urban environments. We will be able to identify the most vulnerable areas to urban heat islands and propose targeted urban design and greening strategies to mitigate these adverse effects. By understanding the complex mechanisms of urban heat islands in Tunisian city centers, we can create more pleasant, energy-efficient, and climate-resilient urban environments. This will contribute to improving the well-being of residents and creating sustainable cities for future generations.

Keywords Urban heat island · City centers · Land surface temperature · Geospatial data · Statistics · Urban landscape metrics · Spatial analysis

1 Introduction

At the urban scale, a diversity of factors, such as the layout of streets, their orientation relative to wind and sun, the presence of vegetation, construction materials and human activities, contribute to creating a mosaic of microclimates complex and distinct. Despite this diversity, conventional weather records fail to fully capture real-world conditions in densely populated neighborhoods, highlighting the urgency of understanding Urban Climate Impact (UCI) at a local scale. The urban heat island (ICU) is the result of a complex interaction between different parameters which vary according to the urban configurations, climate and land uses specific to each city. Within the same urban area, the intensity of UHI can vary considerably. Urban green spaces, such as parks, provide cooler areas, while industrial areas, often devoid of vegetation and generating heat, have higher temperatures [5]. Some mechanisms contribute directly to heating, while others absorb heat through artificial structures [11]. This research aims to deepen the understanding of land surface temperature (LST) in three important Tunisian cities: Tunis, Sousse and Sfax. To do this, data from Landsat 8 satellites were used in conjunction with various geospatial methods. The choice of these three cities was strategic. First of all, Tunis, as the capital, represents the political, economic and cultural center of the country. Sousse, a major tourist destination, is renowned for its urban diversity, ranging from coastal areas to historic districts. Finally, Sfax, as an industrial economic hub, offers a significant example of urban development focused on industry and commerce. The research methodology for this study involves the use of three different software programs to analyze geospatial data: QGIS 3.26, ENVI 5.3 and Fragstats. Each of these software plays a specific role in processing and analyzing data to obtain valuable information on spatial patterns and configuration of urban surfaces.

2 The Emissivity of the Earth's Surface

The emissivity of the Earth's surface, defined as the capacity of a surface to emit thermal radiation relative to an ideal black body, is of crucial importance for understanding the thermal dynamics of our planet. This physical characteristic is essential in various scientific and environmental fields, encompassing the study of climate to the monitoring of ecosystems. Researchers have looked at emissivity to understand surface temperature variations and their specific implications in urban areas. For example, [12] as well as [15] put forward in their previous studies the hypothesis that the composition of the land surface can be estimated using NDVI. They suggested that the area considered bare soil is associated with an NDVI less than 0.2, while vegetation is characterized by an NDVI greater than 0.5. The study by Smith et al. [14] focused on the impact of emissivity on the accuracy of surface temperature measurements in urban contexts, while [3] investigated the influence of emissivity on estimates of heat emissions from different types of soil. The use of satellite data from sensors on board satellites such as Landsat or Sentinel has facilitated the assessment of emissivity over vast areas of land. These data are of crucial importance for establishing precise thermal models, supporting the sustainable management of natural and agricultural resources, as well as for understanding environmental dynamics more precisely. Emissivity assessment often represents a complex process, often requiring field validation to ensure the reliability of the data obtained. However, its essential role in key areas such as climate modeling, environmental monitoring and strategic decisions in land management remains indisputable. Emissivity was estimated from satellite data collected by the Landsat8 satellite. This estimation was carried out using specialized software such as ENVI or QGIS. The maximum and minimum emissivity values thus obtained are recorded in Table 1.

a. **Plant surface**: Green spaces in urban areas play a crucial role in creating more pleasant environments, promoting evapotranspiration and providing areas of freshness. Their development and preservation are key elements of urban development plans (UAP) for a sustainable city [16, 17]. The analysis of these green spaces is facilitated by remote sensing, which makes it possible to interpret vegetation indices from available satellite data. The NDVI (Normalized Difference Vegetation Index) is a remote sensing technique developed in the 1970s. In 1973, the scientist Rouse proposed the concept of using light reflectance in the visible and near-infrared (NIR) wavelengths to assess vegetation health and quantity [13]. The Normalized Vegetation Index (NDVI), calculated from the red (R) and near infrared (NIR) spectral bands, highlights differences in plant density.

Table 1 Emissivities in Tunisian urban centers

Place	Date	Minimum	Maximum
Downtown Tunis	04-05-2023	0.986	0.988351
Sousse city center	27-10-2023	0.986	1.02017
Downtown Sfax	09-05-2023	0.986	0.989307

The formula NDVI = (PIR−R)/(PIR + R) allows these variations to be evaluated. In the context of downtown Tunis, the NDVI varies between −0.0639535 and 0.523876, with negative values near Lake Tunis, where the red reflectance exceeds that of the near infrared. Intervals from 0.03517 to 0.1043 correspond to unvegetated areas, such as built-up areas or bare soil. As for positive values, from 0.1043 to 0.5238, they identify densely vegetated areas. For the city of Sousse, the NDVI varies between −0.251701 and 0.513616, with specific ranges from −0.093409 to 0.177455 for areas near the seafront and constructions. Regarding Sfax, the NDVI ranges from −0.327179 to 0.499616, with ranges from 0.00029 to 0.10279 for areas adjacent to the sea and built-up spaces (see Fig. 1).

This precise analysis makes it possible to distinguish the characteristics of vegetated areas from built-up areas, providing key information for urban planning, green space management and the preservation of the urban environment in these cities.

b. **Water surface**: The presence of water in urban areas has a significant impact on the local climate, influencing temperature and air quality [4]. Two main phenomena result: the evaporation of water and the cooling of the air on contact with the latter. To map water resources in the study region, we used a specific index: the Normalized Difference Water Index (NDWI), which is based on near-infrared (NIR) and infrared wavelengths short (SWIR) according to the Gao method dating from 1996. The NDWI (Normalized Difference Water Index) is a remote sensing technique developed specifically to identify and monitor water bodies or wet areas from satellite images.

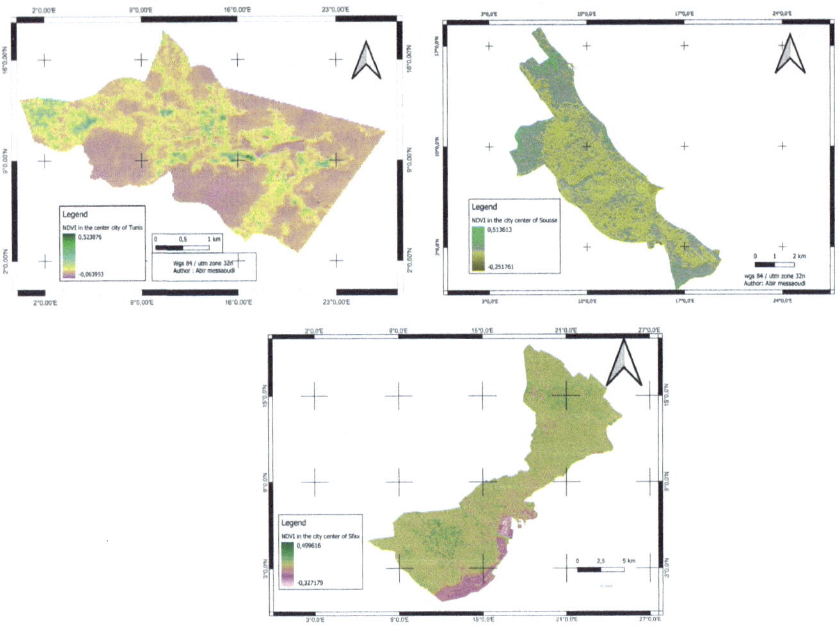

Fig. 1 Standardized vegetation indices in the three Tunisian urban centers

This method relies on the analysis of spectral responses of surfaces observed in the images, thus enabling the distinction between water bodies and non-water areas [10]. The NDWI (Normalized Difference Water Index), a standardized water difference index, is an essential tool for assessing the water presence of different surfaces. Its calculation is based on the comparison of the NIR and SWIR values according to the traditional formula: $NDWI = (\mathrm{PIR} + SWIR)/(\mathrm{PIR} - SWIR)$. The relationship between NDWI and land surface temperature is not direct as in the case of NDBI. However, in certain circumstances, the presence of water can influence variations in ground surface temperature. Water areas generally have a higher heat capacity than surrounding dry areas. As a result, near aquatic areas, land surface temperatures may be slightly lower due to the cooling effect of the water. The Standardized Water Index (NDWI) varies considerably between the cities of Tunis, Sfax and Sousse. These values, ranging from −0.637 to 0.087 for Tunis, from −0.458 to 0.255 for Sfax, and from −0.451 to 0.208 for Sousse, are indicators of the water content in the different areas studied (see Fig. 2).

Negative NDWI values generally indicate surfaces such as dense urban areas or non-reflective materials, which have lower water content. For Tunis, extremely negative values of the index could be associated with highly urbanized areas, paved surfaces or concrete where the presence of water is reduced or absent. However, positive NDWI values suggest the presence of water or reflective surfaces. The higher values of Sfax and Sousse potentially indicate areas with a greater presence of water, such as coastal areas, or more abundant green spaces.

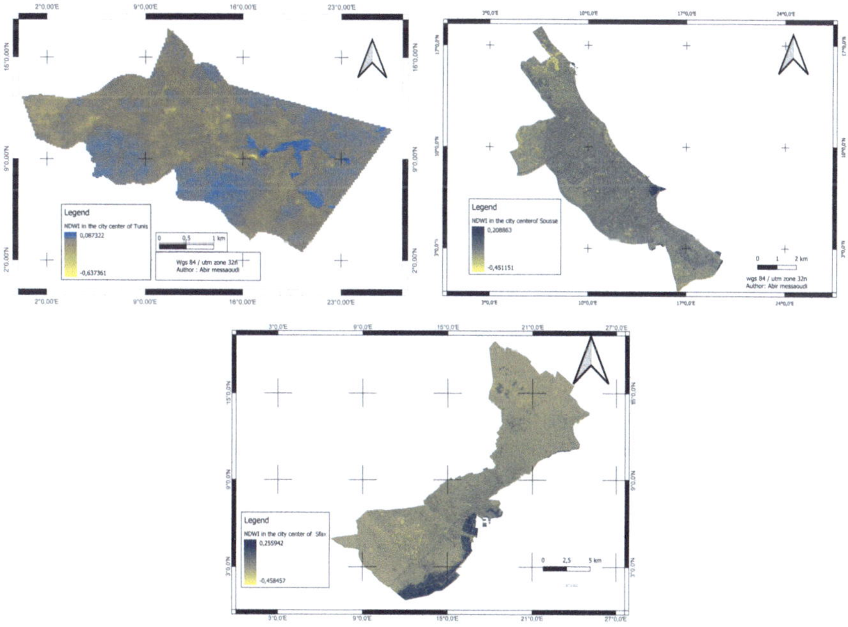

Fig. 2 Standardized water indices in the three Tunisian urban centers

c. **Urban morphology**: Urban morphology is a crucial component of city planning, and to evaluate it, the Normalized Difference Building Index (NDBI) represents an essential tool. This index helps map urbanized areas and urban density within a specific region [18]. The NDBI (Normalized Difference Built-up Index) is a remote sensing technique developed specifically to identify and monitor urban or built-up areas from satellite images. This method relies on the analysis of spectral responses of surfaces observed in the images, thus enabling the distinction between urbanized and non-urbanized areas. Negative values of the NDBI generally indicate the presence of non-built-up surfaces, while positive values indicate the likely presence of built-up surfaces such as roads, buildings, paved roads, etc. [20]. Its calculation formula is defined as follows: $NDBI = (\text{SWIR} + PIR)/(\text{SWIR} - PIR)$, where SWIR represents shortwave infrared and PIR corresponds to near-infrared wavelengths. This measurement makes it possible to detect variations in the density of buildings and urbanized areas. Higher NDBI values are associated with high urban density, highlighting areas where built structures are abundant, while lower values may indicate less urbanized areas or less dense land areas. This approach has been widely used in the literature to assess urban density and urban sprawl in various urban regions. For example, in a study conducted by Wang et al. [19], the Normalized Difference Building Index was applied to analyze urban sprawl and building density in a growing Chinese city. Likewise, in a research work carried out by Li et al. [7], the use of the NDBI made it possible to quantify urban density in metropolitan areas in Europe. The Normalized Difference Building Index (NDBI) is often correlated with the land surface temperature (LST - Land Surface Temperature) in urban studies. This correlation arises from the fact that the density and configuration of urban constructions influence the thermal characteristics of the earth's surface. In general, urban areas with higher building density, as detected by higher NDBI values, tend to absorb more heat due to the presence of building materials like concrete, asphalt, and roofs, which have a greater capacity to store heat. Therefore, these dense urban areas often have higher surface temperatures than less urbanized or less dense areas. The Normalized Difference Building Index (NDBI) reveals significant differences between the urban characteristics of Tunis, Sousse and Sfax. These values, varying from 0.279 to $-$ 0.214 for Tunis, from -0.329 to 0.161 for Sousse, and from -0.329 to 0.1611 for Sfax, provide indications on the extent and density of urbanization in these specific areas (see Fig. 3).

Positive values of the NDBI index indicate a greater concentration of built-up areas. In Tunis, these positive values can mean more urbanized areas or significant construction developments such as amenities and polyfunctional zones. It corresponds to economic, commercial or residential sectors with a high density of buildings and urban infrastructure. On the other hand, negative values of the NDBI index could represent less urbanized areas or unbuilt spaces. For Sousse and part of Sfax, these negative values could indicate less urbanized areas.

d. **Construction materials**: The thermal properties and radiative behavior of construction materials play a crucial role in the thermal regulation of urban environments [6]. Two key parameters influence these properties: albedo and emissivity,

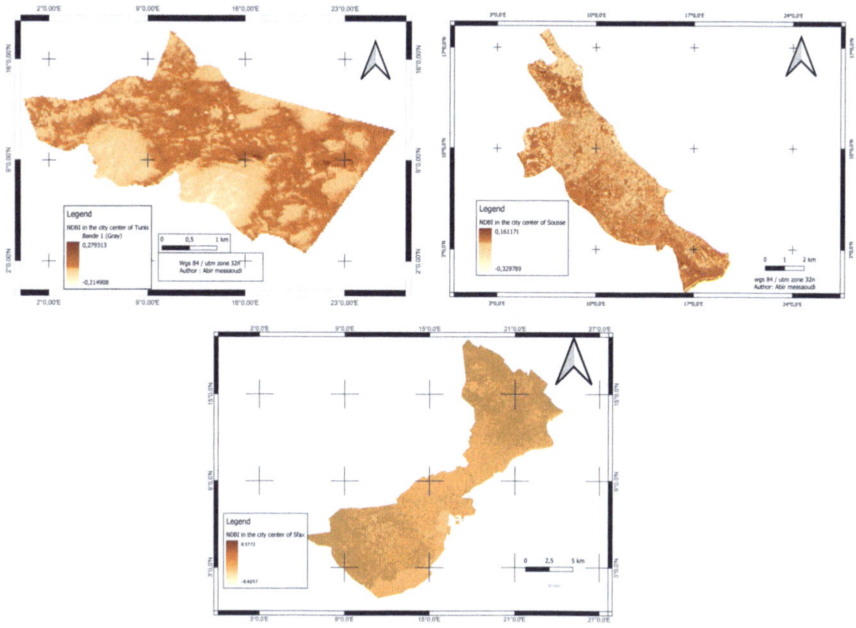

Fig. 3 Standardized construction indices in the three Tunisian urban centers

which are significant indicators of the radiative behavior of materials exposed to solar energy. Albedo, a fundamental parameter indicating the reflectivity of a surface, plays an essential role in regulating the temperature of the Earth's surface. On a scale of 0 to 1, representing the level of reflection of sunlight, a body absorbing all light would have an albedo of zero, while a completely reflective surface would be rated at one. Materials used in urban construction exhibit a variety of albedos, influencing the amount of heat absorbed or reflected. Work by Asaeda and his team [1] examined the thermal conductivity of materials. They found that when the temperature difference between the ground surface and the air is small, the infrared radiation emitted by the ground (L↑) is almost balanced by the descending infrared radiation from the atmosphere (L↓). On the other hand, when this temperature difference is notable, especially for heated surfaces, the radiation emitted by the ground (L↑) becomes significantly greater than the downward radiation (L↓). This difference can contribute to the warming of the lower atmosphere and thus accentuate the phenomenon of urban warming.

For our analysis, a key parameter for assessing surface temperatures is the land temperature index, known as Land Surface Temperature (LST). Calculating this index is complex and involves several steps, including taking into account emissivity and reflectance values. In our study, LST values vary considerably, showing notable differences between maximum and minimum temperatures. In Tunis, in May 2023, we observed temperatures up to 63.6097 °C as the maximum value, while the minimum value was 46.6155 °C. In Sfax, temperatures were even higher, with maximum values reaching 71.095 °C and minimum values of 52.3732 °C. Finally, in Sousse, towards the end of October 2023, temperatures were lower, ranging from 34.84 °C as a maximum to 22.34 °C as a minimum. These temperature differences are closely linked to the nature of the areas studied. Areas with high urban density (for the three cases of medinas in the city center) or with an industrial vocation (for the case of Sfax) present higher temperatures, often attributed to the accumulation of heat in highly urbanized environments. Furthermore, green areas, areas with low urban density and regulated residential areas tend to have lower temperatures (see Fig. 4).

Thus, cities characterized by strong urbanization in urban centrality often have high building density. This contributes to the intensification of urban heat islands. Indeed, the lack of green spaces reduces the city's ability to absorb heat and regulate temperature, while the lack of setbacks between buildings and streets promotes heat retention, creating hot environments and confined. Additionally, anthropogenic activities such as industrial zones, heavy road traffic, and widespread use of air conditioning produce a significant amount of heat. These additional heat emissions amplify the phenomenon of urban heat islands in already densely built and poorly planned areas.

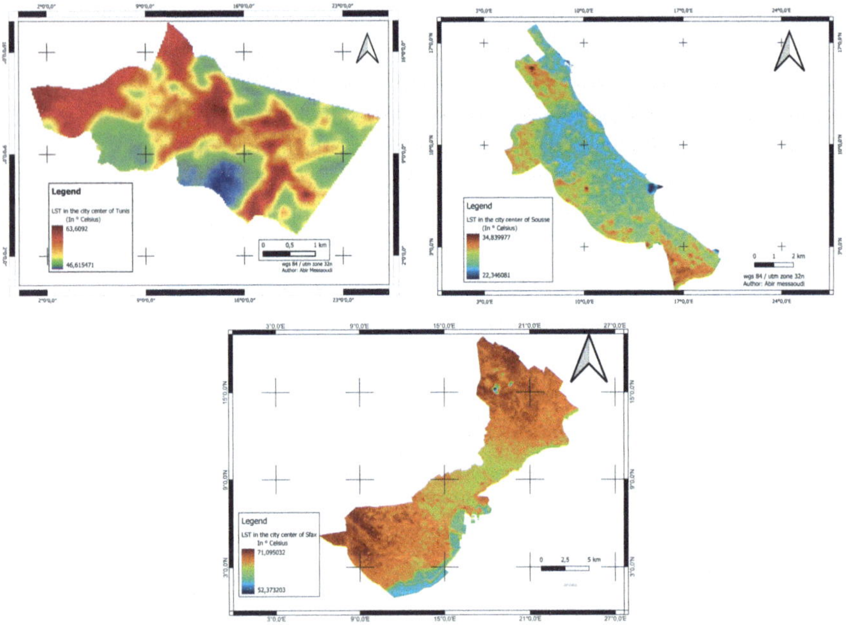

Fig. 4 LST in the three Tunisian urban centers

3 Statistical Analysis

The correlation coefficient (r) of −0.5532 between the NDVI (Normalized Difference Vegetation Index) and the NDBI (Normalized Difference Built -up Index) in Tunis reveals a moderate to moderately strong, but inverse, correlation between these two specific indices (Table 2). This negative correlation suggests that there exists, in certain urban areas of Tunis, an inverse relationship between the density of vegetation and the density of built areas. In other words, where vegetation appears to be less abundant or dense, we find a more pronounced presence of built-up and urbanized areas, and vice versa. This observation may be associated with several factors linked to the evolution of urbanization in the region:

- **Urban expansion**: City expansion at the expense of green spaces can lead to a reduction in vegetation density, thereby promoting the growth of built-up areas.
- **Urban densification**: A greater concentration of buildings, roads, urban infrastructure can be accompanied by a reduction in natural vegetation in certain parts of the city.
- **Territorial planning policies**: Political orientations and urban planning strategies that favor urban development, political decisions regarding zoning, building regulations and land use can have an impact on the distribution between built-up areas and green spaces, sometimes favoring urban growth to the detriment of the preservation of vegetation.

However, the coefficient of determination (R^2) of 0.306 for the NDVI and NDBI data in Tunis indicates that the linear relationship between these two indices explains approximately 30.6% of the variance observed in the density of built areas compared to the density of vegetation. In other words, although this relationship is statistically significant, other factors not included in this model also contribute to the variability observed in the density of built areas in Tunis. Notably, population growth, population density and migration patterns, economic dynamics, employment opportunities and economic development can influence urban growth. The correlation coefficient (r) for Sfax between the values of NDVI and NDBI is approximately −0.695. However, although the coefficient of determination is relatively high (0.483), this means that 48.3% of the variation in building density can be explained by the variation in vegetation density. This leaves 51.7% of the variation unexplained by this relationship, suggesting that other factors may influence building density in Sfax aside from vegetation density. Comparing these values to those of Tunis, we see differences

Table 2 The correlation between NDVI and NDBI in Tunisian urban centers

	NDVI min	NDVI max	NDBI min	NDBI max	R	R^2
Tunis	−0.06	0.52	−0.21	0.27	−0.553	0.306
Sousse	−0.25	0.51	−0.32	0.16	0.362	0.131
Sfax	−0.32	0.49	−0.42	0.57	0.695	0.483

in the strength and nature of the relationship between building density and vegetation density. This difference in the values of R^2 and r between Tunis and Sfax may suggest distinct urban characteristics and development trends between the two cities. A stronger and inverse relationship between building density and vegetation density in Sfax could indicate a different urban planning, where urban expansion or densification occurs at the expense of green spaces or vegetated areas. This may result from faster urbanization or an urban development strategy that neglects the preservation of green areas in favor of real estate development. On the other hand, in Tunis, although the correlation between building density and vegetation density is inverse, it is less strong and indicates a lesser capacity to explain the variation in building density by vegetation density. This may reflect an urban policy or planning that better integrates the preservation of green spaces and a more balanced urban expansion between construction and vegetation. The correlation coefficient (r*)* for Sousse, equal to approximately 0.362, indicates a moderate positive correlation between the Normalized Vegetation Index (NDVI) and the Difference Normalized Building Index (NDBI). This suggests a relationship where increasing NDVI is associated with moderate increasing NDBI and vice versa. However, this level of correlation is not very strong, which may indicate some variability or diversity in the relationship between vegetation density and building density in Sousse. It is possible that urban composition in Sousse is characterized by less uniform patterns or less direct interactions between building density and vegetation density compared to Tunis. The coefficient of determination (R^2) *of* 0.131 for Sousse suggests that 13.1% of the observed variation in the Difference Normalized Building Index (NDBI) can be explained by the corresponding variation in the Normalized Vegetation Index (NDVI). Compared to Tunis, the proportion of the variance in building density explained by vegetation density is significantly lower in Sousse. This could indicate a less direct or less significant relationship between vegetation density and building density in Sousse compared to Tunis. Urban composition in Sousse is characterized by more diverse factors or more complex patterns that influence building density and which are not fully captured by the NDVI.

For the city of Tunis, the value of the correlation coefficient (r) $\approx$0.881 indicates a relatively strong positive correlation between the NDVI (Normalized Difference Vegetation Index) and the NDWI. The coefficient of determination (*R^2), which represents the proportion of the variance in the dependent variable (NDWI) explained by* the independent variable (NDVI), is approximately 0.776 for the city center (Table 3). Thus, for the city of Sousse, the correlation coefficients (r) and determination coefficients (R^2) between the NDVI and the NDWI are 0.927 and 0.859329, respectively. A correlation coefficient close to 1 indicates a strong and positive linear relationship between the two variables, in this case, between the NDVI (Normalized Vegetation Index) and the NDWI (Normalized Difference Water Index). For the city of Sfax, the correlation coefficient between the NDVI (Normalized Difference Vegetation Index) and the NDWI (Normalized Difference Water Index) amounts to approximately $r \approx 0.916$, suggesting a relatively strong positive correlation between these two environmental indices. By calculating the coefficient of determination (R^2), which represents the proportion of the variance of the NDWI explained by the NDVI, we obtain $R^2 \approx 0.839$. This suggests that, for the three urban centers of Tunis, Sousse and Sfax, there is a strong association between the quantity of vegetation (NDVI) and the presence of water (NDWI). In other words, where water is present, there is likely to be an increase in vegetation, and conversely, where vegetation is denser, there might be a greater presence of water or humidity in the soils.

The calculated value $r = -0.995$ indicates a moderate negative correlation between the NDBI and NDWI data for the city of Tunis. This suggests that there is an inverse linear relationship between these two variables. In other words, when NDBI values increase, NDWI values tend to decrease although it is moderate in this case. Which means that 99.2% of the variance of the NDWI variable can be explained by the variance of the NDBI variable using a simple linear model. This indicates a strong inverse linear relationship between NDBI and NDWI for the city of Tunis in this specific context.

For the city of Tunis, the correlation index of approximately 0.94 indicates a strong and positive correlation between the NDWI and LST data. There is an observable trend showing that as NDWI increases, LST also tends to increase. As for the determining coefficient (R^2), it indicates that only approximately 88.36% of the variance in the LST variable can be explained by the variation in the NDWI variable for the city of Tunis (Table 4). This means that NDWI explains part of the variation in LST, but there still remains a part of the variance in LST that is not explained by NDWI for the city of Tunis. The very high Pearson correlation coefficient (r), approximately 0.9767, indicates a strong positive correlation between the NDWI and LST data for Sousse. This means that there is a strong linear relationship between the variation of NDWI and that of LST. The determining coefficient (R^2) indicates that approximately 95.39% of the variance observed in the LST variable can be explained by the variation in the NDWI variable for the city of Sousse. The very high Pearson correlation coefficient (r), around 0.9879, suggests an extremely strong and almost linear correlation between the NDWI and LST data for Sfax. This indicates a very close relationship where the variation in NDWI is almost directly associated with the variation in LST. The determining coefficient (R^2) of approximately 97.59% is very high, which means that approximately 97.59% of the variance observed in the LST variable can be explained by the variation in the NDWI variable for the city of Sfax. This suggests that NDWI has extremely high predictive ability on LST in this city, explaining the vast majority of the variance observed in LST. An increase in surface temperature can stimulate evaporation processes. When heat increases, it tends to accelerate the evaporation of water present in soils, bodies of water and surrounding

Table 3 The correlation between NDVI and NDWI in Tunisian urban centers

	NDVI min	NDVI max	NDWI min	NDWI max	R	R^2
Tunis	−0.06	0.52	−0.63	0.08	0.881	0.776
Sousse	−0.25	0.51	−0.45	0.20	0.927	0.859
Sfax	−0.32	0.49	−0.45	0.25	0.916	0.839

vegetation. This increased evaporation leads to an increase in the amount of water vapor present in the atmosphere.

The correlation coefficients measured for the new surface temperature (LST) and vegetation index (NDVI) data in Tunis, Sfax and Sousse reveal a very strong inverse relationship between these variables (Table 5). The Pearson correlation coefficient respectively (r ≈ −0.94), (r ≈ −0.9711) and (r ≈ −0.9298), indicates an almost perfect negative correlation. This implies that as the surface temperature increases, the vegetation index tends to decrease very consistently, and vice versa. Furthermore, the determining correlation coefficient ($R^2 \approx 0.8836$ or 0.9431 or 0.86452) highlights that almost all of the variance in surface temperature or vegetation index can be explained by the variance in the other variable, thus indicating a strong inverse interdependence between these data in city centers.

The slightly weaker correlation between land surface temperature (LST) and normalized difference vegetation index (NDVI) in Sousse compared to Tunis and Sfax may be due to several factors: Environmental and geographic conditions specific to each city could also influence this correlation. Throughout downtown Sousse, we observe a diversity of green space ratios, which could be linked to this variability. International hotels in Sousse demonstrate a diversity of approaches to landscaping. Some establishments seem less attentive to their local environment, favoring developments centered on large swimming pools occupying a large surface area. These pools act as focal points of the landscape, offering panoramic sea views, often surrounded by low, orderly vegetation to complement the visual aesthetic. On the other hand, other establishments adopt a different approach, highlighting the vernacular model and drawing inspiration from local architecture. These hotels take a neo-vernacular approach, combining elements of local traditional architecture with international landscaping practices. This diversification in exterior design demonstrates the importance of the landscape on the scale of the city and the district, implicitly reflecting the typology of the hotels. These variations demonstrate the significant impact of landscape design on the overall visitor experience and on the integration of hotels

Table 4 The correlation between NDWI and LST in Tunisian urban centers

	LST min	LST max	NDWI min	NDWI max	R	R^2
Tunis	46.61	63.60	−0.63	0.08	0.94	0.8836
Sousse	22.34	34.83	−0.45	0.20	0.9767	0.9539
Sfax	52.37	71.09	−0.45	0.25	0.9879	0.9759

Table 5 The correlation between NDVI and LST in Tunisian urban centers

	LST min	LST max	NDVI min	NDVI max	R	R^2
Tunis	46.61	63.60	−0.06	0.52	−0.94	0.8836
Sousse	22.34	34.83	−0.25	0.51	−0.9298	0.9431
Sfax	52.37	71.09	−0.32	0.49	−0.9711	0.86452

Table 6 The correlation between LST and NDBI in Tunisian urban centers

	LST min	LST max	NDBI min	NDBI max	R	R^2
Tunis	46.61	63.60	−0.21	0.27	0.97914	0.958
Sousse	22.34	34.83	−0.32	0.16	0.97571	0.952
Sfax	52.37	71.09	−0.42	0.57	0.97532	0.951

into their local environment in Sousse, which could also contribute to the variability in green space ratios observed in the downtown. The results for the three Tunisian urban centers, namely Tunis, Sousse and Sfax, reveal marked similarities in terms of relationships between the normalized building index (NDBI) and surface temperature (LST).

These findings demonstrate that highly urbanized areas have maximum surface temperatures. Therefore, it was anticipated that areas where urbanization is predominant would induce significant variations in surface temperature.

The correlation established between the NDBI and LST shows a strong positive relationship, which was observed consistently in each season, as indicated by the correlation coefficients ($r \approx 0.97914$, $r \approx 0.97571$, $r \approx 0.97532$ and the coefficients of determination high $R^2 = 0.958, 0.952$ and 0.951 for Tunis, Sousse and Sfax respectively) (Table 6). This positive correlation highlights that built-up areas contribute significantly to the observed variations in surface temperature, thus reinforcing the urban heat island phenomenon. In contrast, healthy vegetation cover appears to play a crucial role in regulating and lowering surface temperatures. This observation highlights the positive impact of vegetation in moderating local temperatures, suggesting that the presence of dense vegetation could help mitigate the effects of urbanization on surface temperature variations in these urban areas. Correlations between land surface temperature (LST) and urban landscape parameters remained constant in the different study areas, namely Tunis, Sousse and Sfax. These correlations revealed strong positive relationships for some variables, particularly for impervious surfaces, while other variables showed no apparent relationships with LST.

4 Influence of Spatial Landscape Configuration on LST

In the study, we identified five specific spatial metrics at the class level: patch density (PD), average patch size (AREA_MN), average shape index (SHAPE_MN), average fractal complexity (FRAC_MN) and the aggregation index (AI). These metrics were calculated at the class level using Fragstats software (version 4.2v). To assess relationships, we used the 8-cell rule to define patch neighbors. This landscape metric approach was adopted to understand the potential influence of characteristics such as composition, shape, complexity and spatial arrangement of impervious surfaces on the spatial distribution of land surface temperature (LST) in each of the zones of study [2, 8, 9]. It aims to provide valuable insight into how the composition, shape and spatial configuration of landscape features could potentially impact the spatial distribution of land surface temperature in these areas. The use of metrics describing the urban landscape proves essential to understand the spatial variability of LST by taking into consideration the arrangement of landscape characteristics. In all cities studied (Tunis, Sousse and Sfax), compositional variables, such as average surface area (AREA_MN), presented significant positive correlations with the LST of impervious surfaces, reaching a correlation coefficient of 0.9. This suggests a strong and consistent relationship between the average size of impervious surfaces and surface temperature, indicating that in these cities an increase in the size of impervious surfaces is strongly associated with higher surface temperatures. However, for complexity variables such as average shape (SHAPE_MN) and average fractal complexity (FRAC_MN), no significant relationships were observed with the LST of impervious surfaces in the three cities studied. This suggests that, in these specific urban contexts, the shape or complexity of impervious surfaces does not have a measurable impact on surface temperature. The spatial layout variable (AI) could refer to an index or measure that assesses how spatial characteristics, such as the distribution, density, layout or configuration of urban elements (buildings, streets, parks, etc.), affect the distribution of heat on the earth's surface. However, it showed significant negative correlations with the average impervious surface density LST in all cities, with a correlation coefficient of −0.9. This implies that an increase in spatial layout (AI) is associated with a decrease in average surface temperature density on impervious surfaces in these urban environments.

5 Prospects for Improvement for the Urban Context

Adapting to the challenges posed by the urban heat island represents a major challenge for improving the quality of life in the urban centers of Tunis, Sfax and Sousse. A thorough understanding of area-specific climatic characteristics is imperative for designing effective strategies to mitigate the adverse effects of urban heat. These climate data collected by remote sensing methods provide valuable insights to better understand these local climate specificities. To this end, development initiatives are implemented in these cities to respond to the challenges of the urban heat island. In order to increase the normalized vegetation index (NDVI), actions are undertaken to promote the creation and preservation of green spaces, encourage the revegetation of private land, facades and roofs of buildings. Similarly, increasing the Normalized Water Index (NDWI) is being considered through projects to reintroduce water into these urban areas, helping to cool and regulate local temperatures. To reduce the ground temperature index (LST), several measures are recommended, in particular the improvement of the overall emissivity, the use of construction materials with high reflectivity to limit heat absorption, as well as the promotion light shades to increase the reflectivity of surfaces. Adjustments in the orientation of streets and maximization of the sky view coefficient are envisaged to naturally promote shading and the circulation of fresh air. Housing planning and design are being redesigned to take advantage of local climatic conditions. Optimization of natural lighting is favored in architectural design to reduce dependence on artificial lighting. Furthermore, reducing the use of air conditioning is envisaged thanks to natural ventilation and innovative systems such as climatic wells, exploiting the thermal stability of the ground to moderate the interior temperature of homes. These holistic initiatives also include complementary measures such as the insulation of exterior walls, the integration of solar water heaters, the use of renewable energies such as photovoltaic plants, as well as the reduction of asphalt surfaces with permeable coatings, spaces greens and promoting vertical urban development: This approach favors vertical rather than horizontal growth of urban areas in order to preserve the space available for landscaping and urban greening These integrated strategies aim to reduce the impacts of the heat island urban environment and to promote more resilient, comfortable and environmentally friendly urban environments for the inhabitants of Tunis, Sfax and Sousse.

6 Concluding Remarks

The urban centers of Tunis, Sfax and Sousse face major challenges linked to the urban heat island, caused by rapid growth and increased sensitivity to high temperatures. To develop resilience strategies adapted to these dynamic urban environments, a careful assessment of relevant indicators is essential, notably using satellite image processing such as Landsat 8. The analyzes carried out throughout this study reveal a direct correlation between the normalized building index (NDBI), the normalized vegetation index (NDVI), the NDWI (Normalized Difference Water Index) and the intensity of the urban heat island. These findings, based on in-depth cartographic and statistical analyses, allowed a segmentation of the centers of these cities. This approach aims to build urban planning policies that take into account local specificities rather than being based solely on averages which could lack representativeness. According to a prospective vision, modern urban planning had to find methodological and practical tips for better urban adaptation in the face of climate change. However, the resilience of vulnerable urban spaces remains an essential concern for urban planners. Finding a climatic balance) for such a territory requires a double simultaneous intervention, one regulatory and the other operational. Following this assessment, targeted actions are selected to meet the specific needs of these urban areas. Inspired by good global practices, these actions aim to transform vulnerable neighborhoods into resilient spaces, capable of reducing their energy footprint and offering a pleasant living environment, thanks to adapted and flexible developments. This approach demonstrates that strategic urban interventions, based on simple but economically viable and environmentally friendly actions, can transform the challenges of the urban heat island into positive opportunities to strengthen the resilience of the urban centers of Tunis, Sfax and Sousse, while improving the quality of life of their inhabitants.

References

1. Asaeda T, Ca VT, Wake A (1996) Heat storage of pavement and its effect on the lower atmosphere. Atmos Environ 30(3):413–427
2. Ashley NM, Timothy H (2017) River effects on the heat island of a small urban area. Urban Clim 21:262–277
3. Chen F, Yang S, Wang K (2016) Effect of emissivity uncertainty on surface temperature retrieval over urban areas: investigations based on spectral libraries. ISPRS J Photogramm Remote Sens China 114:53–65
4. Gober P, Brazel A, Quay R, Myint S, Grossman-Clarke S, Miller A, Rossi S (2010) Using watered landscapes to manipulate urban heat island effects: how much water will it take to cool Phoenix. J Am Plann Assoc Taylo Francis J 76(1):109–121
5. Hart MA, Sailor DJ (2009) Quantifying the influence of land-use and surface characteristics on spatial variability in the urban heat island. Theor Appl Climatol 95(3):397–406
6. Jeff F (2018) Numerical study of the cooling potential of urban heat island (ICU) reduction techniques in a Mediterranean climate. Civil engineering. INSA de Toulouse 237
7. Li X, Zhou Y, Asrar GR, Marc I, Li X (2018) The surface urban heat island response to urban expansion: a panel analysis for the conterminous United States. Sci Total Environ 605–605:426–435
8. Lucille A (2017) Characterization of urban heat and cool islands in Lyon using satellite, mobile and participatory measurements. Jean Moulin Lyon University, p 135
9. Margarete A, Dubreuil V (2016) Interests of visible and infrared images from the Landsat-8 satellite to model the urban heat island in prudent president– Brazil. In: XXIX th conference of the International Association of Climatology, Besançon, France, pp 95–100
10. McFeeters SK (1996) The use of the Normalized Difference Water Index (NDWI) in the delineation of open water features. Int J Remote Sens 17(7):1425–1432
11. Memon RA, Leung DYC, Liu CH (2008) A review on the generation, determination and mitigation of Urban Heat Island. J Environ Sci 20(1):120–128
13. Rouse JW, Haas RH, Schell JA, Deering DW (1974) Monitoring vegetation systems in the great plains with ERTS. NASA. Goddard Space Flight Center 351:309–317
12. Prasanjit D, Göttsche F-M, Folke SO, Herbert F (2002) Land surface temperature and emissivity estimation from passive sensor data: theory and practice; current trends. Int J Remote Sens 23(13):203–221
14. Schott John R, Boonmee Marvin , Messinger David W (2010). Land surface temperature and emissivity retrieval from thermal infrared hyperspectral imagery. Algorithms and Technologies for Multispectral, Hyperspectral, and Ultraspectral Imagery XII. 6233:1–10
15. Sobrino JA, Jiménez-Muñoz JC, Paolini L (2004) Land surface temperature retrieval from LANDSAT TM 5. Remote Sens Environ 90(4):434–440
16. Tzoulas K, Korpela K, Venn S, Yli-Pelkonen V, Kaźmierczak A, Niemelä J, James P (2007) Promoting ecosystem and human health in urban areas using green infrastructure: a literature review. Landsc Urban Plan 81(3):167–178
17. Wolch JR, Byrne J, Newell JP (2014) Urban green space, public health, and environmental justice: the challenge of making cities 'just green enough'. Landsc Urban Plan 125:234–244
18. Xu H (2008) A new index for delineating built-up land features in satellite imagery. Int J Remote Sens 29(14):4269–4276
19. Yao X, Chen Y, Zhang Q, Mou Z (2016) Assessment of the urban expansion and its impact on the eco-environment—a case study of Hefei municipal area. Sustainability 14(7):10613
20. Zha Y, Gao J, Ni S (2003) Use of normalized difference built-up index in automatically mapping urban areas from TM imagery. Int J Remote Sens 24(3):583–594

The Rehabilitation of Riverbanks in Beijing's Urban Area in the Light of an Ambitious Policy to Enhance Nature and Quality of Life for Residents

Stéphanie Boufflet and Vassilia Pirierros

Abstract Faced with the urgency of the climate issue, awareness has grown and substantial progress has been made in recent years in the design of green and blue spaces in the megalopolis of Beijing. This trend intended to be in favour of biodiversity and the well-being of city residents continues as a further milestone in a long-term strategy for balanced development between man and the living world in the city. We contextualise our research with regard to the place of nature in Chinese ancestral culture and the green and functionalist policies implemented since the 2000s. The 2016–2035 Urban Master Plan is an extension of this, with more ambitious ecological and humanist objectives that we see as an echo of COP 15. Three field studies of recently developed riverbanks in this framework, all located in the densely populated built area of Beijing, allow us to suggest that China has moved towards a contextualization of landscape project, even if attention to the living world is still mainly ornamental. Nevertheless, these developments have enabled residents to appropriate these spaces in their own way. Three years of restricted mobility as part of the global pandemic have undoubtedly had a hand in these new forms of expression in public spaces. They reflect new links between citizens and nature in the urban area, which we believe will be decisive in the coming decades.

Keywords Beijing · Riverbanks · Renaturation · Urban renewal · Urban nature · Urbanity

S. Boufflet (✉)
Department of Theories and Practices of Architectural and Urban Design (TPCAU), National School of Architecture of Paris Val-de-Seine, Paris, France
e-mail: stephanie.boufflet@paris-valdeseine.archi.fr

V. Pirierros
School of Architecture, Tianjin University, Tianjin, China
e-mail: vassilia_pirierros@tju.edu.cn

B. He et al. (eds.), *Urban Climate and Urban Design*, Urban Sustainability,
https://doi.org/10.1007/978-981-96-1521-6_12

1 Introduction

Faced with the problem of global warming, the People's Republic of China has pledged to take urgent measures through international agreements adopting the UN's 2030 Agenda for Sustainable Development in 2015, and signing the Kunming-Montreal Global Framework for Biodiversity, adopted at the 15th meeting of the Conference of the Parties to the United Nations Convention on Biological Diversity in December [6]. Its theme was "Ecological Civilization: Building a Shared Future for All Life on Earth" and Point 12 of this framework focuses on urban and densely populated areas and promotes urban planning that takes biodiversity into account and improves people's health and well-being.

Research has already been carried out, proving the effectiveness of water bodies in cooling the urban climate, particularly in Beijing [4]. Our research further examines the recent development of the riverbanks in Beijing in terms of their potential to support a Chinese urban planning model that is both ecological and humanist, and that has an impact on enhancing the value of nature and the appropriation of spaces. We will consider renaturation here as the act of compensating for the anthropisation of riverbanks. To be as exhaustive as possible, we have crossed three methodological tools; the field study, the interview of local actors and the analysis of regulatory texts.

Beijing's 2016–2035 Urban Master Plan, set in the context of the environmental challenges of climate change and biodiversity loss mentioned in the COP 15, provided the framework for our thinking, which interviews with local players helped us to decipher in terms of the strategies envisaged to protect ecosystems and quality of life in the city. At the same time, a field study enabled us to make a factual assessment of projects recently completed in the city. A comparative analysis of these prototypes will enable us not only to identify their respective weaknesses and merits, but above all to understand the issues at stake in a complex megalopolis. Firstly, we will present the historical and cultural context of Beijing's development and the policies currently in place with regard to the water network. Secondly, we will look at the role that nature can play in a megalopolis. We will analyse the specific features and morphologies of the city's riverside areas. Finally, we will characterise the social needs, and the generic or singular uses that give meaning to each space and its urbanity. This study may serve as a reference for future riverbank regeneration in Beijing and in other cities to provide a balanced response between the interests of Man and Nature.

2 Historical Background to Planning Policies

2.1 An Ancient Culture of Nature in Harmony with the Site

In the culture of Chinese spatial planning, the elements of the greater territory determine the choice of location for any human habitat, whether it's a house or a capital city. Since its origins more than 1000 years BC, the choice of the Beijing site has benefited from a beneficial geographical context in line with the principles of Chinese spatial planning. These principles advocate a search for harmony between mountains and water, between protective hills and abundant water. In this respect, the city of Beijing is ideally situated on a plain protected by mountain ranges to the north and west and watered by the springs of two rivers that flow from north-west to south-east, towards the Gulf of Bohai. This is why so many emperors throughout history have chosen it as their political base. Even though the city has been destroyed several times, the geographical site has always been preserved for the reconstruction of a new capital.

On the scale of the Beijing plain, the two rivers, Chaobai and Yongding, and their multitude of tributaries, give rise to fears of water. Loaded with alluvial deposits, they push back the coastline, creating over the centuries areas of marshland that are more or less unusable for farming. For the city as a whole, the lakes in the imperial gardens at the foot of the hills acted as reservoirs and flowed into a network of water channels, ensuring the mobility of people and goods as well as a principle of regulation. Beijing's ponds and lakes also played a role in regulating heat and were known as summer comfort zones for the emperor and the elite.

2.2 A Maoist War Against Nature

While the heritage of this exemplary urban composition in China, based on a sacralised nature, was maintained for hundreds of years, it was brutally supplanted in the mid-twentieth century by the Maoist ideal, which ushered in a far-reaching urban and environmental revolution. The cultural pairing of "mountains and water" was banished in its entirety, as it inevitably led back to traditional culture. Massive deforestation, the levelling out of mountains and the draining of marshes constituted a veritable war against nature. The anchoring in the land, linked to the search for harmony with the cosmos, is replaced by the human ideal driven by industrial fervour. Water was still feared, however, and it was not until the second half of the twentieth century that water retention infrastructures were built, such as the vast Miyun reservoir to the north of Beijing.

The pressure of Deng Xiaoping's new economic reform policy on land was to cause water to become increasingly scarce throughout the Chinese capital. The phenomenon of grey infrastructure has only exacerbated both surface and underground water shortage problems [9]. However the authorities responded in 2014 with the huge project for the south-north bypass canal (南水北调, Nánshuǐ Běidiào), which will carry water over a thousand kilometers to the northern plain.

2.3 An Urban Dream Challenged by Environmental Awareness

Environmental awareness really took off in the 1990s, in the wake of the economic reforms of the 1980s. In 1993, the Olympic Committee's refusal to allow Beijing to host the 2000 Olympic Games for unofficially political, but officially environmental reasons, undoubtedly contributed to the authorities taking this issue on board and relaunching the Great Green Wall project. This covers more than 20% of China's territory and is divided into green belts in urban areas [3]. In Beijing, these are part of the new environmental policy introduced as part of the 2004–2020 Master Plan. However, this new planning is being overtaken by the current urban dream. In the meantime, Beijing is validating its bid to host the 2008 Olympic Games, one of whose slogans will become the Green Olympic Games (绿色奥运, Lǜsè Àoyùn). In 2006, all of the eligible land in the 2004–2020 Master Plan had already been urbanised.

In terms of the water system, the large majority of the canals in the urban area are being cleaned out. Many of them were used as open-air waste channels. Most of these cleaned canals have been landscaped to a basic but efficient standard. The same types of fast-growing trees from peri-urban nurseries are being planted, regardless of the context. This new blue-green network, initiated in the olden days, has been given a new lease of life and is inaugurating a new relationship between man and nature in the city. This is the starting point for the second phase of development.

3 Current Urban Policy Issues

3.1 Objectives of the 2022 COP 15

In terms of setting Chinese public policy in a larger context, two major events on the global environmental scene have been the main sources of our research. Due to the global pandemic, COP 15 was organised in two parts, part one was held in Kunming in October 2021 and part two in Montreal in December 2022. Despite this, the conference marks China's entry onto the environmental diplomacy stage. Ami Dahan, historian of the politics of climate change noticed "that China has always been very reluctant to accept the rhetoric of "sustainable development", which is fundamentally associated with a social-democratic project. Chinese leaders or officials have preferred to evoke a "low-carbon society", in which the technological dimension is central." [8]. However, between 2012 and 2015, China's position evolved under pressure from the population faced with major waves of pollution and under the influence of many philanthropic financial organizations, think-tanks, and NGOs. According to Terry Townshend, advisor to the Paulson Institute on Chinese environmental issues, the importance of COP 15 part one in Kunming in 2021 and its impact on local governments and civil society in China are undeniable "When China hosted COP 15 part 1 in Kunming, every domestic newspaper, every TV channel led on Biodiversity. In the first two three pages. We cannot overestimate the impact that had in terms of building knowledge and awareness of the importance of biodiversity, among Chinese people."

Objective 12 of the Kunming-Montreal Global Biodiversity Framework sets the following targets for densely urbanised areas "increase the area and quality and connectivity of, access to, and benefits from green and blue spaces". It sets out to "contributing to inclusive and sustainable urbanization" and "improving human health and well-being".

3.2 The Urban Master Plan 2016–2035

The Beijing Urban Master Plan 2016–2035 [7] was issued in September 2017. It is designed as a continuity of the previous one, with the central concern being to control the expansion of the city and protect natural areas. However, it appears to be more ambitious in terms of its green policy project: "we will build a multifaceted and synergistic ecological and environmental governance model, cultivate an ecological culture, enhance the awareness of ecological civilisation among the entire population, and achieve a green transformation of lifestyles and consumption patterns". The Master Plan is built on the territory of the Beijing municipality and is structured (Beijing Urban Master Plan 2016–2035 Fig. 5. Beijing Green Space Structure) around a natural ecological barrier, three green belts, five rivers and nine ecological radials. In the urban area, the aim is to create a network of green and blue spaces (Beijing Urban Master Plan 2016–2035 Fig. 14. Core Area Green Network) that will help to ensure that 95% of built-up areas are within 500 m of a green space. The Plan promotes coordinated development of water and the city. The desire "to restore aquatic systems […] and improve the hydrophilicity of rivers to meet the public's needs for recreation, entertainment, pleasure and experience" is stated. However, it is also attested that safety, flood control and clean water remain top priorities. This Master Plan is part of an ecological and humanist development model that is in line with the objectives of COP15. Under the slogan "a harmonious and liveable city of the first rank in the world", it refers to the promotion of a harmonious society advocated by Hu Jintao since 2007 and prefigures China's desire to appear on the international environmental scene.

3.3 Implementation of Operational Tools

In 2021, the Beijing Municipal Water Resources Administration has published a guide for decision-makers, developers and residents to improve practices [1]. The aim is to bring about a transformation from "waterfront fragmentation" (水岸割裂, shuǐ àn gēliè) to "water city integration" (水城共融, shuǐchéng gòngróng). This involves a far-reaching change in the vision and management system for water resources, with 4 guidelines: 1- the river must be considered beyond its unique and historic function of flood control and drainage, towards a diversification of functions integrating ecology and leisure. 2- the water services are encouraged to coordinate with the city's other services, such as green spaces, roads, etc. 3- upkeep and maintenance tasks are added to the basic construction contracts 4- overall management at the municipal level, above the districts, is put in place. With this in mind, since 2017 the municipality has been promoting major coordinated "ecological restoration" projects, particularly for the Wenyu and Yongding rivers. On an urban scale, numerous micro-projects are being launched throughout the city to enable the renaturation and maintenance of rivers, to create continuity in pedestrian and cycle routes, and even to include facilities such as toilets, leisure areas, etc.

3.4 Comparative Study, Presentation of the Sites

As part of our research, we have selected three sites in dense residential areas that meet the need for citizens to be close to natural areas, as mentioned in the Urban Master Plan and echoing the prescriptions of COP 15. Nevertheless, each of these three sites remains singular and enables us to illustrate the diversity of socio-economic contexts and environmental and urban issues in Beijing. We will consider the space given to Man (M) and to Nature (N) and evaluate it according to the quantities and qualities that characterise them.

Nanchanghe (Haidian District) is a canal of historic value. It was first dug out during the Sui and Tang dynasties (sixth to tenth centuries), but its current course dates from the twelfth century under the Jin dynasty. Its purpose was to transport water from the north-western mountains, stored at Kunming Lake, to the imperial capital. Today, the canal is famous for its boat journeys between the Zoo and the Summer Palace. Various development projects have been carried out since the 2000s. We are looking at the residential section between the Wanshou temple and the Jingmi diversion canal.

Liangmahe (Chaoyang District) is a river dating back to the Yuan dynasty (thirteenth century). It is a bypass of the Bahe River from the city's northern moat. The canal has been cleaned out several times since the 2000s, and a major riverside redevelopment project was completed in 2019. The river runs alongside Chaoyang Park, a major shopping centre and the embassy district, all of which are part of an international and economically prosperous context. We are interested in the residential section to the west between Sanlitun Street and Hujiayuan Street.

Qinghe (Changping and Chaoyang District) is a river dug out during the Yuan dynasty (thirteenth century) which flows for 20 km north of Beijing between the Summer Palace and the Wenyu River. A comprehensive renaturation project is currently underway, and the first part has been completed at the river's eastern mouth. We are analysing the section between the Anli Bridge and Beiyuandong Street, which is bordered to the north by the working-class district of Tiantongyuan, launched in 1999, and to the south by Gated Communities with wealthy populations.

4 The Place of Nature

4.1 A Sacred Nature

This research into the redevelopment of Beijing's water system in the second decade of the twenty-first century is based on history and, first and foremost, culture. In traditional Chinese thought, man is an integral part of his environment, unlike Western thought where nature is external of man. The word nature in Chinese includes the notion of man in his natural environment, 自 Zì, oneself and 然 Rán, a context that we could define as the living world. These two characters together echo traditional Chinese culture, in which man was considered to be part of the living world. The political era of the twentieth century progressively severed the link between man and nature, and advocated human control over the entire territory. China has gone from an inhabited nature to a subjugated nature. It is now a constructed nature, organised around living spaces built for the greatest number of people in sprawling megacities. Nature has become an object of decoration and recreation in the urban space, but it is nonetheless an object of concern.

4.2 Quality and Biodiversity

Water quality is the subject of constant attention from the public and local authorities, who publish results every month [2]. This significant improvement can be seen at http://www.ipe.org.cn/. The rivers studied, which were in class V in 2012, are now in class II, meaning that the water is drinkable and suitable for fishing and swimming. However, these activities are officially prohibited.

Beijing's continental climate, with very marked winter and summer seasons, −5 − 10 °C and dry in winter, 35 40 °C and humid in summer, is a limiting environment for the diversity of flora. According to our observations, the following species are planted in large numbers: acacia, sophora, pine, Spanish mulberry, azerolier, boxwood, wild rose, sea buckthorn, etc. The autumn and spring seasons are very short, but Beijingers attach great importance to the beauty of the vegetation at these times [5]. As a result, landscapers choose ornamental species that flower in spring or are very colourful in autumn (maple, ginko, prunus, peach, deutzia, hydrangea, Japanese corete, viburnum, lavender, rudbeckia etc.). Finally, along the rivers, weeping willows are omnipresent, with occasional reeds, purple loosestrife and water lilies adding to the scenery. Along the Qinghe, the riverbanks are natural, with some pockets more or less left to their natural state. The planting along the banks of Nanchanghe is repetitive and decorative. Along Liangmahe, the landscape design is sequenced and varied, but nature is controlled with alignments and ornamental species. As citizens want to enjoy nature without its drawbacks, anti-insect treatments are regularly used in the city. We didn't see any insects, just sparrows, magpies and spotted dove along Nanchanghe and Liangmahe. Insects were much more common on Qinghe, as were wild birds and ducks (great egret, grey heron, little grebe, northern pintail, eurasian coot, falcated duck, etc.). Beijing is a central location on the migratory bird route between the Arctic and Australasia named the East Asian-Australiasan Flyway. Educational panels have been installed and groups of amateur photographers meet up at migration times.

In conclusion, nature in the city is controlled and recreational to satisfy city inhabitants. A professor we interviewed, believes that biodiversity is not the priority in the city and that the river's function needs to be clarified first: "Is it for the collection and flow of water or is it for recreation? [...] only 10% of the rivers are in the urban area. Maybe the concrete bank is the right consideration. In the urban area the safety of the river is the most important thing. If the river area can be limited, more places can be given to the people."

5 Spatiality

5.1 Definition and Features

The riverbank is a complex space that defines the meeting point between two systems, the linear system of the river flowing from upstream to downstream and the system of the adjacent urban fabric on a neighbourhood scale. Except by boat or swim, the river is inaccessible to humans, who can only experience it from the side. The landscaping takes account of variations in water level in the event of rain, but the banks are public spaces and accessible to all, at all times. Finally, local urban planners have developed an ability to design projects in constantly changing urban contexts, and Yang Yifan, chief planner at the China Urban Development Planning and Design Consulting Company, believes that urban policies have made it possible to create green spaces in quantity, and that quality will follow. To which he adds "we need flexibility [...] It is good to slow down not occupying all the land, keep some strategic green space. We don't know the correct value for the present, but the next generation will be smarter than us."

5.2 Continuity and Accessibility

Along the Qinghe River, on the north bank, the Tiantongyuan residences are fenced off, with access roads more than 500 m from the river and an average of 700 m apart.

Fig. 1 The point at which humans come into contact with water. *Source* Photos taken by the authors

On the South Bank, luxury gated communities have been built with a physical separation strip of private green space, fencing and roads. In contrast, along Nanchanghe, there is a high degree of porosity with the adjacent residential area. There are many residential entrances facing the river. The parks are not enclosed. The urban fabric is more closely meshed, with a road every 100–150 m.

Two routes are systematically distinguished: a route at the height of the city and a route close to the water. The high route, designed for accessibility and traffic flow, is often shared with a road. A movement is underway to transform these roads into soft traffic lanes, particularly along Qinghe. City residents attach great importance to the sight of water, and the development of these areas allows them to get close to it (see Fig. 1). Along Nanchanghe, the path is 50 cm from the river and separated by a 50 cm sculpted stone balustrade. Along Liangmahe, the bank is 1 m above the water level with no guardrail, but a system of distancing is provided by an alley of azerola and bushes. Along Qinghe, the path fluctuates in altitude and distance from the water. Between the high and low paths, there is a difference in level of between 10 and 30 steps, which raises the question of accessibility for all, and ramps for pushchairs and wheelchairs are not yet systematic. In addition, continuity of routes remains a major problem, as highlighted by the study of the Beijing waterfront ([11], p 115). In the sectors studied, pedestrian crossings have been created under road bridges. At Liangmahe, solutions have been found to pass under extremely low bridge tables, creating an impressive physical and bodily experience where the walker's gaze is at water level.

5.3 *Morphology and Potential of the Space*

The Nanchang, Liangma and Qing rivers are respectively 2.75 m, 3.5 m and 6 m lower than the average level of the surrounding city. Green strips of between 70 and 200 m are the result of previous policies. In the 1970s, setback strips were put in place to protect against flooding, and in the 2000s these were turned into landscaped areas. This is particularly evident along Qinghe, which was urbanised in the early

2000s. Along Nanchanghe, buildings have been demolished to make space for urban parks. On the other hand, the dense urban fabric of Liangmahe makes it impossible to create this green edge.

Despite their narrowness, the banks of Liangmahe (see Fig. 2) are laid out in a variety of architectural proposals: squares of 1,000 m^2, plots of 250–500 m^2, platforms on the water, pontoons, pavilions, children's play areas, etc. Along Nanchanghe (see Fig. 3), staircases are used as plazas and the side parks offer a wide variety of gathering spaces and pavilions. Along Qinghe (see Fig. 4), certain sections are dedicated exclusively to movement, with no possibility of gathering. Recreational and sports facilities (basketball court, inline skating rink, fitness equipment, children's play areas) are provided on an occasional basis. All of these facilities contribute to a better quality of life in the city and make reference to the term "urbanity".

Fig. 2 Liangma river section. *Source* Section drawn by the authors. Space given to Man and Nature: M^+ N^+

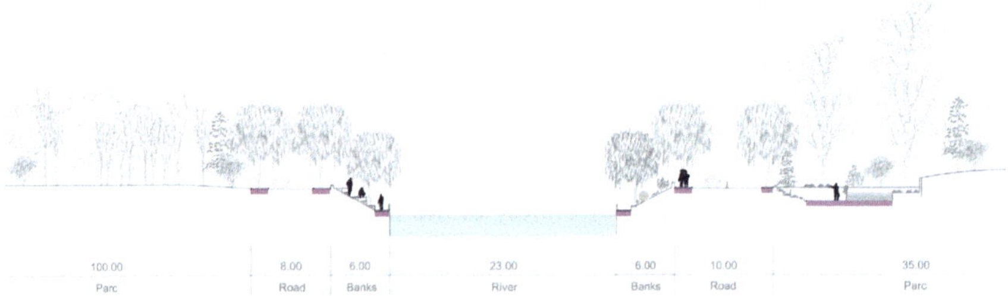

Fig. 3 Nanchang river section. *Source* Section drawn by the authors. Space given to Man and Nature: M^+ N

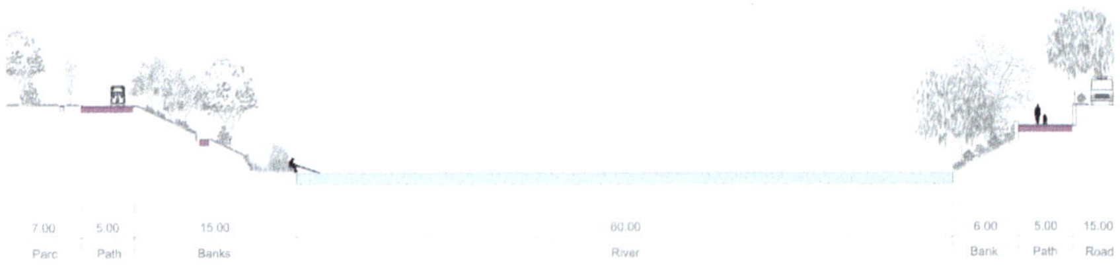

Fig. 4 Qing river section. *Source* Section drawn by the authors. Space given to Man and Nature: M^- N^+

6 Urbanity

6.1 Terminology and Issues

The concept of urbanity refers to the way in which the spaces of a city are receptive to people. Both the Beijing Urban Master Plan and the COP 15 text use terminology that focuses on people and their quality of life. The term 宜居 Yíjū in Chinese can be translated as "livable" or "habitable". The term "well-being" in English relates to people's health but seems restrictive in that it does not include the notion of spatiality. We prefer the term urbanity, which involves a question of balance between urban form and the uses to which residents put it. The challenge of urbanised spaces is to satisfy human needs for sociability and conviviality, which it is interesting here to measure in relation to the mechanisms of collective appropriation of the riverbanks.

6.2 Uses and Social Needs

The banks of Beijing's canals have benefited from a revival of interest that has led to the emergence of new individual, collective and commercial practices. Keeping the water clean was an essential prerequisite for developing these areas as public spaces and enhancing their value. Uses closely linked to water have gradually emerged, such as fishing, open-water swimming and, more recently, water-based activities such as paddling or canoeing. Like green spaces or squares, blue spaces are everyday gathering places. Along Nanchanghe and Liangmahe, informal but regular groups meet up to "do things together" [10] for sports or leisure activities such as taiji, kites, mahjong, dance, music and so on. The more massive and general appropriation of riverside areas is the result of a covid effect. From 2020 to 2023, Beijingers were restricted in their movements outside the municipality and self-limited to the urban area. The usual social venues (shopping centres, restaurants, bars) and the city's parks were intermittently closed, and citizens turned to the rivers, the only accessible places to breathe and interact with nature in the city. During holiday periods or at weekends, residents have taken to setting up their equipment (tents, tables, chairs, etc.) and having picnics, particularly along Liangmahe or Qinhe. Finally, in 2021, the banks of Liangmahe became a place for parties and night-time gatherings, so much so that in May 2022, the lawns were fenced off. On 27 November 2022, this newly appropriated and then banned area was spontaneously chosen by protesters against the zero-covid policy. The riverbanks, a linear space comparable to that of the street, provided a momentary space for expression for the "white sheet movement".

Fig. 5 Seating arrangements. *Source* Photos taken by the authors

6.3 The Bench, a Typology of Contemplation

There are many walkers and joggers along the banks. In addition, the organisation of commercial nautical activities, have created a consumerist relationship with the water space. Stopping in this space becomes an essential issue. Stopping means taking time, paying attention to the people and nature around and, counter-intuitively, being active in the landscape. We would therefore like to focus on the bench as a major piece of public furniture (see Fig. 5). Along Qinghe, benches are few and far between, scattered and anecdotic along the cycle path. Down below, the pontoons and platforms on the water are not systematically equipped with furniture. Along Nanchanghe, we are seeing a proliferation of seating solutions and ways of appropriating the space: a plaza on a step overlooking the river, seating associated with planting beds and equipment (armchairs, chairs, cushions, shelves) brought in by residents. Along Nanchanghe and Liangmahe, the retaining walls of the slopes take the form of continuous benches, providing a valuable space. Finally, everywhere, small folding and transportable stools offer their owners, and in particular fishermen, the possibility of settling freely in the space.

The proximity of a water element is increasingly considered to be conducive to human calm and well-being, in the same way as planted spaces. In this respect, the redevelopment of the riverbanks is enabling residents to find the spatiality of intimacy in the large-scale city, even if it is essentially recreational nature. In the next stage, we need to give more space to the living world, remembering the Chinese scholars of ancient China who advocated humility for man on Earth.

7 Conclusion

Faced with the urgent problem of climate change, the effect of Beijing's green and blue spaces on the local urban climate are better known, and substantial progress has been made in the recent years in the design of these spaces. This dynamic intended to be in favour of biodiversity and the well-being of citizens continues as a further milestone in a long-term strategy of respectful development of the city between fauna, flora and people. In reality, different balances are being established. Renaturation of the riverbanks is underway along the Qinghe, and the long-term potential for ecological restoration needs to be re-evaluated. The banks of the River Nanchang offer a space for appropriation to the local residents. The Liangma river area, meanwhile, is an area of recreational nature in the city and of a leisure society on the scale of the country and its representation to the outside world. Beijing is a very large city, and not all of its citizens can benefit from the regeneration of the riverbanks in their daily lives. However, these spaces are valuable, and may serve as an example for the enhancement of the many existing canals in the capital. In the light of COP15 and ambitious urban policies, Beijing's riverbanks are a laboratory for a Chinese urban planning model that aims to be ecological and humanist, placing nature and people at the heart of a more harmonious development of the city. In practice, both nature and voices remain under control, but the recent renovation of the riverbanks is opening up new prospects for the deployment of living things and for the appropriation of the citizens.

Acknowledgements We would like to thank the people of Beijing, all those who have supported us in this research and have contributed to its success and especially Terry Townshend from the Paulson Institute and Yang Yifan from the China Urban Development Planning & Design Consulting Company.

References

1. Beijing Municipal Commission of Planning and Natural Resources (2021) Beijing River Planning & Design Guidelines. https://ghzrzyw.beijing.gov.cn/en/Notification/202204/P020220411563252736318.pdf
2. Beijing Water Authority. Monthly report on Surface water quality (including water quality of rivers, lakes and large and medium-sized reservoirs). https://nsbd.swj.beijing.gov.cn/dbssz.html
3. Boufflet S (2011) The process of renaturation of the Chinese capital at the dawn of the 21st century: a "green breath" on Beijing? (575 p) Paris Est, University of Orléans. Phd (in French). https://www.theses.fr/2011ORLE1112
4. Cai Z, Guldmann JM, Tang Y, Han G (2022) Does city-water layout matter? Comparing the cooling effects of water bodies across 34 Chinese megacities. J Environ Manag 324:116263. https://doi.org/10.1016/j.jenvman.2022.116263. Epub 2022 Sep 24. PMID: 36166866
5. Cheng S, Zhai Z, Sun W, Wang Y, Yu R, Ge X (2022) Research on the satisfaction of Beijing waterfront green space landscape based on social media data. Land 11(10):1849. https://doi.org/10.3390/land11101849
6. Conference of the parties to the Convention on Biological Diversity Fifteenth meeting (2022). Kunming-Montreal Global biodiversity framework. Target 12, p 10. https://www.cbd.int/doc/c/e6d3/cd1d/daf663719a03902a9b116c34/cop-15-l-25-en.pdf
7. CPC Central Committee and the State Council (2017). Beijing Urban Master Plan (2016–2035) (in Chinese). https://www.beijing.gov.cn/gongkai/guihua/wngh/cqgh/201907/t20190701_100008.html

8. Dahan A (2021) China ecological power, analysis, critiques and perspective. How China's position has evolved in the COPs and on the global climate geopolitical stage, pp 13–18. https://geopolitique.eu/wp-content/uploads/2021/09/GREEN-issue-1-september-EN.pdf
9. Fu W, Yu K, Li D (2018) Water Policy 20, Spatio-temporal relational evaluation of the Beijing water crisis and planning implementation from 1949 to 2013, pp 490–509. https://doi.org/10.2166/wp.2018.107
10. Thireau I (2020) Des lieux en commun. An ethnography of public gatherings in China (386 p.). Cham: Ed. de l'EHESS, Paris (in French)
11. Yang Y, Xu S, Vigano P, Yuan D, Liu H, Lub M, Carvalho Filho L (2020) waterfront analysis and design comparative report of Beijing, Suzhou and Yinchuan. Construct Recognition of the Capital Water System, Beijing. TU Delft Open – BK Books, Cham, pp 92–140

Action Plans and Policy Formulation

Approaches to Urban Climate Change Policymaking: Challenges and Way Forward

Kirti Kusum Joshi and Ajay Chandra Lal

Abstract The world today faces the intertwined challenges of rapid urbanization and climate change. Urban areas, vulnerable to climate impacts, also exacerbate climate change through greenhouse gas emissions. Therefore, it is crucial to address climate change as an urban issue and formulate appropriate policies. Drawing on Nepal's experience as a climate-vulnerable, rapidly urbanizing country, this chapter compares two primary approaches to urban climate change policymaking. The first approach integrates urban concerns into the broader climate change discourse, emphasizing understanding climate change's implications on urban areas and identifying urban challenges within that context. While this approach provides access to global knowledge and international climate funds, there is a risk of overlooking urban nuances amidst broader climate goals. Conversely, the second approach embeds climate concerns within urban policymaking, integrating climate change strategies into urban policies. This approach acknowledges the distinct characteristics of urban areas, allowing targeted interventions. But there is a risk of compartmentalizing climate change as a secondary concern within urban planning, undermining broader climate change efforts. This chapter emphasizes the necessity of an integrated approach that balances urban concerns with climate goals. This entails incorporating climate considerations into urban planning while aligning urban policies with broader climate change efforts.

Keywords Urban area · Climate change · Policymaking · Integrated policy

K. K. Joshi (✉)
Institute of Applied Sciences, Madan Bhandari University of Science and Technology, Chitlang, Nepal
e-mail: kirti.kusum.joshi@mbust.edu.np

A. C. Lal
Institute of Engineering, Tribhuvan University, Lalitpur, Nepal
e-mail: ajay@ioe.edu.np

B. He et al. (eds.), *Urban Climate and Urban Design*, Urban Sustainability,
https://doi.org/10.1007/978-981-96-1521-6_13

1 Introduction

The world today faces the intertwined challenges of rapid urbanization and climate change. Addressing climate change calls for decisive measures in cities and urban areas [9], where more than half of the world's population now reside, and two-thirds will be residing by 2050 [49]. On one hand, cities and urban areas contribute significantly to climate change through greenhouse gas (GHG) emissions [3, 16]. On the other hand, it is in cities and urban areas that the most severe consequences of climate change will be experienced [9, 51]. With the global temperature rising dangerously [27], cities face challenges in terms of both adapting to climate change [1, 7, 25] and mitigating its impacts [26, 31].

Many countries have pledged to take action on climate change, whether through adaptation, mitigation, or both, as part of their commitment to the United Nations Framework Convention on Climate Change (UNFCCC), and have climate policies in place to support this commitment. At the same time, many rapidly urbanizing countries, especially those in the Global South, confront urban development challenges irrespective of the added climate change concerns, and have specific urban policies in place to address these challenges. However, there is scant literature on how countries align their urban policies with their climate policies and vice versa.

Many related previous studies have examined the role of cities or local governments regarding urban climate change policymaking from the perspective of climate governance, focusing on the enablers and barriers to the formulation of such policies [20–22, 28, 30, 51] but do not discuss how cities and local governments integrate urban development policies with climate policies. Certainly, cities and city networks are now increasingly involved in making and implementing decisions on climate change [4, 5, 15, 19, 29, 52]. But like national governments, cities also face the same challenge of integrating policies to address their urban challenges and climate concerns simultaneously.

Without immediate and substantial reductions in GHG emissions across all sectors, global temperatures are projected to surpass pre-industrial levels by 2 °C by around 2050, leading to increased frequency, duration, and severity of extreme weather events [1]. These effects will be particularly exacerbated in cities and urban areas. Urban flooding [32, 43, 46] and urban heat island effects [2, 18, 45] have already emerged on their own as significant threats to urban populations. Moreover, cities are a part of larger geographies, and they cannot address climate change adaptation or mitigation while acting alone. It thus makes sense for countries to formulate specific urban climate change policy at the national level to guide climate actions and urban planning at national, sub-national, and local levels. But the approach to formulating such policy has not garnered much attention, particularly in the case of developing countries.

This chapter aims to fill the gap in research regarding urban climate change policymaking by examining existing approaches to address climate concerns in cities and urban areas. Using the term 'policy' to encompass strategies as well as plans, two key approaches can be identified—(1) one that treats urban concerns as a subset of the

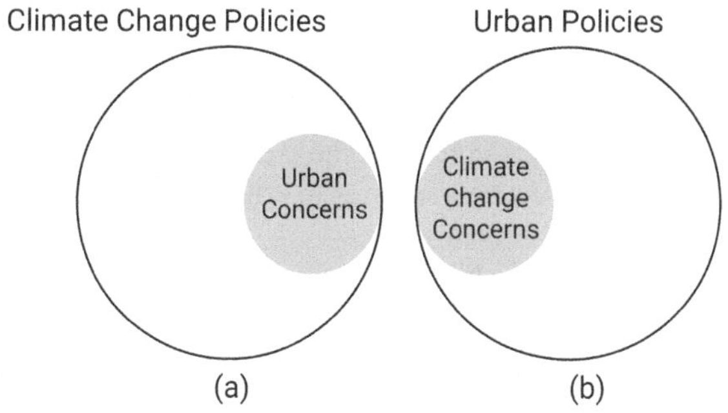

Fig. 1 Primary approaches to integrating urban and climate change concerns in policies

broader climate change policy domain (Fig. 1 Panel a) and (2) another that addresses climate concerns within the domain of urban development policy (Fig. 1 Panel b). This first approach focuses on understanding the overall implications of climate change and identifying specific urban challenges and solutions within that context. It recognizes that climate change affects urban areas differently due to factors such as dense populations and high resource consumption. This approach represents the prevalent practice of climate change policymaking, aligning with the international guidelines and commitments on climate actions.

The second approach to policymaking reflects a growing trend among urban policymakers, where they are now taking into account climate considerations alongside the typical, country-specific urban planning issues. Examples include policies on smart city development, sustainable urbanization, and green development, among others. This approach prioritizes urban development and address climate change concerns as components of urban planning and governance. The emphasis is on integrating climate change mitigation and adaptation measures into urban policies and strategies.

Many urbanizing countries in the Global South rely on international support for their climate actions. In these countries, international development agencies also often provide support to policy entrepreneurs and civil society organizations for policy advocacy [48], which helps make climate change policymaking process consultative and transparent. Treating urban concerns within the overall domain of climate change allows policymakers to benefit from the global knowledge, technical resources, international funds, and other global efforts dedicated to addressing climate change. However, there is a risk that the specific needs and nuances of urban areas may not receive adequate attention and resources compared to broader climate change goals. On the other hand, treating climate concerns within the overall domain of the urban sector enables policymakers to design targeted interventions that align with the immediate needs of urban populations. However, there is a risk of compartmentalizing climate change as a secondary concern within the urban domain, potentially undermining broader climate change efforts.

This chapter compares these two primary policymaking approaches in the case of Nepal, which is one of the world's most climate-vulnerable countries and also a rapidly urbanizing country in the Global South. Nepal's experiences with its climate plans as a part of its international commitments reveal how the climate needs of urban areas can be overlooked in broader climate change discourse. Likewise, Nepal's experiences with urban policymaking reveal how urban policies can fail to address climate change adequately. These experiences offer valuable insights into the policy gaps and corrective measures to create synergy between urban and climate change policies.

This chapter is based on the review of Nepal's key policy documents concerning climate change and urban development, focusing on how urban concerns are addressed in the climate change policies, and vice versa. Figure 2 displays key policies along with their respective publication years. Additionally, the chapter also incorporates insights from key informant interviews with climate policy entrepreneurs, urban policymakers, and practitioners.

2 Climate Change and Urbanization in Nepal

Nepal's vulnerability to climate change is attributed to its fragile topography, from the mountainous cryosphere in the North to the plains in the South, poor economy, and climate-sensitive, subsistence livelihoods of the majority population. In terms of the impacts of weather-related loss events, Nepal ranks as the 12th most affected country for 2019 and 10th most affected country for 2000–2019 in Germanwatch's Climate Risk Index [8]. The direct cost from current climate variability and extreme events, such as floods and landslides, was estimated at 1.5–2% of the country's GDP in 2014, with projections suggesting that the future economic impact of climate change in the country could reach an additional 2–3% of GDP annually by the year 2050 [14, 39].

Temperature in Nepal is projected to rise by 0.92 to 1.07 °C in the medium term (2016–2045) and 1.3 to 1.8 °C in the long term (2036–2065) from the reference period of 1981–2010 [35]. Likewise, annual precipitation is expected to increase by 2–6% in the medium term and 8–12% in the long term. The projections indicate drier winters and wetter monsoon summers, with up to a threefold increase in rainfall. As a result of these changes in climate, extreme weather events (heavy rainfall, snowstorms, high winds, and hailstorms), heat waves, cold waves, floods, landslides, and forest fires are likely to increase in frequency and severity, and chronic or slow onset hazards such as drought, changes in precipitation patterns, snow cover changes, glacier retreat, and GLOFs, are expected to intensify [37].

Along with escalating climate risks, Nepal has also experienced rapid urbanization, with growth rates reaching up to 5% per year [53]. As of 2021, about two-thirds of Nepal's population resides in urban and peri-urban areas, up from 61.5% in 2011 [42]. The rapid urban growth, along with inadequate urban management, has heightened climate vulnerability in urban areas, affecting not only people but also infrastructure. Urban infrastructure, which needs heavy investment, is increasingly at risk from heavy rainfall, floods, and landslides, which are occurring more frequently and

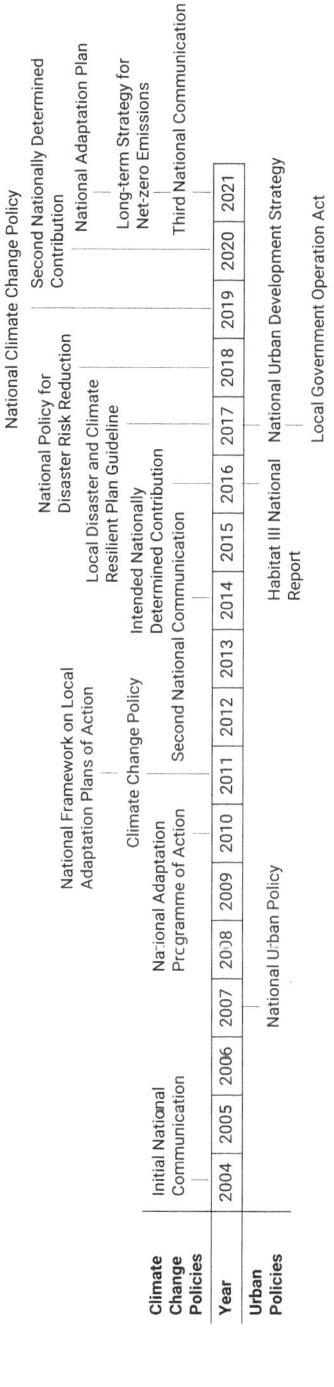

Fig. 2 Nepal's key urban and climate change policies and their respective publication years

intensely. Inadequate considerations in urban planning for climate change adaptation has heightened climate risks in Nepal's urban areas.

Despite negligible contribution to the global annual GHG emission, accounting for only about 0.1% in 2020 (Climate Watch n.d.), Nepal has committed to achieving net-zero greenhouse gas emissions by 2045 [12]. The combination of rapid urbanization, energy-inefficient construction technology, and vehicular growth has gradually increased Nepal's GHG emissions. For instance, the increase in vehicle ownership and usage led the country's transport-sector GHG emissions to more than double between 2013 and 2018 [53]. It is evident that Nepal cannot attain its ambitious emission target without addressing urban emissions.

In summary, Nepal's resilience to climate change hinges significantly on how it manages urbanization from both adaptation and mitigation perspectives. However, the country faces a policy conundrum as it strives to address both climate change and urbanization in an integrated and cohesive manner. The challenge arises from the involvement of multiple government agencies overseeing urban affairs, compounded by a history of poor coordination and collaboration among these entities. This complexity is further exacerbated by the establishment of provincial and local governments as new tiers of government in addition to the federal government following the country's adoption of federal political system in 2015. Nonetheless, the main responsibility and authority to integrate climate change and urbanization policies rest with the federal government.

Nepal does not have a dedicated urban climate change policy—an integrated policy to address climate change in urban areas from both adaptation and mitigation perspectives. The environment line ministry is responsible for formulating climate change policies, whereas there is a dedicated federal ministry to look after urban development policies. These federal ministries strive to integrate concerns of climate change and urban development in their policies in their own way. On the other hand, provincial and local governments have neither financial resource nor technical capacity to formulate climate policies or plans on their own. With the support of development partners, a few local governments have initiated the development of climate action plans, focusing their efforts on addressing the most relevant climate hazards in their territories (see, for e.g., [23, 24, 47]). Provincial governments have not yet made a significant impact in the climate policy arena.

3 Integrating Urban Concerns in Climate Change Policies

3.1 Urban Concerns in Climate Change Adaptation Policies

Although Nepal became a Party to the UNFCCC as early as 1994, Nepal's first urban policy, National Urban Policy (NUP) 2007, mentions nothing about climate change (see [38]). Interestingly, Nepal's Climate Change Policy 2011 also remains silent about the urban dimensions of climate change. This shows the longstanding lack of

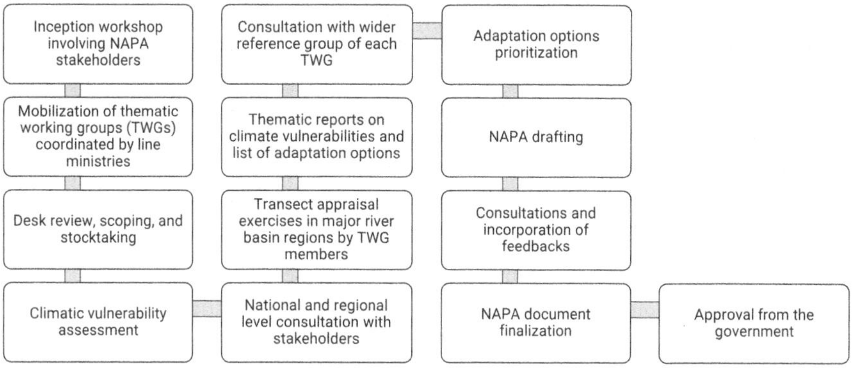

Fig. 3 The NAPA process. *Source* Adapted from MOE [33]

communication and collaboration between the two different agencies responsible for policy formulation on climate change and urban development.

Nepal's first attempt to integrate urban concerns into climate change policies materialized in the form of National Adaptation Programme of Action (NAPA) 2010, which aimed at focusing on the "immediate and urgent adaptation priorities at the national level" [50]. Following prescribed guidelines of the UNFCCC, the NAPA process involved six thematic and two cross-cutting working groups, each led by a different line ministry, with 'urban settlements and infrastructure' as one of the thematic groups [33]. The environment ministry led the overall process. Figure 3 depicts the NAPA process.

The NAPA process identified a long list of adaptation options and then grouped priority activities into nine integrated project concepts including the urban project concept titled 'Promoting climate smart urban settlement'. However, the NAPA's attempt to address adaptation in urban area suffered from three main shortcomings. First, the urban project concept was vaguely defined. For instance, it identified building codes as a tool to promote climate smart settlement but ignored more crucial interventions like land use regulations to promote urban resilience. This showed the NAPA's lack of direction regarding addressing urban climate concerns.

Second, the urban project concept failed to earn ownership from urban decision-makers, suggesting a lack of quality engagement between the environment ministry and its urban counterpart. Moreover, the urban ministry recommended replicating its own ongoing projects, such as eco-cities and healthy cities projects, as a priority adaptation option without sufficient justification or elaboration regarding these projects' objectives, suggesting power dynamics at play rather than a genuine interest in finding solutions to urban climate change problems.

Third, the NAPA failed to produce useful data for local-level analysis since climate change vulnerability assessments were conducted at the district level. Districts are much larger administrative entities than municipalities. The NAPA acknowledged that climate change adaptation varies based on the form, size, and location of settlements, and there is a need to downscale climate change scenarios at the meso-level, but it fell short of fulfilling this need.

Building on the NAPA experience and guided by the Climate Change Policy 2011, Nepal introduced the National Framework on Local Adaptation Plans for Action (LAPA) in 2011 to integrate adaptation options into development policy and planning processes through local- and community-based adaptation plans and pilot projects [44]. Unlike NAPA's top-down approach, the LAPA framework employed both top-down and bottom-up approaches [11] as depicted in Fig. 4. It used a top-down approach to identify climate vulnerability assessment at different geographical scales, such as districts, municipalities, and settlements. At the settlement level, the framework aimed to determine adaptation needs and priorities, leading to a bottom-up approach for integrating adaptation needs into local plans, district plans, and ultimately national plans in that order.

The LAPA framework, which was last updated in 2019, could have been an opportunity to address urban climate adaptation needs but it focussed only on rural areas with about 100 local plans prepared till date. Nevertheless, the LAPA framework offers a valuable guide for creating adaptation plans at local levels, which can also be applied to urban and urbanizing areas.

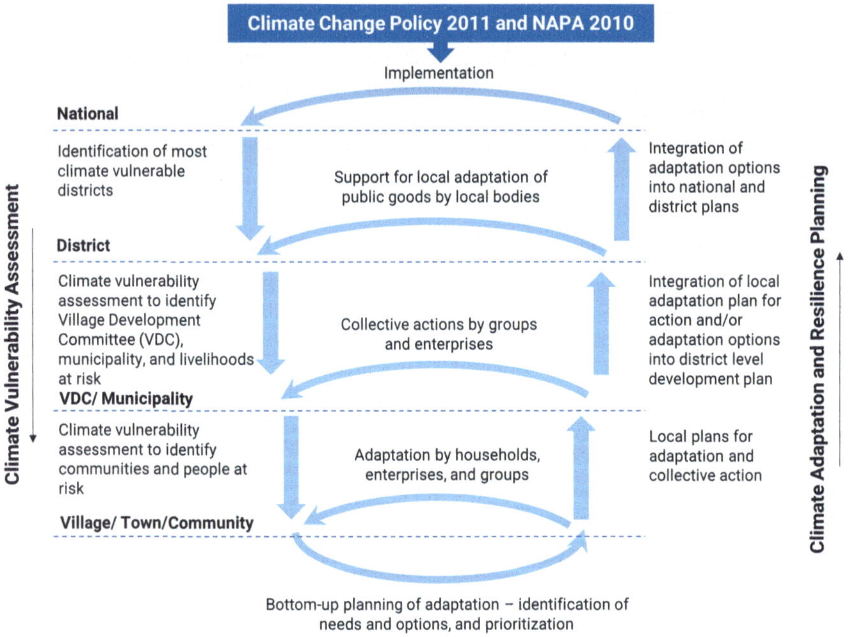

Fig. 4 The LAPA approach to integrating climate change adaptation and resilience into local and national development planning. *Source* Adapted from GON [11]

The National Climate Change Policy 2019, which replaced the 2011 Policy, integrates strategies like climate risk-sensitive land use planning and low-emission technologies in settlement development. However, it treats 'rural and urban settlements' as a single priority area for climate actions, overlooking the distinct roles of urban areas. Similarly, Nepal's National Adaptation Plan (NAP) (2021–2050) follows suit, jointly targeting 'rural and urban settlements' despite initial recognition of urban concerns in its preparatory phase when it incorporated "urban settlements and infrastructure" as a thematic area [34] as in the NAPA. While the NAP addresses urban threats like urban heat island effects, it falls short in fully addressing urban-specific adaptation needs.

Building on the NAPA experience, the NAP process has continued the approach of engaging line ministries through thematic working groups. Figure 5 depicts the NAP process. The NAP provides a better-informed perspective on Nepal's climate change scenarios based on the latest scientific evidence. The NAP also makes an important contribution for local-level adaptation planning by predicting climate extreme events and providing climate risk scenarios at local (municipal) levels for the first time in Nepal. However, the NAP's failure to provide settlement-level analysis was attributed to the lack of data, which the leading ministry had anticipated would be supplied by the urban ministry. This again shows the lack of inter-agency collaboration in policymaking. Compared to the NAPA, the NAP recommended more articulated urban-centric program concepts (Table 1) but the issue of buy-in from the urban ministry remains unclear.

The issue of inter-agency collaboration in policymaking has become even more crucial in Nepal following the enactment of the 2015 Constitution, which established a three-tiered government system comprising federal, provincial, and local governments, each enjoying exclusive and shared constitutional rights. Local governments,

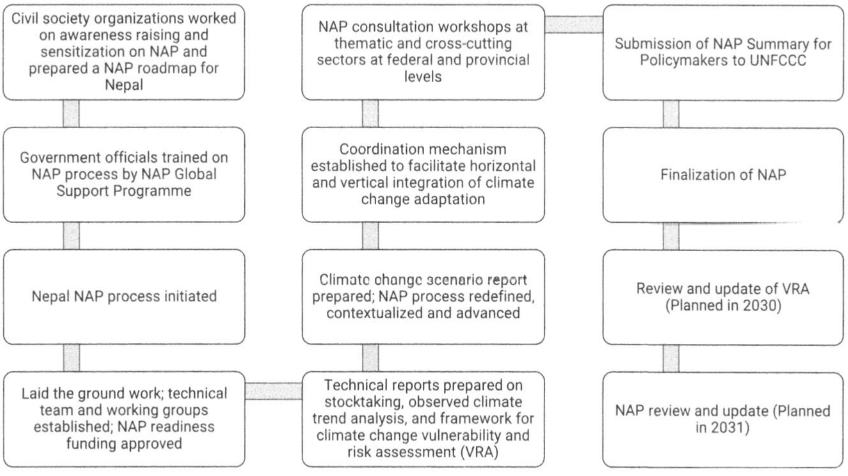

Fig. 5 The NAP process. *Source* Adapted from GON [14]. *Note* VRA: Vulnerability and Risk Assessment

Table 1 Recommended urban-related programs under the NAPA and NAP

Policy initiatives	Urban-related thematic area	Recommended programs	Objectives
NAPA 2010	Urban settlements and infrastructure	Promoting climate smart urban settlement	• Promote climate smart urbanization process in Nepal through national and local capacity building
NAP 2021	Rural and urban settlements	Promoting the circular economy for sustainable urban development	• Pilot climate resilient city planning • Build national capacity for adaptive urban development
		Developing integrated settlement and urbanization models for climate risk reduction and supplying climate adaptation services through nature-based solutions and policy reform	• Resettle/relocate climate and disaster vulnerable populations to safer, serviced areas • Study and identify vulnerable settlements across three ecological zones • Identify safer locations for resettlement and relocation in rural municipal-level strategic spatial plans • Introduce integrated settlement planning for disaster-resilient community development
		Updating and promoting climate resilient building designs, codes, practices and construction technologies and national capacity building to further implementation	• Prepare/revise climate risk-informed urban and rural development plans • Design, pilot, and demonstrate climate and disaster-resilient construction technology • Promote climate resilient building practices • Prepare/revise climate risk-informed urban and rural development plans • Explore and identify environment-friendly building materials and construction technologies • Disseminate information and raise awareness on climate-resilient building practices

Source GON [13], MOE [33]

in particular, are emerging as important players in terms of addressing climate change at community levels. Some local governments have initiated risk-sensitive land use planning, which also incorporates measures for climate change adaptation [17].

Guided by Local Government Operation Act 2017, Local Disaster and Climate Resilient Plan Guideline 2017, and National Policy for Disaster Risk Reduction 2018, the federal government has been incentivising local governments to prepare Local Disaster and Climate Resilient Plan (LDCRP) to mainstream disaster and climate resilience in local development process [10]. However, the LDCRP's efforts to promote ownership and facilitate implementation have faltered due to its reliance on establishing new institutions, allocating financial resources, and navigating complex coordination with municipal, provincial, and federal departments. Most local governments lack financial resources and technical capacity to formulate climate action plans in addition to the regular local development plans.

3.2 Climate Change Mitigation Policies

The need to integrate urban concerns into national climate policy is not limited to adaptation options but also extends to mitigation efforts. Without referring to urban areas, Nepal's Climate Change Policy 2011 emphasized embracing low-carbon socioeconomic development whereas the National Climate Change Policy 2019 has aimed at developing standards for mitigation of GHG emissions in industry and transport sector.

Under the Paris Agreement of the UNFCCC, Nepal submitted its first and second NDC in 2016 and 2020, respectively. The NDC 2020 has established quantifiable activity targets as well as policy targets in key sectors including 'urban settlements'. Table 2 shows targets relevant to urban areas. However, the NDC 2020 does not cover construction and housing sectors even as the Third National Communication (TNC) acknowledges these sectors as among the high carbon emitting sectors in Nepal [36, 37]. It is evident that the NDC 2020 does not adequately acknowledge the urban contributions to the GHG emissions and follows the conventional approach of viewing urban areas from adaptation perspective only.

Nepal's Long-term Strategy for Net-zero Emissions 2021, which builds on the NDC 2020, reiterates and elaborates targets in all other sectors shown in Table 2 except urban settlements. Despite its ambitious target to achieve net-zero emissions by the year 2045, the Strategy is yet another case of downplaying the role of urban areas as significant energy consumers and GHG emitters. Although Nepal's GHG emissions are relatively low at present, they are projected to increase in the near future in proportion with the country's rapid pace of urbanization. The success of Nepal's mitigation efforts critically depends on how urbanization is managed, necessitating the involvement of municipal governments. Without reducing GHG emissions at the local level, achieving the national target is not possible, and the role of municipal governments is crucial in preparing and implementing local mitigation plans.

Table 2 Urban-related policy and quantitative targets in the NDC 2020

Relevant sectors	Targets
Energy	• By 2030, ensure 15% of total energy demand is met by clean energy sources
Transport	• In 2025, ensure electric vehicles (EVs) accounts for 25% of all private passenger vehicle sales (including 2-wheelers) and 20 per cent of all 4-wheeler public passenger vehicle sales • By 2030, increase EV sales to 90% of all private passenger vehicle sales (including 2-wheelers) and 60% of all 4-wheeler public passenger vehicle sales • By 2030, develop a 200 km electric rail network to support public transportation and mass transportation of goods
Cooking	• By 2030, ensure 25% of households use electric stoves as their primary mode of cooking • By 2025, install 500 large scale biogas plants
Waste	• By 2025, treat 380 million litres of wastewater per day and 60,000 cubic metres of faecal sludge per day
Urban settlements	• Adopt national building codes and prepare Integrated Urban Development Plans emphasizing low-carbon and climate-resilient urban settlements in all municipalities • By 2025, revise the urban environment management guidelines to incorporate activities related to promoting low carbon and climate-resilient urban settlements

Source GON [12]

3.3 *Challenges in Integrating Urban Concerns in Climate Change Policies*

Nepal's climate change policymaking has allowed policymakers to benefit from the global knowledge in terms of best practices and efforts dedicated to addressing climate change including access to international funds and technical expertise. For instance, the NAPA Project was supported by the Embassy of Denmark, UK Department for International Development (DFID), Global Environment Facility (GEF), and United Nations Development Programme (UNDP) Nepal [33]. Nepal's NAP process received support from Green Climate Fund (GCF) through a project executed by United Nations Environment Programme (UNEP) [14]. Likewise, Nepal received support from the NDC Partnership as a member for the preparation of its second NDC.

Because of access to international financial and technical resources, the NAPA initiative could produce climate change scenarios and vulnerability assessments at different spatial scales, and the NAP could provide the improved, updated, and upscaled versions of the same. The same resources could have been used to develop knowledge tools and data to better inform urban policies, but the lack of inter-agency collaboration prevented this from happening. Moreover, the perception that the international development agencies are prioritising the climate agenda in Nepal over the country's other development needs could have also affected the prospect of inter-agency collaboration.

Both the NAPA and NAP processes placed an excessive burden on one agency, the environment ministry, to coordinate and gather inputs from partner agencies, attempting to achieve too much in a short period. As a result, other stakeholder agencies became dissatisfied with the process outputs, leading to a loss of interest in owning the process and implementing recommended measures. More specifically, in the urban context, the specific needs and nuances of urban areas could not receive adequate attention and resources compared to broader climate change goals.

4 Integrating Climate Change Concerns in Urban Policies and Its Challenges

Nepal's National Urban Policy (NUP) 2007, which aims to achieve balanced urban growth, enhance urban living standards, and ensure effective urban management, notably refrained from mentioning anything about climate change. But recent urban policies are more mindful about recognizing climate change as an urban risk.

As one of the first instances of integrating climate change into urban policy in the country, Nepal included a plan of action for addressing climate change in its 2016 National Report for the Third United Nations Conference on Housing and Sustainable Urban Development (Habitat III) [40]. The plan of action involves implementing Nepal's climate change commitments, enhancing stakeholder participation and ownership, raising awareness of climate risks, and integrating adaptation into development policies and strategies. The plan notably emphasizes integrating climate change adaptation into urban planning and development, as well as formulating and implementing the NAP with a focus on urban settlements and infrastructure.

The NUDS 2017, first drafted in 2015, has also recognized climate change as an urban threat. It aims to promote innovative urbanization and development strategies considering sustainability, resilience, and climate change challenges. With a vision to achieve a balance and prosperous national urban system by 2031, the NUDS 2017 has identified urban system, urban infrastructure, urban environment, and urban economy as its thematic areas, and urban governance, urban finance, urban investment, and urban land management as mechanisms to realize the vision (Fig. 6). It has embraced sustainability, inclusivity, green asset and technology, efficiency, and resilience as five guiding principles, with 'resilience' broadly addressing disaster and climate risks [41].

The urban ministry's earlier engagements in the NAPA, as well as climate projects, could have influenced the ministry to address climate change in its flagship policy document. However, the NUDS only briefly touched upon the issue of climate change and that also from adaptation perspective. However, it has rightly emphasized the scarcity of information on climate change in urban areas across diverse ecological regions and has identified that the generation of such information as a priority activity,

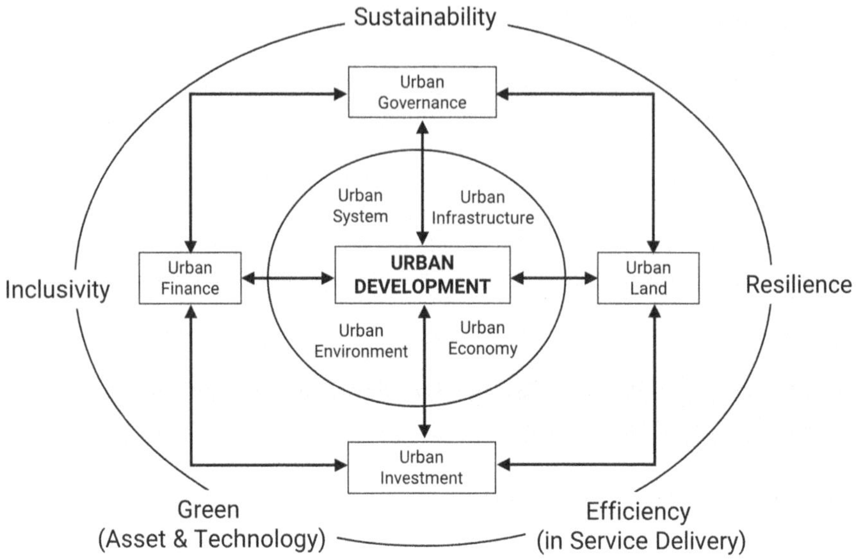

Fig. 6 The NUDS framework. *Source* Adapted from MOUD [41]

which is vital for integrating climate concerns into urban policy. The NAP was one such opportunity but was missed as discussed earlier.

More recently, urban planning tools such as Integrated Urban Development Plans and Municipal Transport Management Plans have emphasized climate-resilient physical planning and infrastructure development. Moreover, it is mandatory for all local governments to prepare medium-term (five-year) periodic plans, which should also address climate change adaptation needs. The latter can be addressed in a more comprehensive manner through the LDCRPs, but the concern is again about the lack of financial and technical capacity of local governments.

Nepal's experience shows that incorporating climate concerns into urban policymaking is not less challenging than integrating urban concerns into climate change policymaking. This is because similar to climate change, urban development is a cross-cutting issue with numerous stakeholder agencies involved in the planning and management of urban and urbanizing areas. The scope of work of the urban ministry overlaps considerably with other agencies. It faces the same problem in gathering feedback from other agencies on urban policies that the environment ministry faces regarding climate change policies.

Integrating climate concerns within the domain of urban planning can provide an opportunity for policymakers to design targeted interventions that address climate change while aligning with urban development goals. However, such opportunity has not been utilized adequately. There are three reasons for this: first, the administrative entities in Nepal that represent cities are urban and urbanizing municipalities. But there is no direct working relationship between the urban ministry and municipal governments for sustained partnership. The complexity involved in interpreting urban policies (responsibility of the urban ministry) with the context of local development (responsibility of the municipal governments) only exacerbates this challenge. Third, lack of settlement-level demographic and climate data poses another hurdle against a meaningful collaboration between the urban ministry and municipal governments. Second, despite possessing constitutional powers, municipal governments lack financial and technical resources for integrating climate change concerns in their urban development policies.

It is also important to note that unlike climate change policymaking, urban policymaking is relatively free from donor influence, which also implies less access to financial resources and international technical support. Resources are vital for the development of tools and data to inform urban climate change policies. There is also a risk of compartmentalizing climate change as a secondary concern within the urban areas, undermining broader climate change efforts.

5 Way Forward

Nepal's experiences show that integrating urban concerns in climate change policies and integrating climate change concerns in urban policies as two separate exercises are counterproductive in tackling the dual challenges of rapid urbanization and climate change because of the individual focus of each approach. Bringing together line ministries involved in formulating urban and climate change policies is crucial to ensure that their efforts complement each other. This can mean sharing of resources, expertise, climate data, and networks (Fig. 7 Panel a). An integrated approach to urban climate change policy is necessary that balances the specific needs of urban areas with broader climate goals (Fig. 7 Panel b).

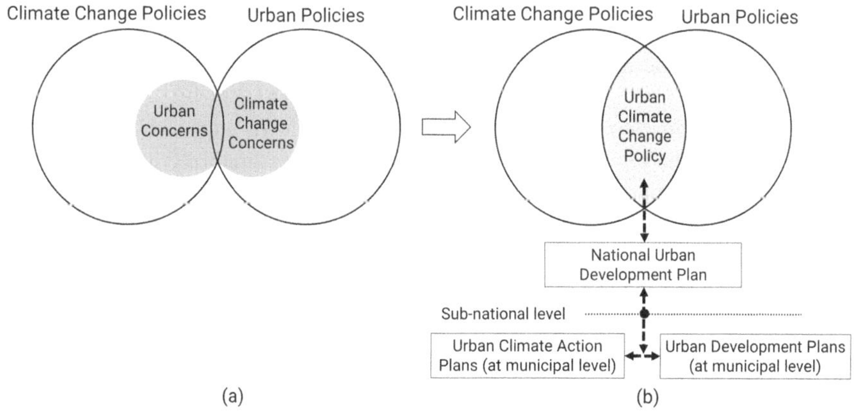

Fig. 7 Way forward for an integrated urban climate change policymaking

Such an integrated approach is also necessary for engaging local governments, as they are best positioned to tackle this dual challenge at local levels, given their constitutional responsibility to address all stages of disaster risk reduction (DRR) at the local level, including climate risks. It is also crucial to involve local governments in climate change initiatives because the impact of climate change is felt locally. Some local governments have initiated actions, often supported by development agencies, as seen in the Nepalgunj Heat Action Plan 2023 [47]. An integrated urban climate change policy should not only define provisions to inform local climate actions through guidelines and technical support but also pave ways for resource sharing and cross-learning opportunities.

The policy should also facilitate integration of urban and climate concerns at national, sub-national (for e.g., provincial), and local (municipal) levels along with vertical integrations between national, sub-national, and local-level policies. This can be achieved through national urban development plan guided by the integrated urban climate change policy, which then can inform local climate action plans and urban development plans (Fig. 7 Panel b).

To ensure the effectiveness of urban climate change policy, it should learn from past shortcomings, particularly regarding institutional roles and engagement. Building on the NAPA, NAP, and NDC experiences and leveraging access to global resources, the environment ministry is best positioned to develop data on climate change trends, climate scenarios, hazard assessment, and GHG emissions, from the national to settlement levels. Likewise, the urban ministry can develop urban settlement-level demographic, socio-economic, and infrastructure-related data. The two ministries should collaborate on urban vulnerability and risk assessment (VRA) and urban emission strategy, with assistance from municipalities for local data collection, field verification, and critical feedback as suggested in the LAPA framework. This would pave way for the urban ministry to formulate national urban development plan integrating urban and climate concerns (adaptation and mitigation). This national plan can then guide municipalities to prepare and implement local climate action plans (or LDCRPs) and urban development plans (or the mandatory periodic plans), for which the municipal governments would require a long-term capacity building support.

Policymaking in Nepal still tends to follow a top-down approach with federal agencies leading the process with consultative workshops organized at sub-national and local levels, primarily for validation. New policy should change this paradigm by incorporating bottom-up policy feedback mechanism, building on the LAPA experience. For instance, the national urban development plan should not only provide guidelines for the local plans but there should be a mechanism for the review of plans, from national to local, and for the allowance of essential adjustments. Access to data, standards, case studies, success stories, and other specific instructions should be easy for all agencies including municipal governments. This can be achieved through a robust information system and knowledge portal.

6 Concluding Remarks

The interconnected challenges of climate change and rapid urbanization demand dedicated and cohesive policymaking. The exploration of two distinct approaches—one integrating urban concerns into the broader climate change domain and the other addressing climate issues within the urban sector—reveals the importance of a balanced and integrated strategy. While the first approach leverages global knowledge, funds, and collaboration, there is a risk of overlooking the specific needs of urban areas. Conversely, the second approach recognizes the unique challenges of urban regions but may risk compartmentalizing climate change within the urban sector. Moving forward, an integrated approach that synergizes the strengths of both strategies is crucial.

Such integrated approach should not only acknowledge the urgency of addressing climate risks at the local level but also position local governments as pivotal actors in forging climate-resilient urbanization. Nepal's efforts on climate change and urban policymaking through separate line ministries have failed to create synergy, resulting into non-optimal use of resources. An integrated urban climate change policy led by the urban ministry in collaboration with stakeholder agencies including local governments will help balance both urban and climate change concerns. Engaging and empowering local governments becomes paramount in this endeavour, as the impacts of climate change are realized locally, and effective mitigation efforts also begin at the local level.

References

1. Adelekan I, Cartwright A, Chow W, Colenbrander S, Dawson R, Garschagen M, Haasnoot M, Hashizume M, Klaus I, Krishnaswamy J, Lemos MF, Ley D, McPhearson T, Pelling M, Kodira PP, Revi A, Sara LM, Simpson NP, Singh C, Solecki W, Thomas A, Trisos C (2022) What the latest science on impacts, adaptation and vulnerability means for cities and urban areas. Indian Institute for Human Settlements. https://doi.org/10.24943/SUPSV209.2022
2. Aghamohammadi N, Ramakreshnan L (2024) Chapter 5—The impact of heat mitigation and adaptation technologies on urban health. In Aghamohammadi N, Santamouris M (eds) Mitigation and adaptation of urban overheating. Elsevier, pp 155–173. https://doi.org/10.1016/B978-0-443-13502-6.00005-1
3. Babiker M, Bazaz A, Bertoldi P, Creutzig F, De Coninck H, De Kleijne K, Dhakal S, Haldar S, Jiang K, Kılkış Ş, Klaus I, Krishnaswamy J, Lwasa S, Niamir L, Pathak M, Pereira JP, Revi A, Roy J, Seto KC, Singh C, Some S, Steg L, Ürge-Vorsatz D (2022) What the latest science on climate change mitigation means for cities and urban areas. Indian Institute for Human Settlements. https://doi.org/10.24943/SUPSV310.2022
4. Bäckstrand K, Kuyper JW, Linnér BO, Lövbrand E (2017) Non-state actors in global climate governance: from Copenhagen to Paris and beyond. Environ Polit 26(4):561–579. https://doi.org/10.1080/09644016.2017.1327485
5. Biesbroek R, Lesnikowski A, Ford JD, Berrang-Ford L, Vink M (2018) Do administrative traditions matter for climate change adaptation policy? A comparative analysis of 32 high-income countries. Rev Policy Res 35(6):881–906. https://doi.org/10.1111/ropr.12309

6. Climate Watch (n.d.) Nepal: GHG emissions. https://www.climatewatchdata.org/countries/NPL. Accessed 14 April 2024
7. Dodman D, Hayward B, Pelling M, Castan Broto V, Chow W, Chu E, Dawson R, Khirfan L, McPhearson T, Prakash A, Zheng Y, Ziervogel G (2022) Cities, settlements and key infrastructure. In: Pörtner H-O, Roberts DC, Tignor M, Poloczanska ES, Mintenbeck K, Alegría A, Craig M, Langsdorf S, Löschke S, Möller V, Okem A, Rama B (eds) Climate change 2022: impacts, adaptation and vulnerability. Contribution of Working Group II to the Sixth Assessment Report of the Intergovernmental Panel on Climate Change. Cambridge University Press, pp 907–1040. https://doi.org/10.1017/9781009325844.008
8. Eckstein D, Künzel V, Schäfer L (2021) Global Climate Risk Index 2021: who suffers most from extreme weather events? Weather-related loss events in 2019 and 2000–2019. Germanwatch
9. Gallardo L, Hamdi R, Islam AKMS, Klaus I, Klimont Z, Krishnaswamy J, Pinto I, Otto F, Raghavan K, Revi A, Sörensson AA, Szopa S (2022) What the latest physical science of climate change means for cities and urban areas. Indian Institute for Human Settlements. https://doi.org/10.24943/SUPSV108.2022
10. GIZ (2020) A guidebook for preparation of local disaster and climate resilient plan. Deutsche Gesellschaft für Internationale Zusammenarbeit (GIZ) GmbH. Kathmandu, Nepal. https://dpnet.org.np/public/uploads/files/LDCRP_Handbook-for-LGs%202021-09-12%2009-54-16.pdf
11. GON (2011) National framework on local adaptation plans for action. Government of Nepal. Kathmandu, Nepal. https://climate.mohp.gov.np/downloads/National_Framework_Local_Adaptation_Plan.pdf
12. GON (2020) Second Nationally Determined Contribution (NDC). Government of Nepal. Kathmandu Nepal. https://climate.mohp.gov.np/attachments/article/167/Second%20Nationally%20Determined%20Contribution%20(NDC)%20-%202020.pdf
13. GON (2021a) National Adaptation Plan (NAP): summary for policymakers. Government of Nepal. Kathmandu Nepal. https://faolex.fao.org/docs/pdf/nep215543.pdf
14. GON (2021b) National Adaptation Plan (NAP) 2021–2050. Government of Nepal. Kathmandu Nepal. https://doenv.gov.np/rules/NAP-2021-2050-1700841577.pdf
15. Gordon DJ, Johnson CA (2018) City-networks, global climate governance, and the road to 1.5°C. Curr Opinion Environ Sustain 30:35–41. https://doi.org/10.1016/j.cosust.2018.02.011
16. Gurney KR, Kılkış Ş, Seto KC, Lwasa S, Moran D, Riahi K, Keller M, Rayner P, Luqman M (2022) Greenhouse gas emissions from global cities under SSP/RCP scenarios, 1990 to 2100. Glob Environ Chang 73:102478. https://doi.org/10.1016/j.gloenvcha.2022.102478
17. Hada C, Shaw R (2022) Integrating disaster and climate change in risk-sensitive land use planning. In Shaw R (ed) Handbook on climate change and disasters. UK, USA: Edward Elgar Publishing. https://doi.org/10.4337/9781800371613
18. Han D, Zhang T, Qin Y, Tan Y, Liu J (2023) A comparative review on the mitigation strategies of urban heat island (UHI): a pathway for sustainable urban development. Climate Dev 15(5):379–403. https://doi.org/10.1080/17565529.2022.2092051
19. Heikkinen M, Karimo A, Klein J, Juhola S, Ylä-Anttila T (2020) Transnational municipal networks and climate change adaptation: a study of 377 cities. J Clean Prod 257:120474. https://doi.org/10.1016/j.jclepro.2020.120474
20. Hughes S (2017) The politics of urban climate change policy: toward a research agenda. Urban Affairs Rev 53(2):362–380. https://doi.org/10.1177/1078087416649756
21. Hughes S, Romero-Lankao P (2014) Science and institution building in urban climate-change policymaking. Environ Polit 23(6):1023–1042. https://doi.org/10.1080/09644016.2014.921459
22. Hughes S, Yordi S, Besco L (2020) The role of pilot projects in urban climate change policy innovation. Policy Stud J 48(2):271–297. https://doi.org/10.1111/psj.12288
23. ICF (2023a) Butwal: building a green, climate-resilient and livable city (2023–2030). SUDRIDH- Nepal Urban Resilience Project, Nepal. https://sudridh-nurp.com/product/green-city-action-plan-2023-2030-butwal-full-report/. Accessed 14 April 2024

24. ICF (2023b) Janakpurdham: building a green, climate-resilient and livable city (2023–2030). SUDRIDH- Nepal Urban Resilience Project, Nepal. https://sudridh-nurp.com/product/green-city-action-plan-2023-2030-janakpurdham-full-report/. Accessed 14 April 2024
25. IPCC (2022a) Summary for policymakers [Pörtner H-O, Roberts DC, Poloczanska ES, Mintenbeck K, Tignor M, Alegría A, Craig M, Langsdorf S, Löschke S, Möller V, Okem A (eds)]. In: Pörtner H-O, Roberts DC, Tignor M, Poloczanska ES, Mintenbeck K, Alegría A, Craig M, Langsdorf S, Löschke S, Möller V, Okem A, Rama B (eds) Climate change 2022: impacts, adaptation and vulnerability. Contribution of Working Group II to the Sixth Assessment Report of the Intergovernmental Panel on Climate Change. Cambridge University Press, pp 3–33. https://doi.org/10.1017/9781009325844.001
26. IPCC (2022b). Summary for policymakers [Shukla PR, Skea J, Reisinger A, Slade R, Fradera R, Pathak M, Al Khourdajie A, Belkacemi M, van Diemen R, Hasija A, Lisboa G, Luz S, Malley J, McCollum D, Some S, Vyas P (eds)]. In: Shukla PR, Skea J, Slade R, Al Khourdajie A, van Diemen R, McCollum D, Pathak M, Some S, Vyas P, Fradera R, Belkacemi M, Hasija A, Lisboa G, Luz S, Malley J (eds) Climate change 2022: mitigation of climate change. Contribution of Working Group III to the Sixth Assessment Report of the Intergovernmental Panel on Climate Change. Cambridge University Press. https://doi.org/10.1017/9781009157926.001
27. IPCC (2023). Summary for policymakers. In: Climate change 2023: synthesis report. Contribution of Working Groups I, II and III to the Sixth Assessment Report of the Intergovernmental Panel on Climate Change [Core Writing Team, Lee H, Romero J (eds)]. IPCC, Geneva, Switzerland, pp 1–34. https://doi.org/10.59327/IPCC/AR6-9789291691647.001
28. Leck H, Roberts D (2015) What lies beneath: Understanding the invisible aspects of municipal climate change governance. Curr Opinion Environ Sustain 13:61–67. https://doi.org/10.1016/j.cosust.2015.02.004
29. Lee T, Jung HY (2018) Mapping city-to-city networks for climate change action: geographic bases, link modalities, functions, and activity. J Clean Prod 182:96–104. https://doi.org/10.1016/j.jclepro.2018.02.034
30. Lee T, Painter J (2015) Comprehensive local climate policy: the role of urban governance. Urban Climate 14(Part 4):566–577. https://doi.org/10.1016/j.uclim.2015.09.003
31. Lwasa S, Seto KC, Bai X, Blanco H, Gurney KR, Kılkış Ş, Lucon O, Murakami J, Pan J, Sharifi A, Yamagata Y (2022) Urban systems and other settlements. In: IPCC (ed) Climate change 2022: mitigation of climate change. Contribution of Working Group III to the Sixth Assessment Report of the Intergovernmental Panel on Climate Change (pp. Chapter 10). Cambridge University Press. https://doi.org/10.1017/9781009157926.010
32. Manawi SMA, Nasir KAM, Shiru MS, Hotaki SF, Sediqi MN (2020) Urban flooding in the northern part of Kabul city: Causes and mitigation. Earth Syst Environ 4:599–610. https://doi.org/10.1007/s41748-020-00165-7
33. MOE (2010) National adaptation programme of action to climate change. Ministry of Environment, Government of Nepal. Kathmandu, Nepal. https://faolex.fao.org/docs/pdf/nep169509.pdf
34. MOFE (2018) Nepal's National Adaptation Plan (NAP) process: reflecting on lessons learned and the way forward. Ministry of Forest and Environment, Government of Nepal. Kathmandu, Nepal. https://napglobalnetwork.org/wp-content/uploads/2018/07/napgn-en-2018-nepal-nap-process.pdf
35. MOFE (2019) Climate change scenarios for Nepal for national adaptation plan. Ministry of Forest and Environment, Government of Nepal. Kathmandu, Nepal. https://lib.icimod.org/record/34554
36. MOFE (2021a) Nepal's third national communication to the United Nations Framework Convention on Climate Change (UNFCCC). Kathmandu, Nepal
37. MOFE (2021b) Vulnerability and risk assessment and identifying adaptation options: summary for policymakers. Ministry of Forest and Environment, Government of Nepal. Kathmandu, Nepal.
38. MOPPW (2007) National urban policy 2007. Ministry of Physical Planning and Works, Government of Nepal. Kathmandu, Nepal

39. MOSTE (2014) Economic impact assessment of climate change in key sectors in Nepal. Ministry of Science, Technology and Environment, Government of Nepal. Kathmandu: IDS Nepal, Global Climate Adaptation Partnership, and Practical Action
40. MOUD (2016) Third United Nations conference on housing and sustainable urban development (Habitat III): Nepal National Report. Ministry of Urban Development, Government of Nepal. https://moud.gov.np/storage/listies/July2019/HIII-Nepal-National-Report_Printed-version.pdf
41. MOUD (2017) National urban development strategy 2017, Part A: Main Document. Ministry of Urban Development, Government of Nepal. Kathmandu, Nepal.
42. NSO (2024) Degree of urbanization (DEGURBA) in Nepal. National Statistics Office, Government of Nepal. Kathmandu, Nepal
43. Rainey JL, Brody SD, Galloway GE, Highfield WE (2021) Assessment of the growing threat of urban flooding: a case study of a national survey. Urban Water J 18(5):375–381. https://doi.org/10.1080/1573062X.2021.1893356
44. Regmi BR, Star C, Leal Filho W (2016) Effectiveness of the local adaptation plan of action to support climate change adaptation in Nepal. Mitig Adapt Strat Glob Change 21:461–478. https://doi.org/10.1007/s11027-014-9610-3
45. Shahfahad, Talukdar S, Rihan M, Hang HT, Bhaskaran S, Rahman A (2022) Modelling urban heat island (UHI) and thermal field variation and their relationship with land use indices over Delhi and Mumbai metro cities. Environ Dev Sustain 24:3762–3790. https://doi.org/10.1007/s10668-021-01587-7
46. Sohn W, Kim J-H, Li M-H, Brown RD, Jaber FH (2020) How does increasing impervious surfaces affect urban flooding in response to climate variability? Ecol Ind 118:106774. https://doi.org/10.1016/j.ecolind.2020.106774
47. Subedi A, Khan R, Singh R (2023) Nepalgunj heat action plan 2023. Red Cross Red Crescent Climate Centre
48. Tanner T, Zaman RU, Acharya S (2018) Influencing adaptation policy: the role of policy entrepreneurs in securing ownership and climate action in South Asia. Learning Paper. Oxford Policy Management
49. UN DESA (2019) World urbanization prospects: the 2018 revision. United Nations, New York
50. UNFCCC (2002) Guidelines for the preparation of National Adaptation Programmes of Action (NAPA). LDC-LEG, Bonn, Germany
51. van der Heijden J (2019) Studying urban climate governance: where to begin, what to look for, and how to make a meaningful contribution to scholarship and practice. Earth Syst Gov 1:100005. https://doi.org/10.1016/j.esg.2019.100005
52. Woodruff SC (2018) City membership in climate change adaptation networks. Environ Sci Policy 84:60–68. https://doi.org/10.1016/j.envsci.2018.03.002
53. World Bank Group (2022) Nepal country climate and development report. CCDR Series. World Bank, Washington, DC

Building Climate-Resilient Cities Through Community Engagement and Participatory Design

Amir Tayyebi

Abstract This chapter investigates community engagement and participatory design as pivotal in forging climate-resilient urban strategies. These approaches involve local communities in the design process, fostering ownership, knowledge enhancement, and social cohesion for sustainable urban resilience. Key concepts of community engagement and participatory design are examined, emphasizing their significance in climate resilience. The chapter outlines integration methods into urban design, such as workshops and collaborative platforms, and the empowerment of communities to input into climate-resilient spaces. Worldwide case studies illustrate the benefits of these approaches, including heightened climate risk awareness, vulnerability identification, social equity, and tailored solutions. Challenges like inclusivity, interest conflicts, and technical-community balance are also discussed. The chapter concludes with guiding principles for practitioners and policymakers to enhance community engagement and participatory design effectiveness, advocating for trust, capacity building, perspective integration, and enduring community-designer-policy partnerships. Embracing these methods, cities can leverage collective wisdom for more sustainable, inclusive, and resilient urban environments against climate change.

Keywords Community engagement · Participatory design · Climate resilience · Urban design strategies · Sustainable cities

1 Introduction

Climate change is a pressing concern of our time [19], presenting a significant threat to cities worldwide [34, 52]. The increasing global temperatures, mainly caused by the emission of greenhouse gases, have led to far-reaching impacts on urban environments [25, 44]. These impacts encompass various ecological, social, and economic dimensions, thus establishing climate change as a multidimensional phenomenon. Urban areas are undergoing notable transformations, such as alterations in weather

A. Tayyebi (✉)
Architecture and Planning Department, Jundi-Shapur University of Technology, Dezful, Iran
e-mail: a_tayyebi@jsu.ac.ir

B. He et al. (eds.), *Urban Climate and Urban Design*, Urban Sustainability,
https://doi.org/10.1007/978-981-96-1521-6_14

patterns, heightened occurrences of extreme events, rising sea levels, and heatwaves [2, 22]. These changes pose substantial risks to infrastructure, human health, and socio-economic stability within urban settings.

Climate change presents significant challenges for urban areas, necessitating the development of urban design strategies that are resilient to climate impacts [24, 50, 66]. By integrating climate resilience into urban design, cities can enhance their capacity to withstand and adapt to the adverse consequences of climate change, thereby safeguarding the safety, well-being, and sustainability of their residents. These strategies encompass a range of measures, such as the incorporation of green infrastructure, adaptive architectural design, efficient energy systems, and sustainable transportation networks, all of which aim to foster the creation of resilient urban environments.

Climate-resilient urban design is an interdisciplinary approach that seeks to create urban environments capable of withstanding and adapting to the various impacts of climate change. This involves the implementation of design strategies that not only address immediate concerns but also anticipate future challenges. **Resilience** in this context refers to the ability of urban systems to absorb disturbances, reorganize while undergoing change, and still retain essentially the same function, structure, and feedbacks [47].

One of the key strategies for climate-resilient urban design is the development of **green infrastructure**. This includes the creation of parks, green roofs, and urban forests, which serve multiple purposes such as reducing the urban heat island effect, managing stormwater, and providing recreational spaces for residents. Green infrastructure helps to mitigate the effects of extreme weather events, such as heatwaves and heavy rainfall, by providing natural cooling and improving water absorption [14].

Another important aspect is the incorporation of **adaptive architectural design**. Buildings and structures are designed or retrofitted to be more energy-efficient, utilize renewable energy sources, and withstand extreme weather conditions. Features such as elevated foundations, flood-resistant materials, and thermal insulation contribute to the durability and sustainability of the built environment [40].

Efficient energy systems are also integral to climate-resilient urban design. The transition to **low-carbon energy sources**, such as solar and wind power, reduces greenhouse gas emissions and dependence on fossil fuels. Additionally, the implementation of smart grids and energy storage solutions ensures a reliable and flexible energy supply that can adapt to changing conditions [48].

Lastly, sustainable transportation networks play a vital role in reducing carbon emissions and enhancing connectivity within the city. The promotion of public transportation, cycling, and walking, along with the development of electric vehicle infrastructure, supports a shift towards more sustainable and resilient urban mobility [48].

Community engagement and participatory design approaches have gained significant recognition and adoption across various disciplines, including urban planning, architecture, and social sciences to facilitate the implementation and realization of these strategies [33, 65]. These approaches are firmly grounded in the belief that active involvement and collaboration of community members are vital for the creation of meaningful and sustainable solutions [11, 32]. By empowering communities to actively engage in decision-making processes, these approaches seek to cultivate a sense of ownership, inclusivity, and social equity. Moreover, community engagement and participatory design approaches prioritize the collective knowledge and expertise of community members, acknowledging their unique perspectives and lived experiences as invaluable assets for problem-solving.

The imperative for community engagement and participatory design in building climate-resilient cities stems from the recognition that climate change disproportionately affects urban populations, particularly those in vulnerable communities. Traditional top-down approaches to urban planning have often overlooked the unique insights and needs of local residents, leading to solutions that may not fully address or even exacerbate existing challenges. The integration of community engagement and participatory design ensures that urban resilience strategies are grounded in the lived experiences and knowledge of those most impacted by climate threats, fostering solutions that are both effective and equitable.

Despite the growing consensus on the value of these methodologies, there remains a significant research gap in understanding how to effectively operationalize community engagement and participatory design in diverse urban contexts. Many studies highlight successful case studies but fall short in providing a comprehensive framework for implementation that is adaptable to different cultural, social, and economic settings. This research seeks to bridge that gap by developing a nuanced understanding of the mechanisms through which community engagement and participatory design can be systematically incorporated into climate resilience planning. By doing so, it aims to provide policymakers, urban planners, and community leaders with actionable insights to create more inclusive and adaptive urban environments in the face of climate change [8].

2 Relevance of Community Engagement and Participatory Design in the Context of Climate Resilience

Community engagement plays a pivotal role in addressing climate resilience, assuming a crucial position in the development of effective and sustainable solutions [46]. Given the escalating impacts of climate change, the active involvement of local communities empowers them to effectively contribute to the identification, prioritisation, and implementation of strategies for adaptation and mitigation [15, 42]. By actively involving community members, their valuable knowledge, experiences, and perspectives can be assimilated into the decision-making process, thereby yielding contextually appropriate and locally tailored solutions (Fig. 1). Furthermore, community engagement cultivates a sense of ownership and shared responsibility, nurturing social capital and fortifying resilience within the community. This form of engagement enhances the efficacy, acceptance, and long-term viability of climate resilience initiatives, ultimately bolstering the community's ability to adapt and prosper in the face of climate challenges.

Participatory design approaches unequivocally position the active involvement of diverse stakeholders, including local communities, as integral throughout the entire design process. This inclusive and collaborative approach astutely acknowledges the inherent complexities and uncertainties associated with climate change, recognizing the imperative for solutions that are context-specific and grounded in principles of social justice [1, 46]. Through the engagement of various stakeholders such as community members, policymakers, scientists, and practitioners, participatory design leverages the collective intelligence, local knowledge, and tangible experiences of those directly impacted by climate change. Such a collaborative process fosters a sense of ownership, empowerment, and co-creation, guaranteeing that solutions effectively address the specific needs, values, and aspirations of the community. Furthermore, participatory design encourages social learning, engenders trust, and amplifies communication and transparency amongst stakeholders, significantly enhancing the prospects for successful implementation and long-term sustainability of climate resilience strategies [45, 49].

Fig. 1 Strength of community engagement

2.1 Understanding Community Engagement and Participatory Design

The concept and principles of community engagement form the bedrock for establishing fruitful partnerships and promoting the active involvement of community members in decision-making processes [54]. Community engagement can be described as a purposeful and inclusive approach that encourages the meaningful participation of individuals, groups, and organizations in identifying, addressing, and resolving matters that directly impact their lives and well-being (Fig. 2) [36, 57]. Several guiding principles underpin the efforts of community engagement. Primarily, inclusivity stands as a vital aspect, ensuring that a diverse range of voices and perspectives are taken into account and given a platform to be heard. This principle recognises and values the inherent diversity within communities, encompassing differences in age, gender, culture, socioeconomic status, and abilities [57]. Another crucial principle is the establishment of mutual respect and trust, creating an environment conducive to open and honest communication between community members and other stakeholders [57]. Transparency serves as an essential pillar, emphasising the need to provide clear and accessible information about the engagement process, its objectives, and potential outcomes. Collaboration and shared decision-making are equally crucial principles, highlighting the significance of working together to devise solutions that genuinely reflect the collective wisdom and aspirations of the community [57]. Lastly, community engagement should strive to build capacity, empowering community members with knowledge, skills, and resources so that they can actively participate in decision-making and take practical steps to address the needs of the community [57].

Engaging with the community is crucial in promoting climate resilience within communities as it brings about a range of benefits [60]. Firstly, community engagement helps to improve knowledge and understanding of the impact of climate change and strategies for adaptation among community members. By having inclusive discussions and educational programs, engagement enables individuals to effectively recognize and respond to climate-related challenges. It also encourages the sharing of knowledge, experiences, and best practices, which can significantly enhance the local capacity to mitigate and adapt to climate risks. Secondly, community engagement creates a sense of ownership and collective responsibility for climate resilience initiatives. By involving community members in decision-making processes, their perspectives and priorities can be incorporated into climate action plans, ensuring relevance, effectiveness, and sustainability. This sense of ownership generates greater motivation and commitment among community members to take action and implement resilience measures. Additionally, community engagement fosters social cohesion and collaboration by bringing together diverse stakeholders, including residents, local authorities, NGOs, and businesses. Such collaboration strengthens social capital, promotes cross-sectoral partnerships, and facilitates resource sharing, all of which contribute to the collective efforts for building climate resilience. Ultimately,

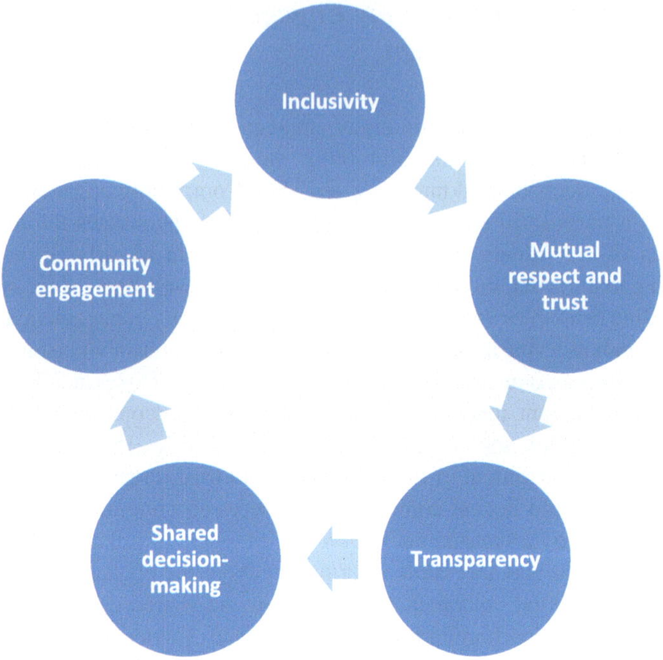

Fig. 2 Guiding principles of participatory design

community engagement cultivates a culture of resilience, empowering communities to adapt, withstand, and thrive in the face of climate change-induced challenges.

Participatory design, also referred to as co-design or collaborative design, is an approach that prioritises the involvement of users and stakeholders in the design process, placing them at the heart of the decision-making process [55]. The aim of this methodology is to create designs that are more user-focused, fair, and responsive to the needs and desires of the intended users or community. The fundamental principles of participatory design include inclusivity and diversity, where a range of perspectives and voices are valued and incorporated into the design process [59]. Empowerment is another crucial principle, as it aims to empower users by enabling them to contribute their insights, ideas, and expertise [68]. Collaboration and co-creation are central tenets, emphasising the importance of working together between designers, users, and stakeholders throughout all stages of the design process. Additionally, iteration and flexibility are highlighted, recognising that participatory design is an iterative process that allows for continuous learning, adaptation, and improvement [63]. Lastly, the principle of reflexivity encourages critical reflection on power dynamics, social norms, and biases that may impact the design process, with the goal of promoting greater equity and inclusivity in the final design outcomes. Through these principles, participatory design aims to encourage more inclusive, democratic, and user-driven design processes that result in products or services that better meet the needs and aspirations of the intended users or community [64].

The connection between participatory design and climate resilience is increasingly being acknowledged as a crucial approach to tackle the challenges posed by climate change [43]. Participatory design methods offer a platform for engaging stakeholders, including local communities, in the process of creating climate-resilient solutions. By incorporating a range of perspectives, knowledge, and insights, participatory design approaches help to improve the understanding of the specific vulnerabilities and priorities of communities in the face of climate change impacts. This involvement encourages ownership and empowerment among stakeholders, as they actively contribute to the identification and development of adaptation strategies that are specific to their context. Through meaningful engagement, participatory design promotes the integration of traditional and local knowledge, ensuring that climate resilience solutions are culturally appropriate and sustainable in the long term. Additionally, participatory design processes facilitate the integration of social, economic, and environmental aspects, enabling comprehensive and holistic approaches to climate resilience. By emphasising collaboration and co-creation, participatory design harnesses collective intelligence and fosters community resilience, allowing for the development of innovative and contextually relevant solutions to climate-related challenges. By combining participatory design with climate resilience initiatives, decision-makers can ensure that climate adaptation measures are inclusive, equitable, and locally embedded, ultimately promoting more robust and effective responses to the complexities of climate change [58].

3 Case Studies and Best Practices

The selection of Copenhagen, Medellín, Portland, Curitiba, and Malmö as case studies for successful community engagement and participatory design in climate-resilient urban strategies was based on a multi-criteria analysis. This analysis considered factors such as the cities' commitment to sustainable development, the implementation of innovative climate adaptation measures, and the active involvement of communities in urban planning processes. These cities have been recognized by various international bodies, including the C40 Cities Climate Leadership Group and the United Nations, for their pioneering efforts in integrating resilience into urban design, which is a testament to their climate-resilient status.

The best practices were collected through a comprehensive review of academic literature, city reports, and policy documents, supplemented by interviews with urban planners and community leaders involved in the projects. This triangulation method ensured a robust collection of data, providing a holistic understanding of the strategies employed by each city. The practices were deemed 'best' based on their effectiveness in enhancing urban resilience, their scalability, and their ability to be replicated in different urban contexts, thereby providing valuable insights for other cities aiming to bolster their climate resilience.

Here are some examples of successful community engagement and participatory design in developing climate-resilient urban design strategies from around the world:

1. **Copenhagen, Denmark**: The city of Copenhagen is a great example of how involving citizens in decision-making processes can lead to sustainable transportation solutions. By engaging the community, Copenhagen was able to create a bicycle-friendly infrastructure, with a wide network of cycling lanes and eco-friendly modes of transportation [9]. This collaborative approach has improved the quality of life for residents and helped address the challenges posed by climate change.

Community engagement and participatory design are integral components of Copenhagen's approach to developing climate-resilient urban design strategies. These processes work by actively involving citizens, local businesses, and other stakeholders in the co-creation and co-design of urban spaces, ensuring that the solutions are not only technically sound but also socially inclusive. In Copenhagen, this has been implemented through a variety of initiatives, such as the creation of the Østerbro Climate Quarter, which adapts the city to extreme weather using affordable green infrastructure (www.c40.org). This neighborhood was transformed into the world's first climate-resilient district, with local hills functioning as water basins and public squares collecting rainwater, thereby reducing the load on the sewer system (www.c40.org).

The inclusion of community engagement and participatory design in Copenhagen is evidenced by the city's strategic vision of 'Co-create Copenhagen,' which seeks new ways of collaborating with citizens across various scales, from community gardens to urban transformation zones [62]. The city's climate adaptation strategy relies on a process of public–private 'co-creation',the participation of citizens, small businesses, and other nonpublic actors in designated areas to remodel the cityscape, adapting squares, streets, lakes, and parks to better cope with future cloudbursts [17]. This strategy has led to more than 10,000 people taking part in the project's 170 citizen-led initiatives (www.c40.org).

Proof of the effectiveness of these methods comes from the tangible outcomes observed in the city. The green and surface-based climate adaptations in the Østerbro Climate Quarter have not only reduced the cost of damage from cloudbursts but are also cheaper to implement than sewer expansions (www.c40.org). Moreover, the project has resulted in environmental benefits like improved air quality and biodiversity, social benefits including enhanced quality of life for citizens, and economic benefits by reducing the financial impact of extreme weather events (www.c40.org). These successes demonstrate the value of community engagement and participatory design in creating sustainable, resilient, and inclusive urban environments.

The extensive community engagement in Copenhagen's urban design strategies has left an indelible impact. By prioritizing sustainable transportation modes, such as cycling, the city has witnessed a remarkable reduction in carbon emissions, contributing to its position as one of the world's most environmentally conscious cities. Additionally, the emphasis on cycling infrastructure has empowered residents to adopt healthier lifestyles, fostering physical fitness and well-being within the community [9]. The participatory nature of the design process has also cultivated a sense of ownership among citizens, making them stakeholders in the city's sustainable future and fostering social cohesion.

Furthermore, Copenhagen's approach to participatory design has greatly improved the city's ability to withstand the impacts of climate change. The emphasis on cycling as a primary mode of transportation has not only reduced carbon emissions but also eased traffic congestion, making the city more adaptable and reducing the potential risks associated with climate-related disruptions. By prioritising the needs and preferences of the community, the urban design strategies implemented in Copenhagen have successfully created a climate-resilient urban environment that safeguards the well-being of its residents and the planet [23].

2. **Medellín, Colombia**: Medellín, a city that previously faced significant social and environmental challenges, has become an impressive illustration of how community engagement and participatory design can strengthen neighbourhoods, improve climate resilience, and promote social transformation [10, 20]. By working collaboratively with local communities, Medellín has successfully rejuvenated disadvantaged areas, effectively addressing issues such as crime, poverty, and susceptibility to landslides. Through the inclusive design process, the city has not only enhanced its physical infrastructure but also instilled a sense of pride, ownership, and unity among its inhabitants [10, 20].

In Medellín, community engagement and participatory design are central to the city's strategy for developing climate-resilient urban design. These processes involve the active participation of local communities in the planning and implementation of projects aimed at enhancing the city's resilience to climate change. For instance, the **Green Corridors** initiative, implemented between 2016 and 2019, is a prime example of participatory design in action (Fig. 3). It involved transforming the city's roads, waterways, and underpasses with a network of trees and plants, leading to a significant reduction in ambient air temperatures and an increase in shade for cyclists and pedestrians (www.ucl.ac.uk). This initiative not only addressed climate resilience by mitigating the urban heat island effect but also improved air quality and increased biodiversity (www.ucl.ac.uk).

The inclusion of community engagement in Medellín's urban design is evident in the city's approach to integrating informal settlements into the urban fabric. The **Metrocable** project, for example, reflects a long-term commitment to engaging with local stakeholders to target spatial interventions effectively. By drawing on lessons learned from past experiences, project teams were able to design an approach that overcame decades of distrust towards formal institutions [18]. This engagement has been instrumental in shaping the city's transformation and ensuring that the interventions meet the needs and expectations of the community.

Evidence of the success of these participatory processes can be found in the tangible outcomes they have produced. The Green Corridors project has been associated with a 2 °C fall in temperatures, a dramatic rise in cycling due to the creation of new bike paths, and a significant drop in morbidity from acute respiratory infections (www.ucl.ac.uk). Furthermore, the Metrocable has been credited with reducing

Fig. 3 Medellín's Green Corridors project. *Source* Photo taken by Peter Yeung

homicide and poverty rates around station areas, demonstrating the positive impact of community engagement on social outcomes [18].

Medellín's participatory design initiatives have led to the creation of lively green spaces that act as crucial social connectors and environmentally sustainable sanctuaries [56]. By engaging community members in the decision-making process, the city has been able to recognise and tackle the specific requirements of different neighbourhoods, promoting fairness and social inclusivity. These green spaces not only offer recreational areas for locals but also aid climate resilience by reducing the urban heat island effect, enhancing air quality, and encouraging biodiversity. Medellín's dedication to participatory design has converted once-neglected areas into thriving, nature-infused environments that encourage social interaction, community well-being, and climate adaptability [56].

Furthermore, Medellín's approach to participatory design goes beyond just creating green spaces. The city has also focused on improving transportation and architectural design [10]. By investing in efficient and accessible public transportation systems, Medellín has not only reduced traffic congestion and air pollution but also provided its residents with more mobility options. Working together with local communities, Medellín has incorporated innovative architectural designs, such as cable car systems and escalators in hilly neighbourhoods, to ensure that everyone has equal access to services and opportunities [38]. These unique design solutions not only demonstrate the city's creativity but also enhance its ability to withstand natural disasters by enabling safer evacuation during landslides and providing strong infrastructural support.

3. **Portland, United States**: Portland is a great example of a city that prioritises community engagement and participatory design when it comes to sustainable urban design practices [61]. By taking a proactive approach, Portland has managed to create a culture of collaboration and inclusivity, which encourages residents to get involved in the decision-making processes that shape their city. This has led to the development of creative green infrastructure solutions that help to reduce the effects of climate change, increase biodiversity, and promote ecological resilience.

Portland has managed to prioritize sustainability and promote environmental stewardship by involving the community in the planning and execution of green infrastructure projects. For example, they have implemented initiatives like bioswales, green roofs, and permeable pavements to effectively manage stormwater runoff, reduce the burden on the city's drainage systems, and lower the risk of flooding. Additionally, incorporating more trees and vegetation in urban areas helps to combat heat island effects, enhance air quality, and provide shade and cooling for residents. These efforts not only improve the city's overall climate resilience but also create pleasant and livable neighborhoods for its inhabitants [4].

Portland's participatory design approach has led to the promotion of walkability and the expansion of public transit systems, in addition to green infrastructure. The city acknowledges the significance of active transportation and transit-oriented development in building sustainable and resilient communities. Through engagement with residents, Portland has identified opportunities to enhance pedestrian and cycling infrastructure, making it more convenient and safer for people to choose alternative modes of transportation. Furthermore, the expansion and improvement of public transit systems have given residents efficient and accessible options, reducing reliance on private vehicles and contributing to lower carbon emissions [4].

Portland's approach to participatory design has been instrumental in promoting walkability and expanding public transit systems, which are key components of its climate-resilient urban design strategy. The city's Urban Design team actively engages with the community and agency partners to explore citywide design issues and opportunities within the built environment. This collaborative process is aimed at enriching the quality, health, and livability of the city by developing concepts and ideas that illustrate future physical changes as Portland continues to grow (www.portland.gov). By involving design professionals, institutions, and community groups, the Urban Design team ensures that the projects reflect the community's needs and priorities (www.portland.gov).

The promotion of walkability in Portland is part of a broader initiative to support 90% of all residents with more walkable and bikeable infrastructure by 20,302. This initiative is included in the city's Climate Plan and aligns with the concept of the "15-Minute City," where all essential goods and services are accessible within a 15-min walk or bike ride from any given point in the city (www.resilience.org). The emphasis on multi-modal transportation options is designed to prioritize pedestrians and cyclists over cars, thereby reducing carbon emissions and enhancing the city's resilience to climate change (www.resilience.org).

In terms of public transit, Portland has made significant strides in expanding its systems to provide safe, direct connections linking various places within the city.

The expansion of public transit is a testament to the city's commitment to reducing reliance on personal vehicles, which is further evidenced by the fact that Portland residents drive significantly fewer miles per person each day compared to the national average (cityobservatory.org). This reduction in driving contributes to what is known as Portland's "green dividend," allowing residents to save on fuel and vehicle costs, which in turn stimulates the local economy (cityobservatory.org).

Moreover, Portland's green infrastructure plays a crucial role in its climate-resilient urban design. The city has implemented a range of green infrastructure projects to mitigate stormwater flooding, protect water resources, and alleviate urban heat island effects (storymaps.arcgis.com). These projects are designed to be both functional and aesthetically pleasing, contributing to the overall sustainability and resilience of the urban environment. The success of these initiatives is a direct result of Portland's participatory design approach, which ensures that the voices of the community are heard and integrated into the planning and implementation of urban design strategies (www.portland.gov).

4. **Curitiba, Brazil**: Curitiba is a city that is well-known for its pioneering urban design strategies, which demonstrate the advantages of community engagement and participatory design in creating urban environments that are resilient to climate change. By actively engaging its residents in the planning and decision-making processes, Curitiba has effectively established sustainable public transportation systems that address the challenges posed by climate change [12, 35].

Curitiba's reputation as a model for sustainable urban planning is largely attributed to its innovative and inclusive approach to community engagement and participatory design. The city's success in developing climate-resilient urban design strategies stems from its commitment to actively involving residents in the planning and decision-making processes. This participatory approach is evident in the city's transformation of urban voids into community gardens, which not only enhances food security and urban vegetation cover but also boosts resilience to climate change (www.c40.org). By repurposing land previously used as garbage disposal sites for urban agriculture, Curitiba has demonstrated a tangible example of community-driven sustainable development (www.c40.org).

The evidence of Curitiba's effective participatory design is also reflected in its long-standing history of urban planning and governance. Since the 1960s, Curitiba has been implementing a continuous planning process that has resulted in efficient public transit, historic and cultural preservation, and a pedestrian-friendly downtown [70]. These achievements are the outcomes of a political project that accommodated diverse interests and facilitated the development of Curitiba's planning process through the active involvement of its citizens [70]. The city's planning model, which emphasizes citizen participation, has been crucial in legitimizing the decision-making process and ensuring that urban design projects meet the needs and expectations of the community.

Furthermore, Curitiba's approach to urban design and planning has been academically recognized for its holistic integration of sustainable design, inclusive urbanism, and community engagement [3]. This what–who–how relationship has significantly contributed to improving urban plans and projects, with additional benefits to the community, public administration, and even the private sector [3]. The city's methodology, which involves comparing data concerning plans and projects with new stakeholder types and engagement tools, has proven to be an effective strategy for achieving truly sustainable design and planning that satisfies the diverse needs of current and future generations [3].

Curitiba's highly acclaimed bus rapid transit (BRT) system was designed and implemented through extensive citizen engagement. The city recognized the significance of affordable and accessible public transportation in reducing traffic congestion and carbon emissions. By involving the community in the design and development of the BRT system, Curitiba ensured that it met the needs and preferences of its residents. As a result, the public transport network is well-utilized, seamlessly integrating with the urban fabric, and offering convenience, reliability, and reduced travel times. This collaborative approach not only enhances transportation efficiency but also contributes to Curitiba's climate resilience goals by reducing reliance on private vehicles and minimizing greenhouse gas emissions [21, 39].

Curitiba has made environmental conservation a priority alongside its focus on public transportation. The city has created green spaces, preserved natural areas, and implemented effective waste management strategies. By using a participatory design approach, Curitiba has been able to identify and protect natural ecosystems and areas rich in biodiversity. These green spaces act as buffers against climate change impacts, improve air quality, and provide opportunities for recreation and leisure. Additionally, Curitiba's waste management strategies, such as recycling initiatives and waste-to-energy programs, demonstrate the city's commitment to sustainable development and environmental stewardship. By actively involving residents in these efforts, Curitiba has fostered a sense of ownership and responsibility, resulting in strong community support and long-term sustainability [35].

Curitiba's participatory design approach has been central to its success in identifying and protecting natural ecosystems and areas rich in biodiversity. This approach is rooted in the city's innovative urban transformation that began in the 1960s, which diverged from modernist planning in favor of a humanistic approach. The city's planning process has incorporated efficient and low-cost solutions for public transportation, land use, and environmental programs, leading to its recognition as Brazil's greenest city and the UN's most ecological city title in 1992 [69]. The participatory nature of Curitiba's urban planning has fostered local stewardship and ownership over natural areas, ensuring their protection and the maintenance of biodiversity.

Evidence of Curitiba's effective protection of ecosystems through participatory design is reflected in its Green Plan and subsequent park system, which are examples of regenerative design and biophilic urbanism with impacts on the regional ecosystem [69]. These initiatives have been successful due to the city's commitment to social-ecological approaches, which integrate the community in the planning process, thereby creating a sense of civic pride and responsibility towards the environment. The participatory scenarios used in Curitiba's ecological restoration have allowed for collaborative solutions that are best suited for the local culture and society, ensuring the preservation of biodiversity and the sustainability of natural ecosystems [16].

5. **Malmo, Sweden**: Malmo, a Swedish city, provides a convincing example of how community engagement and participatory design can lead to the creation of urban environments that are resilient to climate change. By converting a former industrial zone into a sustainable neighbourhood known as the Western Harbour, Malmo managed to encourage cooperation and participation from its inhabitants, which resulted in the construction of energy-efficient structures, the integration of renewable energy sources, and the implementation of sustainable water management systems. This achievement highlights the importance of community involvement in the development of climate-resilient cities [13].

The participatory design process implemented in Malmo's Western Harbour has played a crucial role in promoting environmental sustainability and fostering innovation. By actively engaging the community in decision-making, residents were able to share their diverse perspectives, knowledge, and ideas [13]. This inclusive approach not only empowered the residents but also led to the emergence of creative and sustainable solutions to tackle the challenges posed by climate change. As a result, the neighborhood now showcases energy efficiency, incorporating green building practices, sustainable materials, and smart technologies to minimize energy consumption and reduce greenhouse gas emissions.

The city's authorities took control of the recovery with clear goals and careful planning, ensuring multisectoral work and comprehensive urban programs and policies. The Western Harbour's transformation is a testament to the success of participatory design in achieving urban renewal and sustainable development (smartcitysweden.com). The inclusion of green elements, such as the green space factor and scoring system in construction projects, has been effective and adopted in other parts of Malmö and beyond (smartcitysweden.com).

Furthermore, the Western Harbour's design emphasizes sustainability in the broadest sense, with systems for energy supply and waste treatment that minimize car traffic, thereby enhancing urban mobility. Following the European Housing Expo of 2001, the Bo01 district was designed with a focus on sustainability, featuring green roofs and extensive rainwater run-off systems. The Aktern heat pump plant, local wind power, and solar collectors are integral to the renewable energy system, contributing significantly to the district's heating requirements and electricity generation (smartcitysweden.com). These innovative solutions provide concrete evidence of the participatory design process's role in creating an environmentally sustainable and innovative urban area.

Moreover, the participatory design process implemented in Malmo's Western Harbour has brought about social advantages too. By actively involving residents and stakeholders in the decision-making process, the community has witnessed a boost in social cohesion and a feeling of ownership over the development of their neighborhood [41, 53]. This collaborative approach has helped establish trust, nurtured a strong sense of belonging, and encouraged the adoption of sustainable behaviors and practices among residents. As a result, these social connections and shared responsibility play a crucial role in enhancing the overall resilience and long-term sustainability of the urban district.

3.1 Comparing the Actions in These Five Cities

In the quest for climate-resilient urban design, cities around the world are adopting innovative strategies that involve community engagement and participatory design. The cities of Malmö, Curitiba, Portland, Medellín, and Copenhagen have each taken unique yet comparable paths towards this goal. This subsection compares the actions taken by these cities to develop their climate-resilient urban design strategies.

Malmö, Sweden's third-largest city, has transformed its industrial Western Harbour into a model of sustainability. The city's participatory design process has led to the creation of a district that is not only carbon–neutral but also rich in biodiversity. Malmö's approach emphasizes the integration of green spaces and sustainable transportation, with a focus on renewable energy sources and waste management systems that contribute to a reduced ecological footprint.

Curitiba, Brazil, often cited as one of the most sustainable cities in the world, has long been a pioneer in participatory urban planning. The city's Bus Rapid Transit (BRT) system, parks, and recycling initiatives were all developed through active citizen participation. Curitiba's strategy has been to create an inclusive urban environment that prioritizes green spaces and efficient public transportation, fostering a sense of community ownership and responsibility towards the city's ecological well-being.

Portland, Oregon, is known for its green infrastructure and extensive public transit system, which were developed through a participatory design approach. The city's commitment to walkability and cycling has led to a significant reduction in car dependency, contributing to Portland's reputation as one of America's most sustainable cities. The city's climate action plan involves the community in the decision-making process, ensuring that urban design strategies align with the residents' vision for a resilient future.

Medellín, Colombia, has made remarkable strides in transforming its urban landscape through participatory design. The city's focus on social urbanism has led to the development of public spaces that are accessible and inclusive, such as the famous escalators in the Comuna 13 neighborhood. Medellín's approach to urban design is centered on improving the quality of life for its residents, with a strong emphasis on social equity and environmental sustainability.

Lastly, **Copenhagen**, Denmark, stands out for its climate-adapted urban spaces and commitment to becoming carbon–neutral by 2025. The city's participatory design process involves citizens in the development of green infrastructure and sustainable transportation solutions. Copenhagen's approach to urban resilience includes the integration of nature-based solutions and the promotion of a compact city model, where all essential services are within cycling or walking distance.

4 Guiding Principles and Recommendations for Practitioners and Policymakers

Fostering trust and collaboration: Creating cities that are resilient to climate change requires a comprehensive approach that involves community engagement, participatory design, and building trust and collaboration within local communities [67]. It is crucial to involve communities in the planning and design process to ensure that climate resilience strategies are tailored to meet the specific needs and aspirations of those who will be directly impacted. By actively participating, community members can share their experiences, knowledge, and priorities, providing valuable insights that can improve the effectiveness and longevity of climate-resilient initiatives. Furthermore, involving community members in decision-making processes fosters a sense of ownership and empowerment, encouraging shared responsibility for implementing and sustaining climate-resilient measures.

Promoting capacity building and education: When it comes to creating cities that can withstand the effects of climate change, involving the community in the design process is crucial. However, it's also important to provide education and training to help people contribute effectively. By giving communities the knowledge, skills, and resources they need, they can take an active role in decision-making and help make climate-resilient initiatives a success. Capacity building initiatives might include technical training, educational resources, and practical tools to help people understand climate change, its impacts, and how to adapt [5]. By promoting education and awareness, communities can develop a deeper understanding of the complex relationship between urban development, climate change, and resilience. This knowledge empowers them to make informed decisions and take proactive steps towards building a more climate-resilient future.

Integrating diverse perspectives: When it comes to building cities that can withstand the effects of climate change, it's really important to involve lots of different people and listen to their ideas. Climate resilience is a complex issue that needs input from a wide range of people with different backgrounds, skills, and experiences [6, 37]. By bringing together lots of different perspectives, cities can benefit from everyone's knowledge and come up with new and creative solutions to the challenges posed by climate change. It's important to include everyone in the process so that we can get a full picture of what's going on in each community and create strategies that work for everyone.

Establishing long-term partnerships between communities, designers, and decision-makers: If we want to create cities that can cope with the effects of climate change, it's really important that we work together [7, 51]. This means building strong, long-term relationships between communities, designers, and decision-makers. When we work together like this, we can build trust and make sure that everyone is accountable for what they're doing. This is really important if we want to create solutions that will work in the long term. When we involve communities in the planning and decision-making processes, we can make sure that we're taking into account their local knowledge, needs, and hopes for the future. By working together in this way, we can create solutions that are more than just short-term fixes. We can create solutions that are really integrated and that take into account all the different challenges that climate change presents.

The guiding principles and recommendations proposed in this paper are derived from a comprehensive review of interdisciplinary literature, case studies, and best practices in urban resilience and participatory design. They are grounded in the understanding that climate resilience in urban settings is a multifaceted challenge that requires a collaborative approach, integrating knowledge from various stakeholders, including local communities, urban planners, environmental scientists, and policymakers.

These principles are part of a larger strategic framework that aligns with the identified challenges of urban climate resilience. This framework emphasizes the iterative process of engagement, where feedback loops between practitioners, policymakers, and communities lead to continuous learning and adaptation. The connection between the principles and the challenges lies in their ability to address the complexities of urban systems, ensuring that resilience efforts are proactive, inclusive, and sustainable. This framework serves as a solid foundation for practitioners and policymakers to implement strategies that are not only theoretically sound but also practically viable in building climate-resilient cities.

5 Conclusions

In conclusion, it's evident that cities are increasingly recognizing the importance of involving local communities in urban planning and design. By looking at case studies of Copenhagen, Medellín, Portland, Curitiba, and Malmö, we see the meaningful impacts of this approach, such as better transportation, improved neighborhoods, and more green spaces. The guiding principles mentioned in the paper highlight the need for trust, capacity building, diverse perspectives, and long-term partnerships. These principles are essential for tackling the complex issues surrounding urban climate resilience. By actively engaging communities in the decision-making process, cities can tap into a wealth of knowledge to effectively address climate-related challenges and work towards a more sustainable and resilient future.

References

1. Adger WN (ed) (2006) Fairness in adaptation to climate change. MIT Press
2. Al S (2018) Adapting cities to sea level rise: green and gray strategies. Island Press, Washington, DC
3. Belausteguigoitia J (2019) Inclusive urbanism, sustainable design and community engagement: a holistic approach. WIT Trans Ecol Environ 238:131–141

4. Benedict MA, McMahon ET (2012) Green infrastructure: linking landscapes and communities. Island Press
5. Blair M (2008) Community environmental education as a model for effective environmental programmes. Aust J Environ Educ 24:45–53
6. Bulkeley H, Betsill M (2005) Rethinking sustainable cities: multilevel governance and the 'urban' politics of climate change. Environ Polit 14(1):42–63
7. Bulkeley H, Castán Broto V (2013) Government by experiment? global cities and the governing of climate change. Trans Inst Br Geogr 38(3):361–375
8. Carmen E, Fazey I, Ross H, Bedinger M, Smith FM, Prager K, ... Morrison D (2022) Building community resilience in a context of climate change: the role of social capital. Ambio 51(6):1371–1387
9. Colville-Andersen M (2018) Copenhagenize: the definitive guide to global bicycle urbanism. Island Press.
10. Corburn J, Asari MR, Pérez Jamarillo J, Gaviria A (2020) The transformation of Medellin into a 'City for Life': insights for healthy cities. Cities Health 4(1):13–24
11. Cornwall A (2008) Unpacking 'Participation': models, meanings and practices. Commun Dev J 43(3):269–283
12. de Freitas Miranda H, da Silva ANR (2012) Benchmarking sustainable urban mobility: the case of Curitiba, Brazil. Transp Policy 21:141–151
13. De Rosa SP, de Moor J, Dabaieh M (2022) Vulnerability and activism in urban climate politics: an actor-centered approach to transformational adaptation in Malmö (Sweden). Cities 130:103848
14. Dhar TK, Khirfan L (2023) A sixfold urban design framework to assess climate resilience: generative transformation in Negril, Jamaica. Plos one 18(6):e0287364
15. Dodman D, Bicknell J, Satterthwaite D (eds) (2012) Adapting cities to climate change: understanding and addressing the development challenges. Routledge
16. Durrant E, Howson P, Puttick B, Potts S, Shennan-Farpón Y, Sari N, ... Pfeifer M (2023) Existing evidence on the use of participatory scenarios in ecological restoration: a systematic map. Environ Evidence 12(1):27
17. Engberg LA (2018) Climate adaptation and citizens' participation in Denmark: experiences from Copenhagen. In: Climate change in cities: innovations in multi-level governance, pp 139–161
18. Galvin M, Maassen A (2020) Connecting formal and informal spaces: a long-term and multi-level view of Medellín's metrocable. Urban Transform 2(1):4
19. Garschagen M, Romero-Lankao P (2015) Exploring the relationships between urbanization trends and climate change vulnerability. Clim Change 133:37–52
20. Guerra MI (2014) Regulating neglect: territory, planning, and social transformation in Medellín. University of California, Berkeley Colombia
21. Gustafsson HR, Kelly EA (2016) Developing the sustainable city: Curitiba, Brazil, as a case study. In: How cities will save the world: urban innovation in the face of population flows, climate change, and economic inequality, pp 81–96
22. Hallegatte S (2009) Strategies to adapt to an uncertain climate change. Glob Environ Chang 19(2):240–247
23. Hartz-Karp J, Meister HP (2011) Creating resilient cities through empowered, deliberative participation. In: Resilient cities: cities and adaptation to climate change-proceedings of the global forum 2010. Springer Netherlands, pp 177–185
24. Harvey H, Orvis R, Rissman J, O'Boyle M, Busch C, Aggarwal S (2018) Designing climate solutions: a policy guide for low-carbon energy. Island Press, Washington
25. Hoornweg D (ed) (2011) Cities and climate change: responding to an urgent agenda. World Bank Publications
26. https://cityobservatory.org/fresh-evidence-for-portlands-green-dividend/
27. https://www.c40.org/case-studies/cities100-copenhagen-creating-a-climate-resilient-neighb orhood/
28. https://www.portland.gov/bps/planning/about-urban-design-team

29. https://www.resilience.org/stories/2021-06-14/urban-walkability-gains-a-foothold-in-the-u-s/
30. https://storymaps.arcgis.com/stories/592dc77b7fea4b40a4f38ebff1ad8c27
31. https://www.ucl.ac.uk/bartlett/sustainable/news/2022/apr/medellin-new-chapter-resilience
32. Huxham C, Vangen S (2013) Managing to collaborate: the theory and practice of collaborative advantage. Routledge
33. Innes JE, Booher DE (2010) Planning with complexity: an introduction to collaborative rationality for public policy. Routledge
34. Kondura M (2015) The city and the coming climate: climate change in the places we live
35. Lara FL (2010) Beyond Curitiba: the rise of a participatory model for urban intervention in Brazil. Urban Des Int 15:119–128
36. Larsen AK, Sewpaul V, Hole GO (2014) Participation in community work. International Perspectives
37. Lee KN (2005) Cities and climate change: urban sustainability and global environmental governance. Global Environ Polit 5(4):122–124
38. Levy C, Dávila JD (2017) Planning for mobility and socio-environmental justice: the case of Medellín, Colombia. In: Environ justice and urban resilience in the Global South, pp 37–55
39. Lindau LA, Hidalgo D, Facchini D (2010) Curitiba, the cradle of bus rapid transit. Built Environ 36(3):274–282
40. Lomba-Fernández C, Hernantes J, Labaka L (2019) Guide for climate-resilient cities: an urban critical infrastructures approach. Sustainability 11(17):4727
41. Manzini E (2015) When everybody designs: an introduction to design for social innovation
42. Moser SC, Boykoff MT (eds) (2013) Successful adaptation to climate change: linking science and policy in a rapidly changing world. Routledge
43. Nkoana EM, Verbruggen A, Hugé J (2018) Climate change adaptation tools at the community level: an integrated literature review. Sustainability 10(3):796
44. Nuruzzaman M (2015) Urban heat island: causes, effects and mitigation measures-a review. Int J Environ Monitor Anal 3(2):67–73
45. Pahl-Wostl C (2009) A conceptual framework for analysing adaptive capacity and multi-level learning processes in resource governance regimes. Glob Environ Chang 19(3):354–365
46. Pelling M (2010) Adaptation to climate change: from resilience to transformation. Routledge
47. Raven J (2011) Cooling the public realm: climate-resilient urban design. In: Resilient cities: cities and adaptation to climate change-proceedings of the global forum 2010. Springer Netherlands, pp 451–463
48. Rędzińska K, Piotrkowska M (2020) Urban planning and design for building neighborhood resilience to climate change. Land 9(10):387
49. Reed, MS, Graves A, Dandy N, Posthumus H, Hubacek K, Morris J, ... Stringer LC (2009) Who's in and why? A typology of stakeholder analysis methods for natural resource management. J Environ Manag 90(5):1933–1949
50. Roaf S, Crichton D, Nicol F (2009) Adapting buildings and cities for climate change: a 21st century survival guide. Routledge
51. Romero-Lankao P (2016) Governing carbon and climate in the cities: An overview of policy and planning challenges and options. In: Climate change and sustainable cities, pp 7–26
52. Rosenzweig C, Solecki WD, Romero-Lankao P, Mehrotra S, Dhakal S, Ibrahim SA (eds) (2018) Climate change and cities: second assessment report of the urban climate change research network. Cambridge University Press.
53. Sanders EBN, Stappers PJ (2008) Co-creation and the new landscapes of design. Co-design 4(1):5–18
54. Sanoff H (1999) Community participation methods in design and planning. John Wiley & Sons
55. Schuler D, Namioka A (eds) (1993) Participatory design: principles and practices. CRC Press
56. Schwab E (2013) New public open spaces and old prejudices: public space uses in the Centre of Medellín. In: REAL CORP and CEIT, European Center for Innovation, Planning Times
57. Shaw M, Crowther J (2017) Community engagement: a critical guide for practitioners
58. Sheppard SR, Shaw A, Flanders D, Burch S, Wiek A, Carmichael J, ... Cohen S (2011) Future visioning of local climate change: a framework for community engagement and planning with scenarios and visualisation. Futures 43(4):400–412

59. Simonsen J, Robertson T (eds) (2013) Routledge international handbook of participatory design, vol 711. Routledge, New York
60. Spires M, Shackleton S, Cundill G (2014) Barriers to implementing planned community-based adaptation in developing countries: a systematic literature review. Climate Dev 6(3):277–287
61. Takano T (ed) (2003) Healthy cities and urban policy research. Routledge
62. Tan WGZ (2019) Co-creation and Co-design for community-linked developments in Copenhagen
63. Udoewa V (2022) An introduction to radical participatory design: decolonising participatory design processes. Des Sci 8:e31
64. Van der Velden M, Mörtberg C (2015) Participatory design and design for values. In: Handbook of ethics, values, and technological design: sources, theory, values and application domains, pp 41–66
65. Walljasper J (2007) The great neighborhood book: a do-it-yourself guide to placemaking. New Society Publishers
66. Watson D, Adams M (2010) Design for flooding: architecture, landscape, and urban design for resilience to climate change. John Wiley & Sons
67. Whitmarsh L, Lorenzoni I, O'Neill S (2012) Engaging the public with climate change: behaviour change and communication. Routledge
68. Zhang M (2022, June) Design empowerment: participatory design towards social sustainability. In International conference on human-computer interaction. Springer International Publishing, Cham, pp 274–287
69. Zingoni de Baro ME (2022) Curitiba case study. Regenerating cities: reviving places and planet. Springer International Publishing, Cham, pp 117–162
70. Zurita CEI (2009) Urban design, planning, and the politics of development in Curitiba. In: Contemporary urbanism in Brazil: Beyond Brasilia, pp 202–223

Analysing Policies for Public Green Spaces of Indian Cities

Piyush Pandya, Gagan Jain, and Shweta Prasad

Abstract As urbanization rapidly increases, cities around the world are recognizing the critical role of public urban green spaces (PUGS) in enhancing environmental quality and residents' well-being. Recent studies have shown that green spaces contribute to the economy by increasing individual productivity, attracting investments, mitigating urban heat islands, and making cities climate resilient—thereby making PUGS essential for urban sustainability. In India, however, urban areas face significant challenges in integrating and maintaining these green spaces due to fragmented governance and inadequate policy frameworks. As urban areas fall under the State List in the Indian Constitution's Seventh Schedule, they are primarily governed by Urban Local Bodies (ULBs). This diminishes the role of state and central governments in advisory positions, creating a lack of accountability, and city managers often lack the capacity to understand complex issues. Despite numerous state and central programs, schemes, and policies for PUGS, the absence of logic-based goals with measurable milestones remains a significant challenge. This chapter highlights these challenges within central and state-level policies and significant city-level initiatives. It critically examines Indian urban policies and governance systems at both the union and state levels concerning public urban green spaces (PUGS). The analysis includes a study of best practices from cities in India and abroad. Drawing on global best practices, the findings are supplemented with suggestions for managing urban green spaces in India, considering cultural and behavioural responses, successful financial models, and recent technologies. The study relies on literature surveys and policy reviews, which contribute directly to the development of PUGS in India.

Keywords Urban green public spaces · Urban design and planning · Policy analysis · Sustainable development · Green infrastructure

---

P. Pandya
School of Architecture and Planning, Centre of Excellence-Built Environment and Urban Resilience, Woxsen University, Telangana, India

G. Jain · S. Prasad (✉)
School of Architecture and Planning, Woxsen University, Telangana, India
e-mail: shweta.prasad@woxsen.edu.in

B. He et al. (eds.), *Urban Climate and Urban Design*, Urban Sustainability,
https://doi.org/10.1007/978-981-96-1521-6_15

1 Introduction

United Nations estimates indicate that India has surpassed China in terms of total population [20] . It is also estimated that currently more than 35% of the Indian population lives in cities, which occupy just 6% of the total land area. This percentage is expected to increase to 50% by 2047–50 [21]. While this urbanization trend appears to be a boon for a developing country like India, it inevitably puts tremendous pressure on urban areas, affecting liveability and natural resources. This rapid urbanization also significantly impacts global climate and economic goals. Despite a lower urbanization rate, Asia had 54% of the world's urban population in 2018 [19].

In response to these challenges, the thematic areas of urban management policy (with urban planning approaches), public urban green spaces, and biodiversity emerges as a critical nexus for addressing the environmental impact of urban expansion. This focus pivots on climate resilience, combating urban heat islands, mitigating floods, preserving biodiversity, and even acting as carbon sinks. The chapter also acknowledges the importance of urban green areas from social and economic perspectives [14]. Urban green spaces measurably impact individual mental and physical health, productivity, attraction of investments and workforce, and the direct improvement of the urban environment—a crucial component for the broader canvas of urban sustainability (Fig. 1).

Internationally, cities like Singapore, Copenhagen, and Melbourne have implemented policies that successfully integrate green spaces into urban planning. Singapore's "City in a Garden" vision emphasizes biodiversity and extensive green cover, contributing to urban cooling and increased liveability [17]. Copenhagen's green roof policy and Melbourne's urban forest strategy focus on climate resilience and enhancing public spaces [1, 23]. These international efforts demonstrate that effective policies can significantly mitigate urban challenges related to green spaces, highlighting the importance of strong governance and clear, measurable goals.

Categorically, Target 11.7 of the United Nations' Sustainable Development Goal (SDG) 11 emphasizes urban green and public spaces, in addition to the five themes of SDG 11.3: (i) Urban Planning, Green Cover, and Biodiversity, (ii) Energy and Green Building, (iii) Mobility and Air Quality, (iv) Water Management, and (v) Waste Management.

Therefore, public urban green spaces (PUGS) are not only subjects of botany, ecology, hydrology, and biodiversity but also of urban studies concerning the design, planning, and management of urban areas, urban demographics and sociology, and even urban finance, real estate, and land economics. In this chapter, we refer to gardens, parks, urban forests, and public areas canopied with trees as PUGS. This study is based on the understanding that urban green spaces, which may have a significant positive impact on the urban environment due to their size and continuity and are governed by public authorities, are publicly owned.

Given this outlined importance, Indian cities face unique challenges in addition to the common challenges of rapid urbanization experienced by developing countries, particularly concerning PUGS. Some of these challenges include:

Fig. 1 The social, economic and climatic potential of green spaces. *Source* UNECE. Available at: https://unece.org/sites/default/files/2023-03/Urbanforestpolicybrief_final_0.pdf (Accessed: 13 June 2024)

- Poor local governance
- Weak finances
- Critical infrastructure shortages
- Rapidly deteriorating environment (this leads to diminishing health, productivity, and quality of life)
- Lack of a modern planning framework
- Informality within and outside of governance [15].

Based on SDG 11 and its indicators, along with the challenges for the country and a broader sustainability framework, nine parameters were selected to assess the policies illustrated in Tables 1 and 3, which include:

- Governance and Institutional Frameworks

Table 1 Analysis of national level policies/schemes/programmes/mission/guidelines/acts/rules in India pertaining to UGS [9]

State/national	Policy/scheme name	Focus/parameters								
		Environmental			Social				Economic	
		Directives for green cover/tree cover conservation and plantation	Directives for Physio-spatial integration within natural systems	Directives for diversity of tree species	Public participation in planning, development and maintenance	Directives for equal distribution across the city	Directives for safety, security and access for all	Directives for placemaking as urban insert	Directives for green finance/ financial incentive/ PPP model	Directives for management and maintenance guidelines
National	Nagar Van Yojana, 2020	Y	Y	Y	Y	N	Y	Y	Y	Y
	National Mission for Green India (GIM), 2014 (Under National Action Plan for Climate Change, 2008)	Y	Y	N	Y	N	Y	N	Y	Y
	Urban Green Guidelines (UGG), 2014	Y	Y	Y	N	Y	Y	N	N	Y
	URDPFI Guidelines, 2015	Y	Y	N	N	Y	N	Y	N	N
	Compensatory Afforestation Fund Act, 2016 (CAF Act)	NA	NA	NA	NA	NA	NA	NA	Y	NA

(continued)

Table 1 (continued)

State/national	Policy/scheme name	Focus/parameters								
		Environmental			Social				Economic	
		Directives for green cover/tree cover conservation and plantation	Directives for Physio-spatial integration within natural systems	Directives for diversity of tree species	Public participation in planning, development and maintenance	Directives for equal distribution across the city	Directives for safety, security and access for all	Directives for placemaking as urban insert	Directives for green finance/ financial incentive/ PPP model	Directives for management and maintenance guidelines
	The National Clean Air Programme (NCAP)	Largely following forestated schemes and programmes for tree plantation								
	The Biological Diversity Act, 2002 and Amendment 2023	Y	Y	N	N	N	N	N	Y	N

(continued)

Table 1 (continued)

State/ national	Policy/scheme name	Focus/parameters								
		Environmental			Social				Economic	
		Directives for green cover/tree cover conservation and plantation	Directives for Physio-spatial integration within natural systems	Directives for diversity of tree species	Public participation in planning, development and maintenance	Directives for equal distribution across the city	Directives for safety, security and access for all	Directives for placemaking as urban insert	Directives for green finance/ financial incentive/ PPP model	Directives for management and maintenance guidelines
	Jawaharlal Nehru National Urban Renewal Mission 2005	N	N	N	N	N	N	N	N	N
	Atal Mission for Rejuvenation and Urban Transformation (AMRUT) 2015	Y	N	N	Y	N	Y	Y	Y	N

(continued)

Table 1 (continued)

State/ national	Policy/scheme name	Focus/parameters								
		Environmental			Social				Economic	
		Directives for green cover/tree cover conservation and plantation	Directives for Physio-spatial integration within natural systems	Directives for diversity of tree species	Public participation in planning, development and maintenance	Directives for equal distribution across the city	Directives for safety, security and access for all	Directives for placemaking as urban insert	Directives for green finance/ financial incentive/ PPP model	Directives for management and maintenance guidelines
	Smart City Mission 2015	Y	N	N	Y	N	Y	Y	Y	N
	Heritage City Development and Augmentation Yojana (HRIDAY) 2015	NA	NA	NA	Y	NA	NA	Y	N	NA

- Infrastructure Provision and Maintenance
- Environmental Impact and Biodiversity
- Climate Resilience and Urban Heat Island Mitigation
- Public Health and Well-being
- Social Inclusivity and Accessibility
- Community Engagement and Participation
- Integration with Urban Planning and Land Use Policies
- Financial Mechanisms.

The significance of policy in regulating these challenges cannot be overstated, as effective governance frameworks are crucial for ensuring the sustainability and resilience of urban green spaces. Several countries and cities worldwide have developed policies and strategies to enhance PUGS, addressing challenges such as rapid urbanization, climate change, and social inequality.

a. **Singapore: City in a Garden**

Singapore's "City in a Garden" vision is a prime example of integrating green spaces into urban planning. The city-state has implemented a comprehensive strategy that includes:

Greenery Master Plan: This plan emphasizes increasing green cover through rooftop gardens, vertical greenery, and extensive park networks.

Biodiversity Enhancement: Efforts are made to protect and enhance urban biodiversity by creating nature reserves and biodiversity corridors.

Community Involvement: Programs like the Community in Bloom initiative encourage residents to participate in urban gardening and conservation efforts [17].

These efforts have led to significant improvements in urban liveability, climate resilience, and social cohesion, demonstrating the positive impact of well-planned PUGS.

b. **Copenhagen: Green Roof Policy**

Copenhagen has implemented a green roof policy as part of its broader strategy to combat climate change and enhance urban green spaces. Key aspects of the policy include:

Mandatory Green Roofs: New buildings with flat roofs are required to incorporate green roofs, which help manage stormwater, reduce urban heat islands, and provide habitats for wildlife.

Climate Adaptation: The green roofs contribute to the city's climate adaptation plan by mitigating flooding and reducing energy consumption [23].

Copenhagen's policy showcases how integrating green infrastructure into urban planning can address environmental challenges while promoting sustainability.

c. **Melbourne: Urban Forest Strategy**

Melbourne's Urban Forest Strategy aims to increase the city's tree canopy cover to 40% by 2040. The strategy focuses on:

Tree Planting and Maintenance: Extensive tree planting programs are coupled with maintenance plans to ensure the longevity and health of urban trees.

Community Engagement: Public consultations and educational campaigns are integral to the strategy, encouraging a sense of ownership and participation among residents.

Biodiversity and Resilience: The strategy also aims to enhance urban biodiversity and resilience to climate change by creating diverse and robust green spaces [1].

Melbourne's approach highlights the importance of long-term planning and community involvement in creating sustainable urban green spaces.

Based on insights, the research aims to evaluate existing PUGS policies in Indian cities, identify gaps, and propose actionable recommendations. The significance of policy in regulating these challenges is paramount, as effective governance frameworks are essential for ensuring the sustainability and resilience of urban green spaces. The purpose of the research is to draw from national and international best practices to enhance the management of PUGS in all the other cities in India, ultimately contributing to the broader goals of urban sustainability and liveability.

2 Methodology

2.1 Research Design

This study comprehensively assesses the policies related to public urban green spaces (PUGS) in Indian cities. The research is structured into three main stages: literature review, policy analysis, and case studies.

Stage 1: Literature Review

A thorough literature review was conducted to understand the global and local context of PUGS, including the role of such spaces in urban sustainability, health, and economic productivity. Key sources include academic journals, government reports, and publications by international organizations such as the United Nations and World Bank. The literature review also covers successful international case studies of urban green space policies from cities like Singapore, Copenhagen, and Melbourne. This phase provides a foundation for identifying best practices and understanding the challenges and benefits associated with PUGS.

Stage 2: Policy Analysis

The policy analysis involves evaluating existing urban green space policies at the union, state, and local levels in India. The analysis is guided by the Sustainable Development Goal (SDG) 11 indicators and broader sustainability frameworks.

Data for this phase was collected through document analysis of policy papers, urban planning documents, and government reports. The policies were then evaluated against the selected parameters to identify strengths, gaps, and areas for improvement.

Stage 3: Case Studies

Detailed case studies of select Indian cities were conducted to provide a deeper understanding of the implementation and impact of PUGS policies. The cities were chosen based on criteria such as population size, geographic diversity, and the presence of significant green space initiatives. The case studies aimed to highlight successful practices, challenges, and lessons learned from the ground.

Synthesis and Recommendations

Based on the findings from the policy analysis and case studies, best practices were identified, and actionable recommendations were formulated. These recommendations focus on enhancing the management, governance, and sustainability of PUGS in Indian cities, with an emphasis on policy improvements, innovative funding mechanisms, and community engagement strategies.

Limitations

The primary limitation of this study is its exclusive reliance on secondary data. The study also acknowledges potential limitations, such as the variability in data quality across different cities and the potential biases in self-reported survey data.

3 Planning Approaches for Urban Areas

3.1 Central Environmental Administration, and Urban Green Public Spaces

Among the many challenges with the management of Indian urban areas, one is the environment and the other, physical development management. This analysis examines the preparedness of environmental administration and legislation for management and planning of public urban green areas through revisiting the Environmental Protection (EP) Act, 1986. It focuses on planning tools and planning administration, with the example of the state of Madhya Pradesh representing the other states of India.

The authorities responsible for managing urban growth and fast-paced rural–urban metamorphosis in urban areas often lack expertise, capital, and capacity on many fronts. The situation is so dire that the scheme of capacity buildings states, "Given the lack of overall capacity, the smaller ULBs (Urban Local Bodies) are not in a position to even articulate their demand for training" [8]. At the same time, in the absence of a guard, informal development and reckless formation/conversion of land use on or around environmentally sensitive areas are committed. Similarly, environmental administration and planning in India has its own challenges. They are also fragmented and passive. Central authorities responsible for the environment have no direct say in federal urban planning processes. The cumulative impact of the aforesaid constraints is bound to create morbid cities. It is imperative for planning authorities and individual planning professionals to seek the opinion of experts and/ or research organisations on proposed urban development activities in the context of the environment and development. It has been proven that environmental and cultural protection and promotion contribute to the growth, and planning approaches cannot afford a myopic vision focusing only on development. There is an opportunity for the active participation of existing environmental authorities if a mechanism is provided through policy amendments. The EP Act is a union act. Under Section 5 of this Act, powers were delegated to states.

The Environment (Protection) Act, 1986, amended 1991, and other acts under the EP Act

The environment is a sensitive system with a complex network of interdependencies among different aspects. Urban areas are complex systems, too. Prevention of environmental hazards cannot be achieved by just focusing on one aspect of them. This is an Act to provide for the protection and improvement of the environment and for matters connected therewith.

Urban areas are centres of development and depletion of the environment as well. While environmental legislation focuses largely on industries, theoretically, there should be no hesitation in defining cities as industries where urban green spaces work as coolants.

Extent, Commencement, and Application: This Act extends to the jurisdiction of the Republic of India. It came into force on November 19, 1986. "Environment" as a subject does not figure in the Seventh Schedule in any of the lists; thus, this act becomes important to tie all the loose ends and lacunae of other legislation related to the environment and brings environment as a subject under the purview of the Central Government of the Republic of India. PUGS fall under the purview of urban local bodies and therefore, state jurisdiction. The Act does not specifically address this level, but it impacts these spaces in various ways. Major subjects related to the subject of environment fall under the concurrent list. This means a major chunk of guidelines, rules, laws, and directives are prepared at the central level, and powers are delegated at the state level to implement and execute the legislative structure. The body under whose purview this act comes is the Ministry of Environment, Forest and Climate Change (MoEFCC). Subject to the provisions of this Act, the Central Government shall have the power to take all measures. Under MoEFCC, the Central Pollution Control Board (CPCB) is the national board with oversight powers over state boards. The CPCB, a statutory organisation, was constituted in 1974 under the Water (Prevention and Control of Pollution) Act, 1974 (Fig. 2). Further, CPCB was given more powers and responsibilities under the Air (Prevention and Control of Pollution) Act, 1981. After the commencement of the EP Act, 1986, the Central Pollution Control Board served as a field formation and provided technical services to the MoFCC under the EP Act. It coordinates the activities of the State Pollution Control Boards (SPCBs). It is the apex organisation in the country in the field of pollution control. In addition to that, there are authorities related to different aspects of the environment which provide research and analysis in the environment sector and advises MoEFCC to prepare guidelines and legislations and conducts monitoring.

Limited coordination with the Ministry of Housing and Urban Affairs (MoHUA), state urban development departments, and therefore cities, reduces the roles of CBCB and SPCBs (which are working on the ground and decentralised) to mere spectators at the stage of development plan preparations. The impact of this gap is analysed and visible in the following discussions.

3.2 Management and Protection Approaches for Existing Urban Fabric: Policies, Guidelines, and Schemes Directed by the Central Government

Direct schemes and policies: It is evident that the central level policies do not have a direct say in the urban planning processes of state governments and therefore cities, subsequently having no control over PUGS. Union level policies are either monitoring or advising on environmental subjects in individual cities in the states. This section examines some of those advisory policies, focusing on their impact rather than their origins or premises.

India's diminishing forest and biodiversity resources, focusing on afforestation, wildlife conservation, and sustainable management, is the focus of the Green India Programme. The Compensatory Afforestation Fund Act, 2016 is an agency for green financing emphasizing afforestation and conservation efforts. MoHUA's Urban Greening Guidelines, 2014, provided the first promising policy for incorporating green spaces into urban planning for environmental sustainability on major parameters prior to the Nagar Van Yojana of 2020 (refer Table 1). Other general policies, as highlighted in the table, like The National Mission for Green India (GIM) under the National Action Plan for Climate Change (NAPCC), the National Clean Air Programme (NCAP), and the Biodiversity Management Committee, mandated by the Biological Diversity Act, facilitate their respective focus areas including but not limited to urban areas.

The 2020 Nagar Van Yojana (literally translated as: Urban Forest Scheme) is a promising initiative that aims at creating and maintaining city forests. It integrates the environment into urban landscapes, marking down and developing vacant plots into green pockets for recreation and biodiversity conservation, while curbing the urban blight of unattended empty lots and aligning with larger a framework of sustainable urban development goals. The inaugural phase of the scheme plans to build 400 *Nagar Vans* and 200 *Nagar Vatikas* across the country from 2020–21 to 2024–25 to augment the green cover beyond forested areas in urban and peri-urban regions to promote biodiversity and ecological benefits, subsequently enhancing liveability.

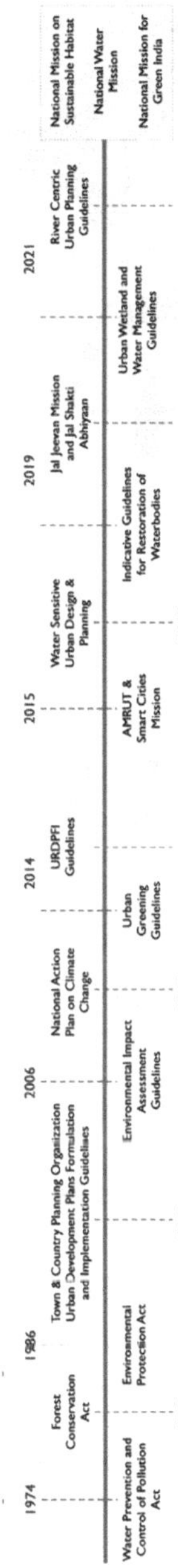

Fig. 2 Evolution of governmental solutions for urban green spaces [7]

As part of other programmes/missions/schemes and policies: The central level missions and schemes on urban development also sparsely focus on PUGS. The assessment of missions' potency requires additional research; however, consideration is appreciated. Primary assessment shows that the Jawaharlal Nehru National Urban and Rural Mission (JnNURM-2005), the Atal Mission for Rejuvenation and Urban Transformation (AMRUT-2015) and the Smart Cities Mission-2015 specifically target sustainability goals, including the preservation and expansion of open spaces at the city level [5]. However, there is an absence of performance indicators or any monitoring or evaluation framework. As iterated earlier under the discussion for planning processes and administration, these special missions also remain ambiguous and amorphous without specifying or mandating PUGS under the guise of freedom and flexibility.

3.3 *State Specific Analysis in India*

India is a federal constitutional republic governed under a parliamentary system consisting of 28 states and 8 union territories. The study of eight union territories is omitted for their smaller areas. Out of 28 states, five of the most urbanized (percentage of their respective population) states are selected for the study, as these states may pose a threat to the environment and subsequently to liveability. This section analyses the selected states within the country based on their policies for the PUGS. It is to be noted that this methodology is not based on the total population living in urban areas but on the percentage of the population living in urban areas. This approach is chosen instead of prioritizing the highest green cover, as green cover also includes forests. Forest and green cover are not the focus of the study, as higher forest cover makes a state less vulnerable. Therefore, if the urbanization level and the tree/forest cover are high in a state, those states will be omitted from the study. This method is employed to scope the study instead of looking at urban green spaces per state due to the limited mapping available in secondary resources.

For urbanization data, this study relies on census data that was collected by the government of India in 2011 in the absence of the latest data. In 2014, the state of Andhra Pradesh was divided into two states, namely Andhra Pradesh and Telangana. Similarly, in 2019 and 2020, Union Territories (UTs) were reconstituted. However, these changes do not affect the study. Refer to Table 2.

The selected states for the study are Tamil Nadu, Maharashtra, Gujarat, Karnataka, and Punjab. Their population size and land cover are above the state average of the country, making them significant states. Particularly, Gujarat and Karnataka, with their green cover and urbanisation, rely entirely on either their respective ULBs or the central government for guidelines and funds. This flawed approach, due to the central government's limited control over ULBs and the ULBs' limited capacity, may result in grey and deteriorating urban areas. As highlighted in Tables 2 and 3, the five states that urgently need effective, climate-specific policies for their cities are having difficulty integrating their tree-planting plans with social factors, even though

Table 2 State wise urbanisation status and green cover analysis. Urbanisation % [13] and % of the forested area [3]

Name	Urban population (in million)	Urbanisation %	% of the forested area	State/UT
Mizoram	0.56	51.5	84.53%	ST
Tamil Nadu	34.94	48.4	20.31%	ST
Kerala	15.93	47.7	54.70%	ST
Maharashtra	50.82	45.2	16.51%	ST
Gujarat	25.71	42.5	7.61%	ST
Karnataka	23.57	38.5	20.19%	ST
Punjab	10.38	37.4	3.67%	ST
Haryana	8.82	34.7	3.63%	ST
Andhra Pradesh ^	28.35	33.4	18.28%	ST
West Bengal	29.13	31.8	18.96%	ST
Uttarakhand	3.09	30.5	45.44%	ST
Manipur	0.82	30.2	74.34%	ST
Nagaland	0.57	28.9	73.90%	ST
Madhya Pradesh	20.05	27.6	25.14%	ST
Jammu & Kashmir^	3.41	27.2	NA	ST
Tripura	0.96	26.1	73.64%	ST
Sikkim	0.15	24.9	47.08%	ST
Jharkhand	7.92	24	29.76%	ST
Chhattisgarh	5.93	23.2	41.21%	ST
Rajasthan	17.08	22.8	4.87%	ST
Arunachal Pradesh	0.31	22.6	79.33%	ST
Uttar Pradesh	44.47	22.2	6.15%	ST
Meghalaya	0.59	20	76%	ST
Orissa	6.99	16.6	33.50%	ST
Assam	4.38	14	36.09%	ST
Bihar	11.72	11.3	7.84%	ST
Himachal Pradesh	0.68	10	27.73%	ST
Delhi	16.33	97.5	NA	UT
Chandigarh	1.02	97.2	NA	UT
Lakshadweep	0.05	78	NA	UT
Daman & Diu^	0.18	75.1	NA	UT
Pondicherry	0.85	68.3	NA	UT
Goa	0.9	62.1	NA	UT

(continued)

Table 2 (continued)

Name	Urban population (in million)	Urbanisation %	% of the forested area	State/UT
Dadra & Nagar Haveli^	0.15	46.6	NA	UT
Andaman & Nicobar Islands	0.13	35.6	NA	UT

*Average annual during 2001–11
^Reconstituted states/UT after 2011

all states appear committed to urban afforestation. Innovative green financing for the plantation to reduce the burden on the government and post-plantation maintenance and monitoring strategies are also absent in many states.

4 Indian Cities, Global Cities, and Best Practices

4.1 Provisions Related to Town and Country Planning (TNCP) Governance and Processes

The section highlights that ULBs with their limited capacity require support from the state government as the central government has limited intervention capacity to run the regular operations and management of the country's urban areas. The central bodies that work at local levels do not consult city governments when they create environmental atlases or when they monitor pollution, even though pollution sources are not centralized but spread out across regions. However, pollution is just a symptom of the main cause, the cause which the contributed by ULBs due to inadvertent land- use plan. The EP Act does not specifically mention the town and country planning (TNCP) authorities of the respective states. Instead, the focus is on monitoring and curbing the pollution. Whereas generally states' TNCP acts underline the consultation with different authorities and departments of the government without specifying departments of state government or the central government. The authorities related to different aspects of the environment, particularly natural geography, have their own demarcated areas and regions which are generally kept out of the jurisdiction of the state's Directorate of Town and Country Planning. Generally, urban natural geographical features fall under the management of respective ULBs, unless authorised otherwise.

There is no robust study, guideline, or policy provided under the EP Act, 1986, or its implementing system that underlines the complementing land use for environmental hotspots. It is studied on a case-by-case basis by performing an Environmental Impact Assessment (EIA). The environmental impact assessment tool has its own flaws that are criticised by many competent professionals and organisations

Table 3 Analysis of state level policies/schemes/programmes/mission/guidelines/acts/rules in India pertaining to UGS. *Source* Websites of respective states and their departments

State/national	Policy/scheme name	Focus/parameters								
		Environmental			Social				Economic	
		Directives for green cover/tree cover conservation and plantation	Directives for physio-spatial integration within natural systems	Directives for diversity of tree species	Public participation in planning, development and maintenance	Directives for equal distribution across the city	Directives for safety, security and access for all	Directives for placemaking as urban insert	Directives for green finance/financial incentive/PPP model	Directives for management and maintenance guidelines
Tamil Nadu	Chennai Metropolitan Development Authority, Policy Note-2023–24 (for 5904 sq.km. 30 urban areas and 1321 villages)	Y	Y	Y	N	N	N	Y	Y	N
	Environment and Climate Change Policy Note 2023–24 (Tamil Nadu Climate Change Mission)	N	N	N	N	N	N	N	N	N
	Green Tamil Nadu Mission	Y	Y	Y	Y	N	N	Y	Y	Y

(continued)

Table 3 (continued)

State/ national	Policy/scheme name	Focus/parameters								
		Environmental			Social				Economic	
		Directives for green cover/tree cover conservation and plantation	Directives for physio-spatial integration within natural systems	Directives for diversity of tree species	Public participation in planning, development and maintenance	Directives for equal distribution across the city	Directives for safety, security and access for all	Directives for placemaking as urban insert	Directives for green finance/ financial incentive/ PPP model	Directives for management and maintenance guidelines
	Forest Policy Note 2023–24	Y	N	Y	Y	N	N	N	N	Y

(continued)

Table 3 (continued)

State/ national	Policy/scheme name	Focus/parameters								
		Environmental			Social				Economic	
		Directives for green cover/tree cover conservation and plantation	Directives for physio-spatial integration within natural systems	Directives for diversity of tree species	Public participation in planning, development and maintenance	Directives for equal distribution across the city	Directives for safety, security and access for all	Directives for placemaking as urban insert	Directives for green finance/ financial incentive/ PPP model	Directives for management and maintenance guidelines
	Housing and Urban Development Department Policy Note 2023-2024	N	N	N	N	N	N	N	N	N

(continued)

Table 3 (continued)

State/ national	Policy/scheme name	Focus/parameters								
		Environmental			Social				Economic	
		Directives for green cover/tree cover conservation and plantation	Directives for physio-spatial integration within natural systems	Directives for diversity of tree species	Public participation in planning, development and maintenance	Directives for equal distribution across the city	Directives for safety, security and access for all	Directives for placemaking as urban insert	Directives for green finance/ financial incentive/ PPP model	Directives for management and maintenance guidelines
	Municipal Administration and Water Supply Department Policy Note 2023–2024	Y	Y	Y	N	N	N	Y	Y	N

(continued)

Table 3 (continued)

State/national	Policy/scheme name	Focus/parameters								
		Environmental			Social				Economic	
		Directives for green cover/tree cover conservation and plantation	Directives for physio-spatial integration within natural systems	Directives for diversity of tree species	Public participation in planning, development and maintenance	Directives for equal distribution across the city	Directives for safety, security and access for all	Directives for placemaking as urban insert	Directives for green finance/financial incentive/PPP model	Directives for management and maintenance guidelines
	KNMT, Singara Chennai 2.0, Namakku Name - Thittam and AMRUT 2.0	Y	Y	N	N	N	N	Y	N	N

(continued)

Table 3 (continued)

State/national	Policy/scheme name	Focus/parameters								
		Environmental			Social				Economic	
		Directives for green cover/tree cover conservation and plantation	Directives for physio-spatial integration within natural systems	Directives for diversity of tree species	Public participation in planning, development and maintenance	Directives for equal distribution across the city	Directives for safety, security and access for all	Directives for placemaking as urban insert	Directives for green finance/financial incentive/PPP model	Directives for management and maintenance guidelines
Gujarat	Gujarat Urban Development Mission	Y	N	N	Y	N	N	Y	Y	Y

(continued)

Table 3 (continued)

State/ national	Policy/scheme name	Focus/parameters								
		Environmental			Social				Economic	
		Directives for green cover/tree cover conservation and plantation	Directives for physio-spatial integration within natural systems	Directives for diversity of tree species	Public participation in planning, development and maintenance	Directives for equal distribution across the city	Directives for safety, security and access for all	Directives for placemaking as urban insert	Directives for green finance/ financial incentive/ PPP model	Directives for management and maintenance guidelines
	Cultural Forests of Gujarat 2023 (Aapna Sanskrutik Vano)	Y	N	N	Y	N	N	Y	N	Y

(continued)

Table 3 (continued)

State/ national	Policy/scheme name	Focus/parameters								
		Environmental			Social				Economic	
		Directives for green cover/tree cover conservation and plantation	Directives for physio-spatial integration within natural systems	Directives for diversity of tree species	Public participation in planning, development and maintenance	Directives for equal distribution across the city	Directives for safety, security and access for all	Directives for placemaking as urban insert	Directives for green finance/ financial incentive/ PPP model	Directives for management and maintenance guidelines
	First State Action Plan on Climate Change – Gujarat 2014 (Draft 2021 not to be cited)	Y	N	N	Y	N	N	Y	Y	N

(continued)

Table 3 (continued)

State/ national	Policy/scheme name	Focus/parameters								
		Environmental			Social				Economic	
		Directives for green cover/tree cover conservation and plantation	Directives for physio-spatial integration within natural systems	Directives for diversity of tree species	Public participation in planning, development and maintenance	Directives for equal distribution across the city	Directives for safety, security and access for all	Directives for placemaking as urban insert	Directives for green finance/ financial incentive/ PPP model	Directives for management and maintenance guidelines
	State Policies and Programmes for Inclusive Urban Development	N	N	N	N	N	N	N	N	N

(continued)

Table 3 (continued)

State/national	Policy/scheme name	Focus/parameters								
		Environmental			Social				Economic	
		Directives for green cover/tree cover conservation and plantation	Directives for physio-spatial integration within natural systems	Directives for diversity of tree species	Public participation in planning, development and maintenance	Directives for equal distribution across the city	Directives for safety, security and access for all	Directives for placemaking as urban insert	Directives for green finance/financial incentive/PPP model	Directives for management and maintenance guidelines
Karnataka	Karnataka Urban Development Policy	N	N	N	N	N	N	N	N	N

(continued)

Table 3 (continued)

State/ national	Policy/scheme name	Focus/parameters								
		Environmental			Social				Economic	
		Directives for green cover/tree cover conservation and plantation	Directives for physio-spatial integration within natural systems	Directives for diversity of tree species	Public participation in planning, development and maintenance	Directives for equal distribution across the city	Directives for safety, security and access for all	Directives for placemaking as urban insert	Directives for green finance/ financial incentive/ PPP model	Directives for management and maintenance guidelines
	Krushy Aranya Protsaha Yojane (KAPY) and raising of seedlings for public distribution and Maguvigondu Mara Salegondu Vana	Y	N	N	Y	N	N	N	N	N

(continued)

Table 3 (continued)

State/ national	Policy/scheme name	Focus/parameters								
		Environmental			Social				Economic	
		Directives for green cover/tree cover conservation and plantation	Directives for physio-spatial integration within natural systems	Directives for diversity of tree species	Public participation in planning, development and maintenance	Directives for equal distribution across the city	Directives for safety, security and access for all	Directives for placemaking as urban insert	Directives for green finance/ financial incentive/ PPP model	Directives for management and maintenance guidelines
	Tree Park Scheme and DaiviVana Scheme	N	N	N	Y	N	N	Y	N	N

(continued)

Table 3 (continued)

State/ national	Policy/scheme name	Focus/parameters								
		Environmental			Social				Economic	
		Directives for green cover/tree cover conservation and plantation	Directives for physio-spatial integration within natural systems	Directives for diversity of tree species	Public participation in planning, development and maintenance	Directives for equal distribution across the city	Directives for safety, security and access for all	Directives for placemaking as urban insert	Directives for green finance/ financial incentive/ PPP model	Directives for management and maintenance guidelines
	Greening of Urban Area (GUA)	Y	N	N	Y	N	N	Y	N	N

(continued)

Table 3 (continued)

State/ national	Policy/scheme name	Focus/parameters								
		Environmental			Social				Economic	
		Directives for green cover/tree cover conservation and plantation	Directives for physio-spatial integration within natural systems	Directives for diversity of tree species	Public participation in planning, development and maintenance	Directives for equal distribution across the city	Directives for safety, security and access for all	Directives for placemaking as urban insert	Directives for green finance/ financial incentive/ PPP model	Directives for management and maintenance guidelines
	Karnataka State Action Plan on Climate Change Version 2.0 2021	Y	N	N	Y	N	N	Y	Y	N

(continued)

Table 3 (continued)

State/ national	Policy/scheme name	Focus/parameters								
		Environmental			Social				Economic	
		Directives for green cover/tree cover conservation and plantation	Directives for physio-spatial integration within natural systems	Directives for diversity of tree species	Public participation in planning, development and maintenance	Directives for equal distribution across the city	Directives for safety, security and access for all	Directives for placemaking as urban insert	Directives for green finance/ financial incentive/ PPP model	Directives for management and maintenance guidelines
Maharashtra	Maharashtra State Forest Policy 2008	Y	Y	Y	Y	Y	N	N	Y	Y

(continued)

Table 3 (continued)

State/ national	Policy/scheme name	Focus/parameters								
		Environmental			Social				Economic	
		Directives for green cover/tree cover conservation and plantation	Directives for physio-spatial integration within natural systems	Directives for diversity of tree species	Public participation in planning, development and maintenance	Directives for equal distribution across the city	Directives for safety, security and access for all	Directives for placemaking as urban insert	Directives for green finance/ financial incentive/ PPP model	Directives for management and maintenance guidelines
	Maharashtra State New Housing Policy and Action Plan, 2015	Y	N	N	N	N	N	N	N	N

(continued)

Table 3 (continued)

State/ national	Policy/scheme name	Focus/parameters								
		Environmental			Social				Economic	
		Directives for green cover/tree cover conservation and plantation	Directives for physio-spatial integration within natural systems	Directives for diversity of tree species	Public participation in planning, development and maintenance	Directives for equal distribution across the city	Directives for safety, security and access for all	Directives for placemaking as urban insert	Directives for green finance/ financial incentive/ PPP model	Directives for management and maintenance guidelines
	Corporate social responsibility policy of forest development corporation of Maharashtra limited	Y	Y	N	Y	N	N	N	Y	NA

(continued)

Table 3 (continued)

State/ national	Policy/scheme name	Focus/parameters								
		Environmental			Social				Economic	
		Directives for green cover/tree cover conservation and plantation	Directives for physio-spatial integration within natural systems	Directives for diversity of tree species	Public participation in planning, development and maintenance	Directives for equal distribution across the city	Directives for safety, security and access for all	Directives for placemaking as urban insert	Directives for green finance/ financial incentive/ PPP model	Directives for management and maintenance guidelines
	Maharashtra (urban areas) protection and preservation of trees act, 1975 and amendment act 2021	Y	Y	Y	Y	N	N	N	N	Y

(continued)

Table 3 (continued)

State/ national	Policy/scheme name	Focus/parameters								
		Environmental			Social				Economic	
		Directives for green cover/tree cover conservation and plantation	Directives for physio-spatial integration within natural systems	Directives for diversity of tree species	Public participation in planning, development and maintenance	Directives for equal distribution across the city	Directives for safety, security and access for all	Directives for placemaking as urban insert	Directives for green finance/ financial incentive/ PPP model	Directives for management and maintenance guidelines
	Maharashtra State Water Policy - 2019	N	Y	N	Y	N	N	N	Y	Y

(continued)

Table 3 (continued)

State/ national	Policy/scheme name	Focus/parameters								
		Environmental			Social				Economic	
		Directives for green cover/tree cover conservation and plantation	Directives for physio-spatial integration within natural systems	Directives for diversity of tree species	Public participation in planning, development and maintenance	Directives for equal distribution across the city	Directives for safety, security and access for all	Directives for placemaking as urban insert	Directives for green finance/ financial incentive/ PPP model	Directives for management and maintenance guidelines
	Maharashtra water resources regulatory Authority Act 2005 and Amendment Act 2016	N	Y	N	N	Y	N	N	N	N

(continued)

Table 3 (continued)

State/ national	Policy/scheme name	Focus/parameters								
		Environmental			Social				Economic	
		Directives for green cover/tree cover conservation and plantation	Directives for physio-spatial integration within natural systems	Directives for diversity of tree species	Public participation in planning, development and maintenance	Directives for equal distribution across the city	Directives for safety, security and access for all	Directives for placemaking as urban insert	Directives for green finance/ financial incentive/ PPP model	Directives for management and maintenance maintenance guidelines
	Unified development control and promotion regulations for Maharashtra state 2020	Y	Y	N	Y	N	N	N	Y	Y

(continued)

Table 3 (continued)

State/ national	Policy/scheme name	Focus/parameters								
		Environmental			Social				Economic	
		Directives for green cover/tree cover conservation and plantation	Directives for physio-spatial integration within natural systems	Directives for diversity of tree species	Public participation in planning, development and maintenance	Directives for equal distribution across the city	Directives for safety, security and access for all	Directives for placemaking as urban insert	Directives for green finance/ financial incentive/ PPP model	Directives for management and maintenance guidelines
Punjab	The Punjab Urban Planning and Development Building Rules, 2021	Y	Y	As per UGG 2014	N	N	N	N	N	N

(continued)

Table 3 (continued)

State/national	Policy/scheme name	Focus/parameters								
		Environmental			Social				Economic	
		Directives for green cover/tree cover conservation and plantation	Directives for physio-spatial integration within natural systems	Directives for diversity of tree species	Public participation in planning, development and maintenance	Directives for equal distribution across the city	Directives for safety, security and access for all	Directives for placemaking as urban insert	Directives for green finance/financial incentive/PPP model	Directives for management and maintenance guidelines
	Urban Greening and Tree Plantation Strategy and Guidelines 2023	Y	Y	Y	Y	Y	N	N	N	Y
	Punjab State Action Plan on Climate Change 2014	Y	Y	Y	Y	N	N	N	Y	Y

Table 4 Per capita Urban Green Spaces (UGS) in the major metropolitans of India [13, 16]

City	City area sq. km	City population in million (census 2011)	Area of UGS (sq. km)	Per capita UGS (sq. mt. per person)
Pune	484.61	3.12	69.49	0.22
Hyderabad	172	7.75	8.72	0.88
Chennai	176	8.69	11	1.26
Surat	326.5	4.46	11.84	3.32
Ahmedabad	464	5.57	21.8	3.91
Mumbai	735	18.45	122	6.61
Bangalore	709	8.43	150	17.80
Delhi	1484	16.34	324.44	19.84
Chandigarh	114	1.05	50	47.37

[24]. In the context of the chapter, it should be highlighted that EIA only concerns certain projects in their isolation and of a certain scale. While a developmental and augmenting proposal for a city can be seen as a large project as well, that is broken into smaller projects to avoid EIA while putting stress on the environment.

It is important to note that city development plans are not scrutinized through Environmental Impact Assessment, a concern recently highlighted by the Supreme Court of India [2].

Due to the lack of a system and guidelines, a city must rely on the wisdom of the individual decision- makers responsible for the preparation of urban management and planning documents. It cannot be monitored or verified even if TNCP acts of states mandate to have green belts, parks, gardens, etc. as part of all statutory and non-statutory plans like zonal, development, local area plans, and schemes without specifying the details like the required area and complementing land use, etc. The states and concerned urban planners and authorities rely on advisory documents and guidelines prepared by the central government or international competent authorities like WHO. This situation leaves a huge gap that can only be filled by voids i.e., PUGS that are integrated into natural systems and liveability. Yet, there is no direct mandatory mechanism to ensure that in the existing governance.

4.2 Indian and Global Cities and Their Practices for PUGS[1]

As highlighted in 3.1, the sole responsibility to provide desired PUGS based on stated parameters relies on urban planners working for respective ULBs (Table 4). Following are a few highlights of selected city strategies.

[1] The initiatives and data stated for the cities are gathered from their respective city websites.

Chandigarh, India: The city was planned by Le Corbusier with a focus on extensive landscaping and greenery. It stands out for a high quality of life and a clean environment after nearly 60 years. The city prioritises urban greening, afforestation, and green belts with limited post plantation monitoring. The city exceeds recommended open space norms, with 19.69% of the first-phase areas designated as open spaces, and yet the corporation aims to develop one garden or green belt in each sector annually to retain the initial lush greenery, low-rise buildings, and clear vistas threatened by the later phases due to smaller and denser developments. Community open spaces in group housing aim to maintain a connection with nature, incorporating the idea of placemaking.

New Delhi, India: Greening initiatives of the most polluted Indian city include establishment of a Tree Authority and Officer under the Delhi Preservation of Trees Act, 1994, obligating permissions for tree-related activities, and a Tree Help Line for reporting illegal tree felling while recovering costs from non-compliant tree planting, compensatory planting, and careful tree selection ensuring diversity. Focusing on the smallest spaces for trees in otherwise dense fabric, a million trees were planted in the public spaces of the City Plants a Million Tree Campaign in 2011. Green Leap Delhi and the Tree Ambulance Servicehad ensured public participation.

Ahmedabad, India: Government interventions through the City Development Plan and Development Plan focus watershed management, urban poor housing, transportation, and environmental improvement, including projects like the Sabarmati Riverfront. Greening efforts focus on parks, green streets, waterfront development, and urban groves with an innovative PPP model with Anand Milk Unit Limited (AMUL). The other city specific schemes are the Greening of Ahmedabad Programme and the Ahmedabad Green Partnership Project.

Bangalore, India: Previously famous for its gardens, the megacity boasts 20% green cover, with higher plant species diversity. Urbanization from 1999–2014 led to an 184% increase in land use, reducing vegetation, and lakes face issues with encroachment and pollution. Unforeseen migration boosts due to the IT industry mean that the city's gardens are visibly reducing [6].

Pune, India: The city's greening efforts include the Maharashtra (Urban Areas) Protection and Preservation of Trees Act, 1975, requiring written permission from the Tree Authority for tree felling. With public participation, the city and the Tree Public Foundation planted 50,000 trees. The Pune Municipal Corporation engages in green financing through advertisement rights. No efforts can be seen on the part of the maintenance head [12, 18].

Chennai, India: Another megacity with a recurring problem of floods, greening efforts involve the Tamil Nadu Tree Protection Act, requiring permission from the District Committee for tree felling. "Royal trees" on private lands need the Chief Conservator of Forests' approval. Government-led tree plantation schemes planted 6.4 million saplings in 2012 and 6.5 million in 2013. Observations include urban afforestation efforts and the initiation of a tree census (CMDA Chennai, 2023).

London, UK: The global city emphasises on accessibility to gardens and maximises urban greening with the "right place, right tree" principle. The goal is to plant two million trees by 2025. Despite efforts, there's a highlighted lack of public participation [4].

New York City, USA: The global city has a comprehensive Tree Census detailing tree information, Trees Count 2015, and the Million Tree Initiative. PlaNYC committed to building more blue belts and green streets, to incentivize green roofs, to require green parking lots, and to form an inter-agency Best Management Practices Task Force in 2007. Through tree giveaways and Mulch Fest recycling, public participation is ensured [10].

Insights like on-street placemaking contributed by trees, interdepartmental coordination, public participation, biodiversity, intensification of individual green efforts etc. from national and global practices, along with the analysed state and national policy profile, suggest the following policy recommendations for states.

5 Recommendations

Indian states prioritise more on economic and physical development than liveability and the environment [11], often relying solely on afforestation. However, addressing urban land and environmental pressures requires a comprehensive understanding of the behaviour of private, public, and individual entities within cities. Providing access to green finance for cities is crucial in the complex landscape of urban development and the environment. Policies for urban areas must consider inclusivity and physical access for all while also considering age, gender, disability, sexual identity, and marginalized groups based on class and caste, ensuring safety and security for everyone.

Effective urban projects, including Public Urban Green Spaces (PUGS), require cross-departmental, cross-level, multidisciplinary, and cross-stakeholder coordination through re-engineering urban governance. An example that incorporates informal and cultural spaces into green areas, as suggested by the Nagar Van Yojana 2020 to prevent urban decay. Such orchestration of urban strategies, cultivation of community pride, responsibility, and ownership emerge as a pivotal motif disburdening government and enriching the socio-ecological fabric of cultural India.

Lessons that can be taken from international examples like Singapore, Copenhagen, Melbourne, London, and New York City for PUGS in India include:

1. **Integrated Planning**: Incorporating green spaces into broader urban planning strategies.
2. **Community Involvement**: Engage the community in planning and maintaining green spaces to foster ownership and sustainability.
3. **Climate Resilience**: Use green spaces to enhance urban resilience to climate change, mitigating effects like heat islands and flooding.

4. **Biodiversity Enhancement**: Protect and enhance urban biodiversity through well-designed green spaces to improve ecological sustainability and urban living quality.

Urban development plans should be created more swiftly and included in Environmental Impact Assessments (EIA) to fully understand their impacts. These plans should also consider complementary land uses, cultural landscapes, nature-culture linkages, and nature-based solutions while preserving traditional ecological knowledge and techniques. Recent national and state policies in India show promise but still have gaps that need addressing for better implementation. This study could help Indian state governments and urban administrations align with national policies and create effective individual strategies.

References

1. City of Melbourne (2012) Urban forest strategy: making a great city greener 2012–2032
2. Das A (2023, January 11) Make environmental impact assessment mandatory for urban development, recommends supreme court; cites condition of Bengaluru as warning. www.livelaw.in: https://www.livelaw.in/top-stories/make-environmental-impact-assessment-mandatory-for-urban-development-recommends-supreme-court-cites-condition-of-bengaluru-as-warning-218641
3. Forest Survey of India, MoEF (2021) India State of Forest Report 2021. Forest Survey of India, MoEF, Dehradun
4. Greater London Authority, London (2011) The London plan. https://www.london.gov.uk/priorities/planning/publications/the-london-plan. Accessed 28 May 2014
5. Gujar S, Deshmukh A, Gupta R (2022) Review of open space rules and regulations and identification of specificities for plot-level open spaces to facilitate sustainable development: an Indian case. ICSEEGT 2022.
6. Halbe A (2023) Satellite images show green cover of Bangalore reducing alarmingly fast. Research Matters. https://researchmatters.in/article/satellite-images-show-green-cover-bangalore-reducing-alarmingly-fast
7. INDE: Integrated Design (2022) Building India's blue-green network: a strategic framework for Bengaluru. INDE, Banglore
8. Ministry of Urban Development (2015) Capacity building scheme for urban local bodies (CBULB). Ministry of Urban Development
9. MoEFCC, GoI (2023, November 30) Resources. Ministry of Environment, Forest and Climate Change. https://moef.gov.in/moef/resource/publications/index.html
10. New York City Department of Parks & Recreation, NYC (2015) How the tree census works. http://www.nycgovparks.org/trees/treescount/about. Accessed 2 June 2015
11. Panda A, Thakur P (2023, November 30) Liveability of Indian cities: turning urban legend into reality. www.teriin.org: https://www.teriin.org/article/liveability-indian-cities-turning-urban-legend-reality
12. Pune Municipal Corporation (PMC) (2014) Pune Municipal Corporation—Garden department notification. http://www.punecorporation.org/informpdf/Garden/Expression_of_Interest_for_divider_Islands_&_Gardens.pdf. Accessed 18 May 2015
13. Registrar General and Census Commissioner of India (2023, November 28) Population. Census of India. https://censusindia.gov.in/
14. Rodenburg C, Baycan-Levent T, LeeuWen E, Nijkamp P (2001) Urban economic indicators for green development in Cities. Greener Manag Int 36:105–119. http://www.jstor.org/stable/greemanainte.36.105

15. Roy A (2009) Why India cannot plan its cities: informality, insurgence and the idiom of urbanization. Plann Theory 76–87
16. Sangwan A, Saraswat A, Kumar N, Pipralia S, Kumar A (2022) Urban green spaces prospects and retrospect's. In: Castanho RA, Fernández JC (eds) Urban green spaces. IntechOpen. https://doi.org/10.5772/intechopen.102857
17. Tan PY, Wang J, Sia A (2013) Perspectives on five decades of the urban greening of Singapore. Cities 32:24–32
18. Times of India, and Madaan N (2013) NGO completes planting of 50 000 trees in Pune. http://timesofindia.indiatimes.com/city/pune/NGO-completes-planting-of-50000-trees-in-Pune/articleshow/21730046.cms. Accessed 12 Dec 2014
19. United Nations (2023, November 26) 2018 revision of world urbanization prospects. https://www.un.org: https://www.un.org/en/desa/2018-revision-world-urbanization-prospects
20. United Nations Department of Economic and Social Affairs, P. D (2022) World population prospects 2022: summary of results. United Nations, New York
21. United Nations Department of Economic and Social Affairs (2019) World population prospects 2019. United Nations, New York
22. United Nations Economic Commission for Europe (2021) Sustainable urban and peri-urban forestry: an integrative and inclusive nature-based solution for green recovery and sustainable, healthy and resilient cities (Policy Brief, p. 14)
23. Voskamp IM, Van de Ven FH (2015) Planning support system for climate adaptation: composing effective sets of blue-green measures to reduce urban vulnerability to extreme weather events. Build Environ 83:159–167
24. Yadav VS (2018) Environmental Impact assessment: a critique on Indian law and practices. Int J Multidiscip Res Dev 1–5

Teaching Practice: Analyzing and Transmitting the Wisdom of Climate-Adaptive Construction of Traditional Heritage Settlements

Xu Li, Pengcheng Liu, and Baojie He

Abstract Climate change and increasingly prominent energy demands exert significant constraints to urban sustainable development. Energy consumption of buildings is determined by both individual buildings and urban morphology. Therefore, it is imperative to improve environmental quality through urban design. To support this, this study aims to report how the education shapes theoretical and practical components in design courses by the teaching practice of the environmental performance a heritage Ciqikou district in Chongqing, China. By analyzing the construction wisdom inherent in local climate adaptation, this study is expected to provide theoretical guidance for climate-responsive urban design practices. The teaching practice includes two parts: theory part and design part. The theory part consists of the interpretation of spatial characteristics and principles of multi-scale urban form adaptation to climate, and the analysis of research progress of relevant topics all over the world. In the design part, students were expected to analyze the wisdom of traditional settlements adapted to local climate at different scales, and explore the inheritance of the traditional settlements. Through environmental performance simulation and multi-objective optimization based on genetic algorithm, the students explored the spatial patterns and their combinations adapted to the regional wind and thermal environmental conditions, investigated the behavioral patterns of people adapting to the microclimate, and summarized the wisdom of building adapted to the local climate. Overall, the teaching practice is based on the combination of green design and cultural heritage, aiming to cultivate students' ability to comprehensively use the

X. Li · P. Liu · B. He (✉)
Centre for Climate–Resilient and Low–Carbon Cities, School of Architecture and Urban Planning, Key Laboratory of New Technology for Construction of Cities in Mountain Area, Ministry of Education, Chongqing University, Chongqing 400045, China
e-mail: baojie.unsw@gmail.com

B. He
School of Architecture, Design and Planning, The University of Queensland, Brisbane 4072, Australia

Institute for Smart City, Chongqing University, Liyang 213300, Jiangsu, China

CMA Key Open Laboratory of Transforming Climate Resources to Economy, Chongqing 401147, China

B. He et al. (eds.), *Urban Climate and Urban Design*, Urban Sustainability,
https://doi.org/10.1007/978-981-96-1521-6_16

relevant knowledge of meteorology, environmental science, and planning to analyze the climate adaptive characteristics of settlements and apply them to design. After the course, students are expected to initially master the characteristics of climate-adaptive urban form, responsive environmental performance assessment software, the application of climate-adaptive urban design theories and methods, and innovative solutions from different perspectives.

Keywords Pedagogical practice · Climate adaptation · Spatial morphological features · Spatial patterns · Camping wisdom · Heritage · Teaching practice

1 Introduction

Climate change and the increasingly prominent contradiction between energy supply and demand are greatly restricting the development of Chinese cities and towns. The energy consumption of building groups is closely linked to urban form, which can influence the energy consumption of buildings by about 10–30% or more [7]. Therefore, improving environmental comfort by means of urban and rural planning and design, so as to reduce building energy consumption and carbon emissions, is an important topic for research in the field of urban and rural planning in China, as well as one of the important contents of teaching practice.

In the current era of green and low-carbon development, adapting to the regional climate for the reduction of building energy consumption has become a problem that must be considered in spatial planning and design. At present, urban design teaching system in China has mainly focused on spatial form design [1, 2, 4], multi-scale urban spatial system planning [8, 12, 13], as well as their connection with the economy, history and culture, political system, ecological environment, engineering technology, etc. [3, 6, 9–11, 14–16]. However, the teaching system has failed to consider how to realize green energy saving in urban design, and there is limited consideration of the interaction between the city and the climate. As a result, the students have insufficient knowledge and theoretical methods of the basic knowledge and theoretical methods of the climate-adaptive urban design, making them difficult to adapt to the new industry development for positive climate mitigation and adaptation. It is urgent to carry out the teaching of climate-resilient urban design courses.

Considering the complexity of urban issues and the fact that teaching will inevitably involve a wide range of cross-disciplinary knowledge and skills, the urban design teaching group has developed the "Climate Resilient Urban Design Course". The course combines theory and design parts to systematically explore the intersection of urban planning with meteorology, environmental physics and other disciplines. This can potentially cultivate complex planning professionals with knowledge related to green energy saving. In particular, the course incorporates environmental performance simulation and intelligent optimization into the teaching plan, and is closely integrated with the traditional morphological design teaching. The course takes the Chongqing Ciqikou Historical District in Chongqing, China as an example to carry out teaching practice, and guides students to actively explore spatial forms adapted to the regional climate and climate-adapted urban design by analyzing the correlation between the characteristics of the environmental performance and spatial forms.

2 Course Design

The course consists of theory and design parts, with the characteristics of integrating quantitative research into traditional planning and design, combining theory with practice, and deepening students' understanding of climate-resilient urban design. About 60 students were involved in the theory part, and students could choose their own practical topic according to their interests. The classes related to climate resilience were conducted in two weeks, 8 hours in total, during which the lecturer explained the progress of climate resilient design research and related basic knowledge, and assigned students to carry out a literature review on the corresponding topic in groups in order to better understand the current situation and development trend (Table 1).

Students (8–12 students) who choose the Climate Adaptation track would further participate in the design class program. The design classes were conducted in 12 weeks with 48 class hours in total. Teachers were required to (i) explain the main contents, requirements and time schedule of the course, (ii) assign students to work in groups to evaluate the spatial environmental performance of the site and form a research report on the spatial environmental performance evaluation (in groups), (iii) analyze the spatial patterns with climate resilience, (iv) propose optimized designs

Table 1 Course schedule

Progress arrangements		
Serial number	Design (thesis) work content	Duration (weeks from/to)
	Lecture in theoretical classes, completion of literature review in groups	Theoretical course weeks 1 and 2
1	Related lectures, breakout groups	Design class week 1
2	Current situation research and report (PPT) Presentation of the preliminary program of the research design	Design class weeks 2–3
3	Study design determination	Design class weeks 4
4	Measurement, simulation and analysis	Design class weeks 5–8
	Midterm review chart	Design class weeks 9
5	Optimized schematic design/design guideline development proposed for key sites	Design class weeks 10–11
6	Ortho-Commentary	Design class weeks 12

for key sites and carry out simulation and verification (in groups), and (v) propose guidelines for designing spatial patterns with climate resilience. The setting of such classes was to (i) cultivate students' values of combining green and low-carbon design with historical and cultural heritage preservation, (ii) expand the width of cross-disciplinary knowledge and consolidate the depth of knowledge of planning disciplines, and (iii) to enhance their comprehensive ability through theoretical analysis and design practice.

3 Course Introduction and Task Interpretation

3.1 Course Introduction

Through storytelling and examples, students' interests and thoughts were aroused, and discussions were held on what issues were of concern in climate-resilient urban design to introduce the course teaching.

(1) Take the military history case "Battle of Red Cliff" as the starting point to point out that the change of wind direction is one of the key factors for the victory of the Wu army in the Battle of Red Cliff (Fig. 1). Through the analysis of geography, hydrology and climate characteristics, Cao's army stationed in Wulin behind the largest freshwater lake in China at that time "Yunmengze". By utilizing the meteorological principle of "lake wind", the Wu army successfully waited for the southeasterly wind blowing towards Wulin at night to win the war.
(2) Take the case of settlement design as an entry point to illustrate that the form features reflect the adaptation to the climate—should the wide side of the building face south or north in the planning and design of the settlement and why?

As a result, half of the students chose the inaccurate answer, reflecting that some students did not have enough knowledge about climate adaptation, lacked energy-saving consciousness in traditional planning and design, and ignored climate adaptation and energy-saving design in the program. This further demonstrated that previous teaching system failed to deliver climate-adaptive urban design knowledge. At the same time, it also triggered students to think about the importance of green energy saving and climate adaptation.

3.2 Interpretation of the Course Tasks

The course covers the compilation and review of current research, measurement and simulation of spatial environmental performance, analysis of climate-adaptive spatial forms, design guidelines for climate-adaptive spatial forms, and optimization of key

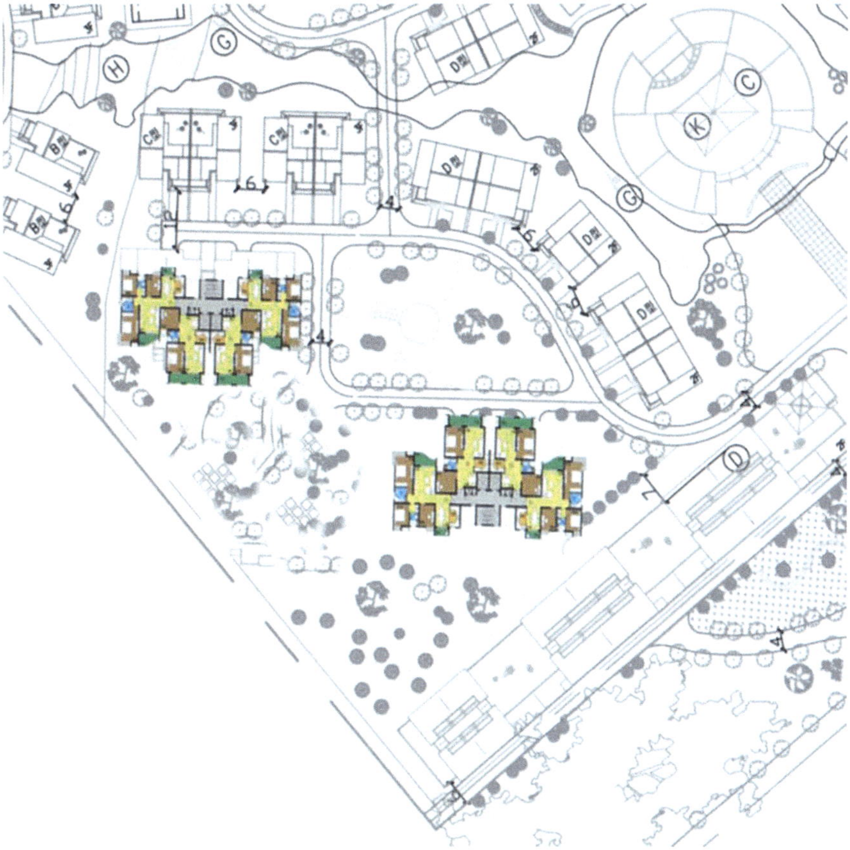

Fig. 1 Examples of settlement design

lots, to form a complete and coherent progressive teaching system. Research-based teaching is integrated into the design, helping students to understand the reasons behind the forms and cultivate their meticulous logical analysis ability (Fig. 2).

4 Overview and Review of the Current State of Research

4.1 Importance and Purpose

As the basic board of the research stage, students were required to master the stage characteristics, methodological paths and hot topics of existing research through the analysis of the current literature, and discover the deficiencies and put forward the countermeasures to provide support for their respective research topics. Through the

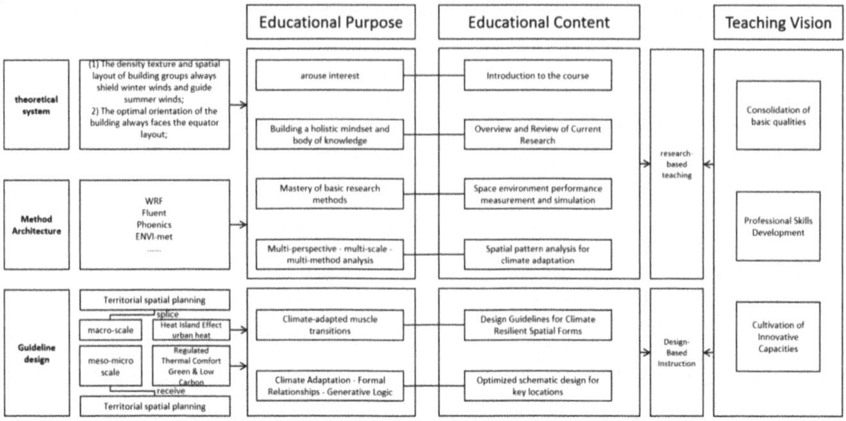

Fig. 2 Teaching system framework

combination of lectures, assignments and discussions, the training of the review was strengthened, and students were trained to construct an overall system of ideas and methods for conducting scientific research.

4.2 *Teaching Content*

It explains the characteristics and principles of climate adaptation under different regions in the current literature from the levels of landscape pattern and settlement site selection, structural form and scale of streets and lanes, architectural layout and components, and green landscape. For example, there are differences in the architectural texture and layout of form to adapt to different regional climates; the density texture and spatial layout of building groups always shield winter winds and guide summer winds; the best orientation of buildings is always oriented towards the equator [5], and so on. In the teaching process, the application of these climate adaptation principles in urban and architectural design was vividly illustrated by combining practical cases.

4.3 *Status of Assignments*

(1) Selection of Topics

Based on the content of the lectures, students would select topics in small groups, collect and review literature from different perspectives such as different climate zones, multi-scale, spatial patterns and microclimate correlation according to their own research interests, and the contents of the review will include research phases, trends and methodologies, and conclusions of the research, etc. The paper was used as the course assignment and would be evaluated in the final grade. The synthesis paper was used as the coursework of the theory class and evaluated at the end of the semester.

(2) Completion and Shortcomings

Each group had completed the review of existing literature from different perspectives, and gained a certain mastery of the results, methods and hot spots of this research direction. However, it is found that the number and depth of reading literature were insufficient, and the learning of basic knowledge of related cross-disciplines and related principles should be strengthened.

The teaching team made a return visit to the students after the first year of the program, and some of them thought that the explanation of the principles of climate adaptation was complete and systematic during the teaching process of literature review, especially the explanation of the processes and paths of microclimate research through practical cases was impressive. However, they also thought that most of the literature review work was concentrated in the first 1–2 weeks, with a lack of follow-up research. It was suggested that the literature research could be carried out throughout the design class, and that students could select topics for themselves and gradually improve the literature review during the research process.

5 Environmental Performance Measurement and Simulation

5.1 Importance and Purpose

Spatial environmental performance assessment and simulation is a necessary research method in traditional settlement climate adaptation research. In the teaching practice of this part, students could use group cooperation to fulfill the requirements of microclimate numerical simulation and field observation and other related tasks according to their interests and research directions.

5.2 Teaching Content

(1) Introduce the mainstream simulation software of current numerical simulation technology, including Fluent, PHEONICS, ENVI-met and so on, as well as explain the operation process of these software, covering meteorological conditions, three-dimensional spatial model construction, computational domain, boundary conditions and physical parameter settings, grid division and computational solutions and other parts.

(2) Introduce the use of meteorological monitoring instruments and precautions during field observation. Later on, we provide specific guidance on the selection of spatial location and time arrangement in conjunction with the direction of the selected topic of the field observation team.

5.3 *Student Work and Teacher Evaluation*

The results of spatial environmental performance evaluation and simulation cover macro-, meso-, and micro-scale.

(1) At the macro level, some students used remote sensing images to invert the Land Surface Temperature (LST), integrated Morphological Spatial Pattern Analysis (MSPA) and connectivity analysis to identify cold islands and heat islands, calculated the cooling rate and constructed the heat diffusion resistance surface based on the surface cover, and used the circuit theory model to identify corridors, obstacles and pinch points. Based on the surface cover, the cooling rate was calculated to construct the heat diffusion resistance surface, and the circuit theory model was used to identify corridors, obstacles, and pinch points to construct the cold island and heat island networks in the central urban area of Chongqing. This work applied circuit theory into the identification of urban cold and heat island networks, excavated the current urban thermal environment problems, identified thermal landscape obstacle points, and provided a reference for the improvement of the thermal environment in the central urban area. The work reflected the thermal environment of the area around the base.
(2) At the level of streets and alleys and their structures, some students used PHEONICS numerical simulation to simulate the wind environment of the settlement, and then evaluated the microclimate adaptation characteristics of the structure, scale, and spatial locations of the ancient town of Ciqikou. The evaluation of the spatial environmental performance further suggested the question "How does the herringbone-like street network structure of Ciqikou affect the ventilation of the settlement?" This is the selected topic. The students demonstrated the effects of the main street orientation, the presence or absence of branch alleys, the form of intersections, the orientation of branch alleys, and the gradient of branch alleys on the overall ventilation performance of the settlement (Table 2). The student excavated the problem through environmental performance simulation and used multiple working condition designs to analyze the climate-adaptive spatial form characteristics (Table 3). The logic of the analysis was well-organized, and the conclusions of the study were of some value.
(3) Some students also used ENVI-met to simulate the microclimate characteristics of different street intersection forms, and validated the simulation results by field observation, and summarized the following types of climate-adapted street intersection forms. As the combination of simulation and field observation is used to measure the spatial environmental performance, the credibility of the results is high.

Table 2 Influence of street network structure on the overall ventilation performance of the settlement

working condition	Variable Control Chart	research purpose	wind direction	research objective
working condition1	Main street orientation  With or without branch alleys 	1. is the current orientation layout of Main Street based on climate adaptation or topographic or traffic adaptation? 2. the importance of branch alleys to overall ventilation performance	south wind	Average summer wind speed (m/s)
working condition2	Intersection form  cross shape T-shape	Effect of intersection form on overall ventilation performance		
working condition3	Branch alley orientation 	Influence of Branch Lane Orientation on Overall Ventilation Performance		
working condition4	Slope of branch alley 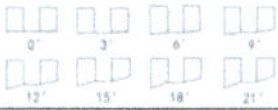	Influence of Branch Lane Slope on Overall Ventilation Performance		

Table 3 Climate adaptive street intersection morphology types

Type of intersection	Typical Example of the Ciqikou Neighborhood	Climate Adaptive Orientation Angle
"T"-shaped intersection		Angle between 30°and 45° to the prevailing wind direction 
"Y"-shaped intersection		Angle between 15°- 30° to the prevailing wind direction 
"Ten"-shape intersection		Angle between 30° 45° to the prevailing wind direction 

During the teaching process, we found that the progress of the actual measurement group was often limited by weather conditions. For example, sometimes a typical day with high summer temperatures was encountered only two weeks after the course, so the progress of the measurement was slow. The simulation group was limited by the computational power of the computer, which was time-consuming, and most of the students had a short time of exposure to the software, so they did not have a good grasp of the fine-grained microclimate simulation in a complex mountainous area.

Most of the students chose PHEONICS and ladybug tools with a higher degree of interface visualization and relatively simple software operation, for the numerical simulation of microclimate. However, some of the students' simulation results contained problems such as anomalies and inconsistency with the actual situation. For example, when simulating a simple model, the thermal environment of ladybug tools could calculate the UTCI value better; however, the wind environment simulated by butterfly showed abnormal results, while PHEONICS could simulate a reasonable wind speed ratio. When simulating a complex mountain model, the wind and heat values simulated by ladybug tools were abnormal, and the wind environment simulated by PHEONICS also showed abnormal results. Overall, the wind environments simulated by PHEONICS were overall we more reasonable. However, when using PHEONICS to simulate UTCI, its value would be above 100 from time to time, not consistent with the actual condition. The students thought that butterfly and PHEONICS could simulate the wind and thermal environments of small-scale sites or buildings with topography more quickly, while only PHEONICS could simulate the wind and thermal environments of large-scale sites or buildings with topography and other complex models, but its accuracy needs to be verified. This reflects the fact that the existing models are based on CFD architecture to simulate urban wind-heat environments, but the results are not consistent across different simulation software and platforms at this stage. Teaching feedback is beneficial to further deepen the requirements of the research methodology and software application. In the future teaching practice, it should be necessary to further (i) improve the accuracy of the underlying architecture of the CFD numerical model, (ii) study the selection of the turbulence model, the boundary conditions and physical parameter settings, (iii) limit the mesh division standard to a reasonable range, and (iv) strengthen the validation of the field observation to ensure the credibility of the simulation results of the wind field, temperature and humidity field.

In terms of field observation, some students reported that there was little actual measurement data, which made it difficult to form valuable and credible research, so the guidance for students should be further strengthened in selecting topics and spatially located measurement points. In addition, some students reflected that due to the weather, the course did not promote the early wind-heat environment measurement enough, so it is suggested that the time of the course should be chosen in the time near winter or summer, so as to better promote the content of the measurement.

6 Spatial Morphological Analysis for Climate Adaptation

6.1 Importance and Purpose

Mining typical spatial patterns of climate adaptation and camping wisdom and experience are the main purpose of this course design. The teaching team focused on guiding students to analyze the climate adaptation characteristics of the heritage area of Ciqikou from multiple scales and different perspectives, and thereby encouraging students to adopt qualitative and quantitative research methods for analysis.

6.2 Teaching Process

In the teaching process of this link, students were required to:

(1) Analyze the spatial patterns (low-level and high-density) adapted to regional wind and heat environmental factors at multiple scales, such as district (colony), block, building group combination and architecture, in the context of the heritage settlement of Ciqikou, and focus on analyzing the generating mechanism of the corresponding spatial patterns.
(2) Combining the results of the survey and evaluation of the spatial environmental performance of the Ciqikou Historical District, evaluate its climate adaptability, and identify the spatial features with climate adaptability; at the same time, conduct a survey on the energy use behavior of the residents (merchants) and the activities of the tourists, and refine the behavioral patterns of climate adaptability.

In this way, a research-based instructional program design was used to guide students in identifying scientific problems and parsing mechanisms. Students closely integrated research and design, built programs on relevant research, and reinforced the logical relationship between prior research and spatial design.

6.3 Student Work and Teacher Evaluation

Students used diverse research methods to analyze spatial patterns of climate adaptation.

(1) Correlation analysis

Some students divided the research units based on the morphological typology of the heritage settlement of Ciqikou and divided them into several categories based on the morphological features of the units to identify the morphology of the ancient town of Ciqikou - a prototype microclimate research unit. Based on the results of unit division of morphological typology, the correlation between unit morphological features and microclimatic factors was further studied, and the factors that could significantly influence the microclimate of Ciqikou were gradually screened out to summarize the relevant regularity texture (Fig. 3). The student understood and analyzed the complexity of the built environment of the city through the idea of comprehensive comparison and classification, summarized the similarities and differences, and conducted research on the characteristics of each unit in different categories. The research methodology used transformed urban morphological characteristics into quantitative data, and used a programming language to implement comprehensive statistical analysis, rapid classification, and iterative computation and clustering of the data. Overall, the research methodology and pathway used by the student was reasonable, but for the microclimate simulation session, the microclimate simulation results obtained by using QGIS calculations alone should be further verified.

(2) Intelligent Optimization

Based on the overall simulation results of the heritage settlement, some students summarized the street intersection prototype and used orthogonal experiments to study the changing law of wind and thermal environment performance of intersection patterns such as flat T-type, climbing T-type and falling T-type. The genetic algorithm, afterwards, was used to explore the optimal values of each spatial morphology parameter of the mountain street intersection and analyze the spatial pattern of the mountain intersection. The student used intelligent optimization algorithm to consider the adaptive characteristics of each spatial morphology parameter of intersections to microclimate in complex mountain environment, and overcame the problems of single-variable-by-variable analysis, artificial coupling, and inefficiency through automatic computer iteration, and found the optimal values of spatial morphology parameters of climate-adaptive intersections, and analyzed the spatial camping mode of their climate adaptability. Overall, it is a new attempt to explore the climate adaptation of traditional heritage settlements (Table 4).

(3) Activity behavior research

In the process of settlement adaptation to climate, in addition to the expression of spatial construction, there is also a certain pattern of climate adaptation in people's activities. We expected to identify the behavioral patterns of people's active adaptation to climate in the field research on the ancient town of Ciqikou (Fig. 4). Some students made preliminary explorations. For example, some students investigated the spatial and temporal distribution of people's activities at important spatial nodes of the town, but they did not show the systematic pattern of the behavior.

In this teaching session, students were active enough to try to adopt new methods for spatial pattern climate adaptation analysis. For example, a student used genetic algorithm to solve the multi-objective optimization problem of climate, and its purpose was to try to get excellent spatial patterns of settlement patterns. However, in the actual study, it was found that it was difficult to find valuable pattern features by setting up spatial patterns extensively to search for the best. In response to this objective problem, we instructed the student to start from the actual case of the Ciqikou

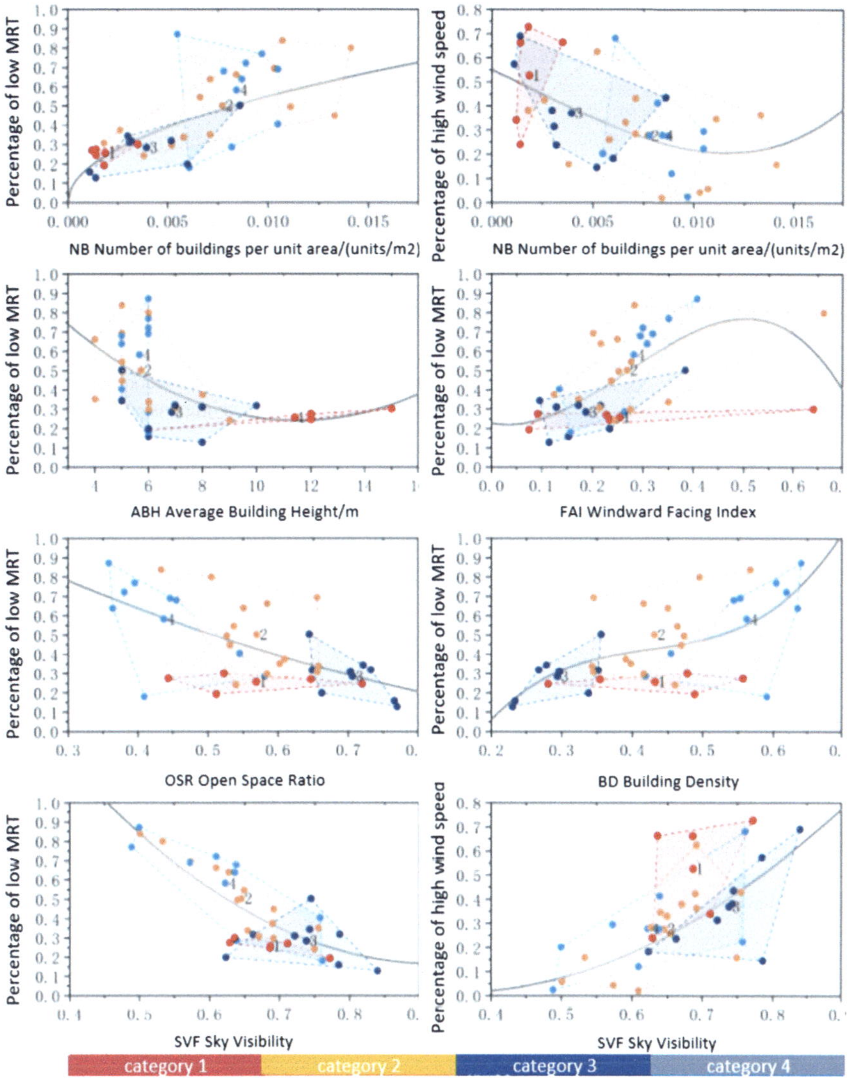

Fig. 3 Correlation study between morphological characteristics of cells and microclimatic factors

Historical District, and from the results of the wind and heat environment obtained from the simulation, to find good morphological elements for the establishment of spatial prototypes, and then to find the range of morphological indicators with good environmental performance in such spatial prototypes through the multi-objective search for optimization. There was also feedback from the students that they should go to the field for more observation to ensure the rationality of the prototype setting.

Table 4 Study of optimal spatial patterns for mountain intersections

Working condition	Variable control chart	Morphological parameter variables	Parameter settings	
Working conditions 1,2 	Macro: Main Street Orientation 	Main Street Orientation	all working conditions	0-360°
		Main Street Length	Working condition 1	50-70m
			Working condition 2	40-60m
	Meso: Intersection structure 	Length of sub-street	Working condition 1	25-50m
			Working condition 2	40-60m
		Street height	all working conditions	3-12m
	Micro:node morphology 	Node length and width	all working conditions	5-15m

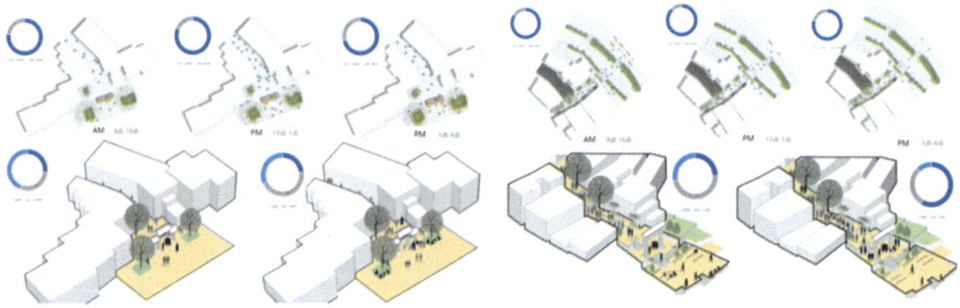

Fig. 4 Behavioral survey of activities in Ciqikou

7 Design Guidelines for Climate-Resilient Spatial Forms

7.1 Importance and Purpose

The purpose of this teaching session is to combine the above research with planning and design practice, to transform the relevant results of research-based teaching into spatial pattern mapping and design guidelines, and to provide planning practitioners with a high degree of visualization and easy-to-apply guidance for the relevant technology.

7.2 Teaching Process

The formulation of design guidelines for climate-adaptive spatial patterns is an important link in the application of research to design practice. Teaching should train students to transform research results into technical guidelines for designers to guide design practice. We combined the relevant technical achievements in the field of architecture to explain the relevant principles of architectural scale and the process of generating guidelines, providing references for students to formulate climate-adaptive urban design guidelines.

7.3 Students' Assignments and Teachers' Evaluation

Combined with the research on the design of related working conditions, some students proposed the morphological design guidelines for low-rise high-density construction areas in four aspects: main street layout, branch alley layout, public space layout and slope construction (Fig. 5). Some students also proposed climate adaptation design guidelines for single form elements from different form elements (Table 5). Some other students proposed optimization guidelines for Ciqikou and the surrounding areas from the levels of street morphology and building components (Table 6). These students' morphological design guidelines have clear semantics and are of some significance to design practice (Tables 7 and 8). However, due to the limited class time in this course, the practical application value and accuracy of these guidelines should be further verified.

Aiming at the problems students had in the formulation of guidelines, the teaching team summarized that the following elements should be deepened or improved:

(1) In-depth understanding of the microclimate of the traditional settlement of Ciqikou: Before planning and designing the guidelines, it is necessary to have

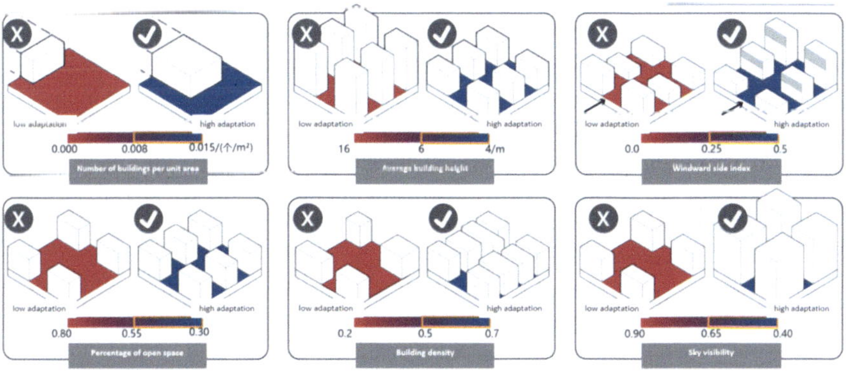

Fig. 5 Operation of the design guidelines for climate-resilient spatial forms

Table 5 Climate adaptability assessment and optimization strategies for Ciqikou and surrounding areas

Morphological unit strategy based on microclimate optimization

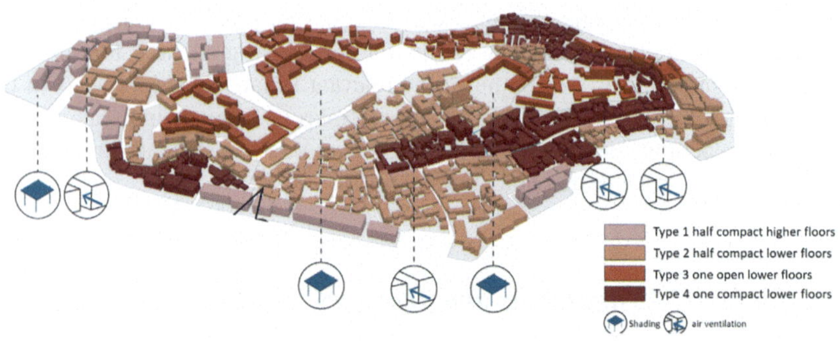

Type 1-Average microclimate adaptation	Type 1 is mostly located in the south side of the periphery and the height is high, should ensure that the ventilation channel between the building is smooth, the formation of a number of air inlets; south-oriented activities site by the radiation temperature is high, can add a number of shade facilities.
Type 2-Good microclimate adaptability	Type 2 constitutes the main part of the block, with a large area, compact along the street and loose internally, which can be appropriately opened up along the street surface to optimize ventilation to evacuate internal heat.
Type 3-Poor microclimate adaptation	Type 3 open space accounted for a large proportion of the building less, subject to high radiation temperature can add a variety of shade facilities, such as open space shade shed building façade of the shade budding components, etc.
Type 4-Good microclimate adaptation	Type 4 is the core protection zone of the neighborhood, which is a compact strip with little open space, mainly utilizing the shade facilities along the street façade, and needs to strictly ensure that the ventilation channels along the street surface are smooth.

an in-depth understanding of the microclimate environment in which the traditional settlement of Ciqikou is situated, including climatic conditions, topography, geomorphology, vegetation, and other factors, as well as the impact of these factors on the spatial layout of the settlement. This will help to formulate more reasonable planning and design guidelines.

(2) Respect and protect the original street texture: In the planning and design guidelines, the original street texture should be respected and protected, including the shape and direction of the street, as well as the relationship with the surrounding buildings. On this basis, the adjustment of open space function and building function. It is definitely not feasible to improve the microclimate by destroying the traditional features such as large-scale demolition and construction and planting trees in the main street.

Table 6 Morphological design guidelines for low-rise high-density construction areas

Spatial Scale	Guideline Strategy
Main Street Layout	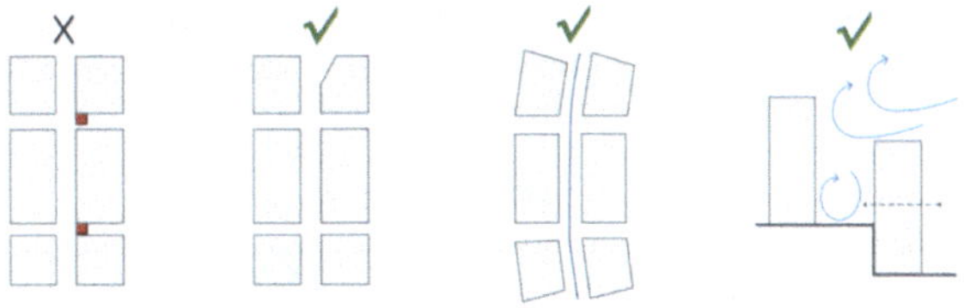 1. Prohibition of additional obstructions at intersections 2. Main Street opens into a flared shape 3. Main Street retains a certain curvature 4. Lower buildings on the windward side of both sides of Main Street to enhance ventilation in the Church
Layout of the branch alleys	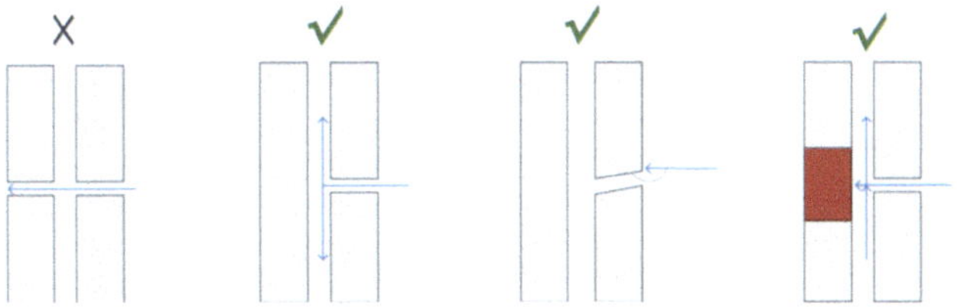 1. Branch alley crosses Main Street 2. Branch Alley T-intersects Main Street 3. Branch alley at an angle to the prevailing wind direction 4. Important building arrangement: at the end of the branch alley
Layout of the public space	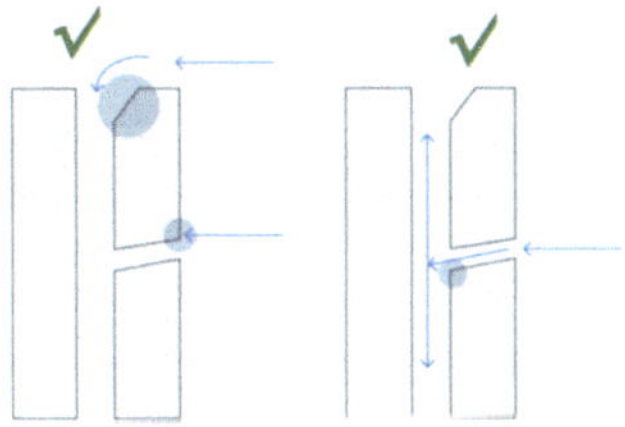 1. Public spaces are arranged at street openings 2. Public spaces are arranged at T-intersections
Slope construction	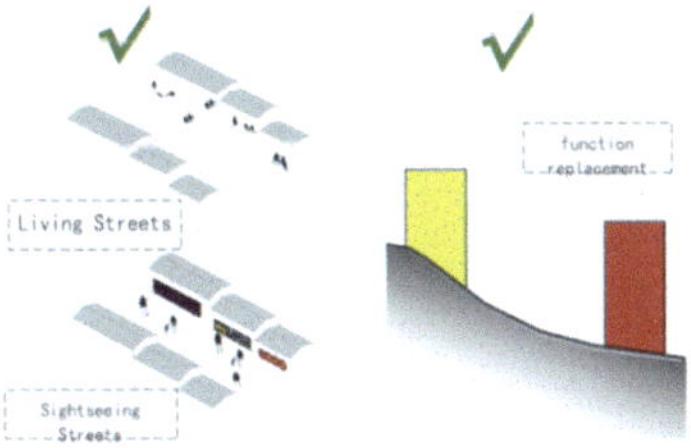

Table 7 Climate adaptability design guidelines for street morphology

scale		Conclusions of the study: reach a verdict	icon: domi nance	icon: inf erior	Optimiz ation guide lines	Application Sche matic: New construction and optimization
Nakag anmatsudo Street	Widths	1.1-5m UTCI better performance 2. Appr oaching 5m to reach bett er level 3.5-10 m show elev ated UTCI 4. Above 10m recov ered lower U TCI	 widt hs : 6.042m UTCI : 36.56	 wi dths : 7.279m UT CI : 43.59	 ≦5m increase outdoor activity space  5-10m additional shading	 Maintain street widths within 5m better  Additional shading for 5-10m street widths
	a spect ratio	1. UTCI performs we ll at <1, lowe st at 0.5 2. UTCI increased at 1-1.5, highest near 1.5 3. UTCI gradually de creases at >1 .5	 aspe ct ratio: 0.757 UTCI : 37.12	 asp ect ratio : 1.118m UT CI: 40.5	 ≦1Savings on shading facilities  1-1.5 Increase in shading facilities	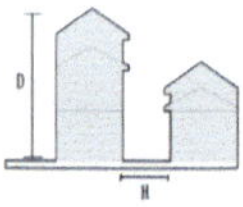 Street aspect ratios should be kept within 1 better 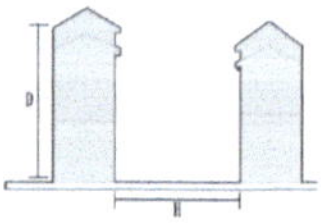 ≦1 Laying out outdoor activities >1 Adding shading

(continued)

Table 7 (continued)

Orientation	1.0-50 degreesCelsius UTCI increases 2. UTCI stays low at 100-180 degrees, optimal at 120-130 degrees	 orientation：124 UTCI：36.56	 orientation：46 UTCI：40.22	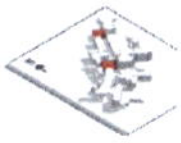 Northeast-southwest facing streets increase outdoor activity space	 Parallel streets for east-west oriented buildings Vertical streets for north-south oriented buildings
				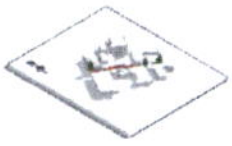 North-west - south-east increased shading	 North-west-south-east increased shading

(3) Practical solutions should be put forward to alleviate the high temperature in summer and guarantee the safety of tourism in the ancient town: From the perspective of climate adaptation, planning cool tourism routes in the ancient town and implementing the planning of facilities along the cool routes through the design guidelines, so as to avoid the human health problems caused by the high temperature and heat, which can, to a certain extent, guarantee the safety of tourism trips during the high temperature period in summer.

(4) Combining carbon neutral strategy and energy saving demand to propose behavioral guidance: It is essential to analyze the energy use behavior of actual merchants, residents and tourists from the perspective of energy saving and propose energy saving-oriented behavioral guidance.

8 Optimized Schematic Design for Key Locations

8.1 Importance and Purpose

The teaching purpose of this session is to apply the typical spatial patterns and climate adaptability principles formed by the research in the above stages to the planning and design practice. Accordingly, it should consider the relationship between form and body, the logic of program generation, etc., focusing on cultivating students' innovative thinking and stimulating their creative ability. In the study of the actual sample of the Ciqikou Ancient District, the teaching team encouraged students to put

Table 8 Climate adaptability design guidelines for building components

scale	Conclusions of the study				optimization gui delines	Application Schematic
		reach a verdict	icon dominance	inferior		New construction and optimization
Micro-scale building components	shield from the sun	1. Sun-shading facilities are effective, as well as convenient and economical. 2. Almost all street-level stores have awnings on the first floor.	 UTCI: 37.69	 UTCI: 39.09	 Retractable Awnings  GreenRattan Rack  Yard and street plantings	 Increase in greenery along the street  Addition of retractable awnings
	air convection	1. Improving building air convection can help improve human thermal comfort. 2. New areas may consider adding courtyards. 3. Optimized areas may improve the door-to-window ratio.	 UTCI: 37.69	 UTCI: 43.60	 courtyard  skylights  Increase window-to-wall ratio	 Addition of an inner courtyard  Increase the ratio of windows to doors
	transpiration of plants	1. A single type of lean-to vertical greening with a combination of open vertical greening is predominant. 2. backsplash buildings and plants provide prolonged shading for the street.	 UTCI: 37.69	 UTCI: 38.25	 Green Rattan Rack  Greening of walls  Green plants along the street	 Increase in greenery along the street  Increase in wall greening

forward specific optimization designs and strategies for areas that were not climate-adaptive.

8.2 Teaching Process

In the actual teaching, some students had insufficient ability to translate theory into design practice, weak articulation between planning and design and climate adaptation research, and some students still had a poor understanding of the basic principles of protection and renewal of historic districts, which should be strengthened in the next stage of teaching practice. For example, some students planted street trees in the streets of ancient towns to alleviate the hot environment, but ignored the current problem that the streets of ancient towns are narrow and there is no space to plant street trees.

8.3 Student Work and Teacher Evaluation

The results of this section were relatively weak due to the limitation of teaching hours and the fact that most of the students focused on the preliminary regularity study. Only some students combined the analysis of simulation and measurement results to propose optimized design strategies to improve the wind and thermal environments in terms of riverbank form, riverfront functional area design, pavement form and structures.

Feedback from teaching and learning:

(1) Strengthen the in-depth research on the key lots, including their historical and cultural background, social assets, etc., to provide a comprehensive analysis for the optimization of the design scheme.
(2) Comparative research can be conducted in the fall semester in conjunction with the spring semester course's related research results on key lots, as well as comparing the differences in the design of optimized solutions for key lots under summer and winter climate adaptation research.

9 Conclusions

In the context of climate change and the tension between energy supply and demand, urban design can not only stay in the relatively traditional teaching practice, but also comply with the strategic needs such as carbon neutrality, break through the established boundaries of disciplines, and strive to realize the cross-discipline, in line with the needs of the era of green and low-carbon development. This course tried to integrate climate-adaptive spatial morphology research into the teaching of urban design course, by combining theoretical classes with design classes, through several teaching sessions, including current research and review, spatial environmental performance measurement and simulation, climate-adaptive spatial morphology analysis, climate-adaptive morphology design guidelines and optimization of the design of key lots. In the process of teaching, we provoked students' thinking and interest by telling stories and analyzing design cases and deepened students' understanding of related knowledge points through interactive discussions. In the course practice process, we enabled students to master the research and design of the spatial morphology characteristics through (i) the literature review writing and prompting students to grasp the theoretical knowledge system related to climate adaptation, (ii) numerical simulation, field observation, field survey and other methods to guide students to analyze and discover the spatial morphology characteristics of climate adaptation and explore its mechanism, and (iii) the typical spatial pattern sorting, guideline development, optimization design to guide students to return to the morphology and design itself. Through the combination of typical spatial patterns, formulation of guidelines, and optimization of design, students were guided to return to the form and design itself. Morover, research-based teaching and design were closely integrated to ensure students can recognize the important influence of microclimate in urban design practice and put forward the spatial morphology control guidelines for climate adaptation. In the context of green and low-carbon development of urban construction, this course can introduce the perspective of climate adaptive design and cultivate the cultivation of composite professionals with the knowledge of climatology, environmental physics and urban and rural planning, thereby providing a reference for the teaching reform of the planning discipline.

Acknowledgements Many thanks go to the permissions of image and table use by Wenxuan Wu, Ying Liu, Kai Zhang, Xuanting Zhou, Rongzhi Jia, Xinhui Zhu, Binxin Zhang, Qianyu Shi and Junjie Zhang at School of Architecture and Urban Planning, Chongqing University, China.

References

1. Belof M, Kryczka P (2024) Between architecture and planning: urban design education in poland against a background of contemporary world trends. J Plann Educ Res 44(3):1030–1047
2. Cao K, Xiao J (2018) From "Imitating Forms" to "Following Rules"—Exploration of urban and rural planning teaching methods focused on form relationships and generation logic. Art Des (Theory) 2(12):144–146
3. Chiaradia AJ, Sieh L, Plimmer F (2017) Values in urban design: a design studio teaching approach. Des Stud 49:66–100
4. Gu K (2020) The teaching of urban design: a morphological approach. J Plan Educ Res 40(4):472–481
5. Han D (2021) Integrated design method for climate adaptive green public buildings. Southeast University Press, Nanjing
6. Harvard University (2019, Nov 25). https://www.harvard.edu/
7. Leng H, Chen X, Ma Y (2020) Research progress and enlightenment on the impact of urban morphology on building energy consumption. J Architect 2:120–126

8. Linovski O, Loukaitiou-Sideris A (2013) Evolution of urban design plans in the United States and Canada: what do the plans tell us about urban design practice? J Plan Educ Res 33(1):66–82
9. Ma X, Lan L, Chen J (2023) Analysis of urban design introductory courses in American architectural schools—A case study of five schools: MIT, Yale, Harvard, Berkeley, and UPenn. Architect 3:32–38
10. MIT - Massachusetts Institute of Technology (2019, Nov 30). http://www.mit.edu/
11. Rong C, Meng G (2013) Enlightenment from urban design teaching in Blekinge Institute of Technology for landscape architecture education in China. J Lands Res 5(10):27–28
12. Sun S (2023) Integration of "Space+"—Teaching reform of urban design course in the senior years at Southeast University. Urban Des 2:42–49
13. Sun T, Xu K, Mei M (2023) Exploration of the capability cultivation system of urban design undergraduate course. Urban Des 2:18–35
14. Szmelter A, Zdunek-Wielgolaska J (2019, September) The history of city planning at the Faculty of Architecture, Warsaw University of Technology: Theory and Practice. In: IOP Conference Series: Materials Science and Engineering, Vol 603, No 4, p 042033. IOP Publishing
15. Yang C, Geng H (2017) Analysis and suggestions on urban design education system—Learning from the urban design education system and core courses in American Universities. J Urban Plann 1:103–110
16. Yang J, Gao Y, Luo J (2011) Research on key training areas and methods in urban design teaching system. Urban Plann 35(8):55–59

The manufacturer's authorised representative in the EU is Springer Nature Customer Service Centre GmbH, Europaplatz 3, 69115 Heidelberg, Germany. If you have any concerns regarding our products, please contact ProductSafety@springernature.com

Printed and bound by CPI Group (UK) Ltd, Croydon, CR0 4YY
01/10/2026
02225941-0006